Brazing

Manufacturing Processes and Materials Series

Series Editors
Professor John Wood
Cripps Professor of Materials Engineering
Department of Engineering Materials and Design
University of Nottingham
Nottingham
NG7 2RD

Diran Apelian
Provost
Worcester Polytechnic Institute
100 Institute Road
Worcester, MA 01609-2280
USA

1. **Advanced Welding Processes**
   J. Norrish
2. **Powder Metallurgy**
   The process and its products
   G. Dowson
3. **Sheet Metal Forming**
   R. Pearce
4. **Hot Isostatic Processing**
   H.V. Atkinson and B.A. Rickinson

Brazing: For the Engineering Technologist

M. Schwartz

Staff Engineer of the Sikorsky Aircraft Division of United Technologies Corp., CT, USA

CHAPMAN & HALL

London · Glasgow · Weinheim · New York · Tokyo · Melbourne · Madras

**Published by Chapman & Hall, 2–6 Boundary Row, London
SE1 8HN, UK**

Chapman & Hall, 2–6 Boundary Row, London SE1 8HN, UK

Blackie Academic & Professional, Wester Cleddens Road, Bishopbriggs,
Glasgow G64 2NZ, UK

Chapman & Hall GmbH, Pappelallee 3, 69469 Weinheim, Germany

Chapman & Hall USA, One Penn Plaza, 41st Floor, New York NY 10119,
USA

Chapman & Hall Japan, ITP-Japan, Kyowa Building, 3F, 2-2-1
Hirakawacho, Chiyoda-ku, Tokyo 102, Japan

Chapman & Hall Australia, Thomas Nelson Australia, 102 Dodds Street,
South Melbourne, Victoria 3205, Australia

Chapman & Hall India, R. Seshadri, 32 Second Main Road, CIT East,
Madras 600 035, India

First edition 1995

© 1995 M. Schwartz

Typeset in 10/12 Palatino by EXPO Holdings, Malaysia
Printed in Great Britain by St Edmundsbury Press, Suffolk.

ISBN 0 412 59510 9 (HB)   0 412 60480 9 (PB)

A catalogue record for this book is available from the British Library

Library of Congress Catalog Card Number: 94-72662

♾ Printed on permanent acid-free text paper, manufactured in
   accordance with ANSI/NISO Z39.48-1992 and ANSI/NISO Z39.48-
   1984 (Permanence of paper).

To Anne-Marie

Contents

Preface

There is a myriad of combinations of materials and processing methods available to the designer of structures. Each individual part of a structural design may be subject to a broad spectrum of requirements and influences. The more complex the structure, the more important is the materials selection process. Such is the case with the design of components and assemblies and their subsequent use.

As the development process begins, the first two questions that should come to the designer's mind are: 'What material can be used to make this?' and 'How can this material be joined?' For the designer, these are not simple questions to answer. Both the materials and the method of joining those materials are crucial to the structural design. To accomplish the selection process effectively, the first requirement is for the design organization to include individuals who are familiar with a variety of related technologies. It is important for the design team to recruit experts from all related areas; only then can all of the possible options be considered. Also, the design engineers must be familiar with each of the basic design parameters and how they interact.

This confidence was recently exhibited in a survey of design and manufacturing engineers. Even though they probably do not have a clearer crystal ball than any other group in today's society their views of the new decade showed the emergence as a competent, confident, concerned crowd, anxious to apply the tools of technology to the problems of the world.

As a result of the above and other major concerns the remaining years of the 20th century will be ones in which most governments seriously address the global challenges of being a highly visible player in the manufacturing (joining) community.

Monies in the eighties were spent heavily on the tools of joining technology and not enough attention was paid to application and maintenance skills that would be essential to the successful application of the rapidly emerging technologies – brazing ceramics to metals; heating methods like microwave and laser processing; application and forms of braze filler materials (CVD, sputtering, RS foils); use of vacuum atmospheres and precise control of heating cycles with sensors and programmers.

Improvements in joining science and technology must keep pace with advances in materials science and technology or else the benefits of these new materials will not be achieved in the marketplace.

The engineering schools of the world increasingly are facing up to the fact that in the future engineers must play a key part in management's strategic planning group. The university system recognizes that tomorrow's manufacturing engineers will have to have enhanced capabilities due to three significant factors: increased product sophistication and variation; a global manufacturing environment; a multitude of social and economic changes.

The education of the 21st century engineer will require a radically different education. They will be a systems integrator and not classified as a metallurgist or a brazing or a manufacturing engineer or a quality engineer or an industrial engineer, for the individual will require talents in each of these fields. The assurance of quality must be built into the manufacturing process and the initial product design. The handling of components to and from manufacturing systems, once the role of the industrial engineer, now will be part of the systems integrator function. The increasing use of advanced materials such as composites and ceramics is rapidly breaking down the separation between material suppliers and manufacturers, because in today's world raw material is not only converted into engineering materials but is also converted into the finished part in the very same process. The manufacturing engineer of the future must be fully conversant with modern materials applications. Those who are involved in developing industrial computer networks must also be familiar with the manufacturing processes themselves.

I hope that this book will assist in furthering the education of the engineers of the future.

M. Schwartz

1

Brazing fundamentals

1.1 INTRODUCTION

Brazing is a method of permanently joining a wide range of materials and has wide application in fabricating components of commercial value. It is distinguished from welding in that the process takes place at temperatures below the melting points of the materials to be joined. Brazing is distinguished from soldering by the temperature of processing; processing above 450°C is classified as brazing, Figure 1.1.

The value in brazing is its capability for forming strong bonds between materials of widely differing composition and properties such as ceramics and metals. Joints that are accessible and parts that may not be joinable at all by other methods often can be joined by brazing. Complicated assemblies comprising thick and thin sections, odd shapes, and differing wrought and cast alloys can be turned into integral components by a single trip through a brazing furnace or a dip pot. Metal as thin as 0.01 mm and as thick as 150 mm can be brazed.

Understanding the fundamentals will provide a basic understanding of the brazing process through a review of the factors fundamental to the process itself. The nature of the interatomic (metallic) bond is such that even a simple joint, when properly designed and made, will have strength equal to or greater than that of the as-brazed parent metal or non-metal. The shapes of a braze fillet are naturally excellent and the meniscus surface formed by the fillet metal as it curves across corners and adjoining sections is ideally shaped to resist fatigue.

Brazing comprises a group of processes (Figure 1.1) in which coalescence is produced by heating to suitable temperatures above 450°C and by using a ferrous and/or nonferrous filler metal that must have a liquidus temperature above 450°C and below the solidus temperature(s) of the base metals(s) or nonmetal(s). The braze filler metal is distributed between the closely fitted surfaces of the joint by capillary attraction.

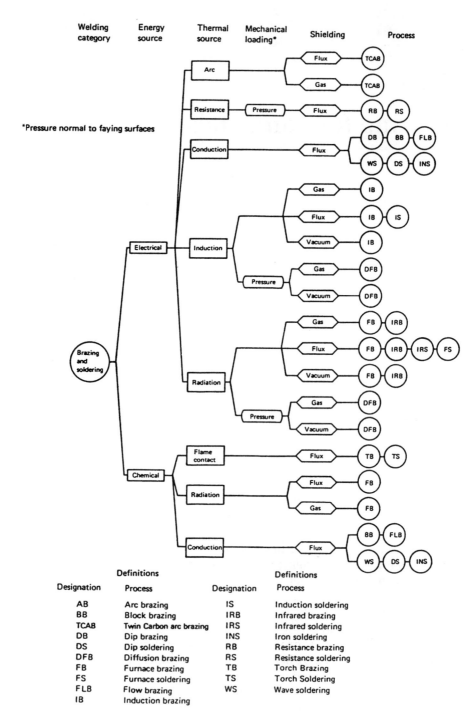

Designation	Process	Designation	Process
AB	Arc brazing	IS	Induction soldering
BB	Block brazing	IRB	Infrared brazing
TCAB	Twin Carbon arc brazing	IRS	Infrared soldering
DB	Dip brazing	INS	Iron soldering
DS	Dip soldering	RB	Resistance brazing
DFB	Diffusion brazing	RS	Resistance soldering
FB	Furnace brazing	TB	Torch Brazing
FS	Furnace soldering	TS	Torch Soldering
FLB	Flow brazing	WS	Wave soldering
IB	Induction brazing		

Fig. 1.1 Brazing and soldering classification diagram.

Brazing has four distinct characteristics:

1. The coalescence, joining, or uniting of an assembly of two or more parts into one structure is achieved by heating the assembly or the region of the parts to be joined to a temperature of 450°C or above.
2. Assembled parts and brazing filler metal are heated to a temperature high enough to melt the filler metal but not the parts.
3. The molten filler metal spreads into the joint and must wet the base metal surfaces.
4. The parts are cooled to freeze the filler metal, which is held in the joint by capillary attraction and anchors the part together.

1.1.1 Historical background

Jewelry fashioned by Egyptian craftsmen approximately 1475 BC has been discovered which illustrates the utility of the joining method from the earliest recorded times.[1] As the catalog of materials expanded, the technology of brazing increased and has been an enabling method for joining, used in the manufacture of many products.

Initially the development of materials progressed from the noble metals, gold and silver, to copper, to the reactive iron and steels, to the current range of superalloys, refractory metals, titanium, ceramics, composites and aluminides, while a parallel development occurred with braze filler metals. Initially filler metals were mixtures of metallic salts and organic reducing agents, however as time progressed, copper and brass were used heralding the filler metals of today and hopefully tomorrow.

Concurrently the existing brazing processes and techniques were improved and new ones developed. These included direct and indirect methods of heating and the capability of controlling the ambient atmospheres which are all employed in the mass production applications of today.

Formal development of the technology has been assisted in this century through the interests of the metallurgical professions in developing a theoretical basis for the process and in teaching brazing in engineering programs at university level.

Brazing generally generates much less thermally-induced distortion or warpage since a part can be uniformly heated to brazing temperature and therefore, minimize any type of distortion.

In most brazing processes, the most important point to the manufacturing engineer is that brazing readily lends itself to mass-production techniques. It is relatively easy to automate because the application of heat does not have to be localized and the application of braze filler metal is less critical, Figure 1.2.[2,3] In fact, given the proper clearance conditions

Brazing fundamentals

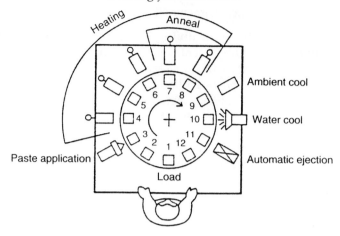

Fig. 1.2 Closely-timed sequence of 12-station rotary machine handles application of brazing filler metal, heating of brazed joints, annealing, and cooling.

and heat, a brazed joint tends to 'make itself' and is not dependent on operator skill.

The attributes of the brazing process are listed below.

- Economical fabrication of complex and multicomponent assemblies
- Simple means for achieving extensive joint area or joint length
- Joint temperature capability approaching that of base metal
- Excellent stress distribution and heat transfer
- Ability to preserve protective metal coating or cladding
- Ability to join cast materials to wrought metals
- Ability to join nonmetals to metals
- Ability to join widely different metal thicknesses
- Ability to join dissimilar metals
- Ability to join porous metal components
- Ability to fabricate large assemblies in stress-free condition
- Ability to preserve special metallurgical characteristics of metals
- Ability to join fiber- and dispersion-strengthened composites
- Capability for precision production tolerance
- Reproducibility and reliable quality control techniques available.

The application of automation is made easy since there are numerous means of heating a joint and/or part as well as producing multiple-braze operations during one heating cycle.

Although during brazing there is no melting of the base metal, the heating can affect the properties of the materials being joined. In the process of brazing, the joint temperature is raised to a point where an intermediate filler metal becomes molten and fills the joint between base materials and forms a metallurgical bond with the base materials being joined. Contained within this simple description of the process are a significant number of basic metallurgical and chemical processes taking place within and on the surfaces of the materials involved. An appreciation of the complexity of the process is a necessity in designing and producing a braze joint with closely controlled physical and chemical properties.

That the process is feasible and can be successfully carried out in a cost-effective manner has been clearly demonstrated by its prominent use in manufacturing products of all types.

In brazing, one is dealing at the most fundamental level with the interfaces between the phases: liquid-solid, vapor-solid, vapor-liquid. It is the processes occurring at these interfaces and within the contacting phases that constitute the process and determine the properties of the resulting joint.

The phenomena of wetting and flow of a liquid on the surface of a solid are basic to most models developed to describe the formation of a braze joint. Wetting of the base materials by the braze filler metal is required to provide the bonding needed and is characterized through the thermodynamic concept of capillarity and the free energy of formation of product phases formed in the reactions occurring during brazing. Flow is required to fill the joint with molten braze filler metal and is characterized kinetically in terms of the resistance against which the capillary driving force must work. Both wetting and flow are strongly influenced by interactions between solids, vapors, and liquids occurring at the interfaces and within the bulk phases. It is the extent of these interactions that determine the properties of the resulting joint.

A discussion and description of the theories developed for wetting and flow as they apply to brazing[4,5] is found in Chapter 2 and the technology required to braze successfully is described in the remaining chapters of this book.

1.2 CORRECTING A POPULAR MISCONCEPTION

The design engineering department of a large furnace brazing firm wished to produce large fillets on several furnace-brazed assemblies. In the past it had been a 'rule of thumb' that large fillets were undesirable while smaller fillets were preferred. Why not produce larger fillets? The design people wanted larger fillets in the hope they could reduce the stress concentration at the brazed joint.

Unfortunately, this is a bad practice since there are extremely few cases where the size of the fillet will have any marked effect on the properties of the joint, particularly fatigue.

1.2.1 The case for small fillets

There are major reasons for not wanting large fillets in a brazed joint. To dispel at the outset any doubts about this fact, some of the reasons are listed below.

1. The larger the fillet, the more porosity encountered. This occurs in the 99.90% pure copper filler metal used in copper brazing. At the brazing temperature, copper picks up iron and thus we no longer have a single melting point. As the copper starts to cool in the joint, some of the material is sucked out of the fillet, leaving porosity at the surface around the crystals (dendrites) of the higher melting material that is already frozen.
2. The porosity in fillets does not provide suitable strength in the fillet, particularly when the parts are subject to fatigue stresses. If high-stressed joints are to be encountered, it is desirable to provide a base metal fillet, which will have a much higher strength, both in tension and in fatigue, as well as a better distribution of stresses through the area.
3. Large fillets cause difficulty in interpreting liquid penetrant inspection. In fact, liquid penetrant inspection should not be used to inspect brazed joint fillets. If there is a crack, it should be optically visible. If there is a large quantity of porosity down the large fillet, most inspectors will erroneously indicate this as a crack because there is an in-line indication.
4. One major problem with specifying fillet size is the position of the joint in the furnace. If the joint is horizontal, such as a piece of flat material with a vertical section brazed to it, you can add a lot of filler metal and get a large fillet; but when you turn that assembly so that the joint is vertical, gravity will pull a free-flowing filler metal out of the fillet area and leave only a very small fillet. Of course, down at the bottom, if the filler metal is not wicked away, there will be an excess of material. This does not give you a uniformly large fillet all the way down, however, only at the bottom. The fillet the rest of the way up is quite small. While it is possible to use a brazing filler metal that has a large liquidus and solidus range and braze within that range to increase the size of the fillet, however, in trying to do this, the quantity of liquid filler metal is limited, and therefore, the joint may or may not be properly filled.
5. Due to the normal blueprint variation in dimensions, the clearance of a joint can vary, thus the volume of a joint varies by several hundred per cent. Thus, if the same quantity of filler metal is used, the fillet sizes will vary as the clearance (i.e. 0.03 to 0.10 mm) or volume of the joint varies, thus presenting more difficulties. If paste materials are used, then it is the operator's judgement as to how much paste to apply to a

given joint to provide a 'fill' of the joint plus the excess material to provide a fillet. Thus, this is a judgement call and is another variable that has to be added to the clearance, etc. (Most of the people who are applying paste filler metal get very adept at applying the proper amount for the joints they are working on.)

6. In copper and silver braze filler metals, the fillet can be harder or softer than the base metal. In copper brazing of steel, the filler metals are considerably softer and do not have the physical properties in the fillet that is obtained with the joint or in the base metal proper. Thus, small fillets give better uniformity and as previously mentioned, if strength is needed, stresses should be moved to the base metal by redesign.

7. The nickel braze filler metals obtain their strength and ductility through diffusion brazing. The essential variables are time, temperature and quantity of filler metal. The rate of diffusion is directly proportional to time and temperature, and inversely proportional to the quantity of filler metal. Thus, large fillets normally do not get adequately diffused so that the hard, not too ductile alloy is in the large fillet while small fillets usually are adequately diffusion brazed, hence an increase in the ductility and remelt temperature. With little or no diffusion brazing, a nickel base filler metal joint will remelt at a low temperature and will be low in ductility. On the other hand, when adequate diffusion brazing takes place, the fillet area becomes soft and ductile and the remelt temperature increases dramatically and can be in excess of 1371°C (melting temperature of nickel braze filler metals).

Therefore, from the above, small fillets are highly desirable.

REFERENCES

1. Brooker and Beatson, *Industrial Heating*, Iliffe and Sons Ltd., London, 1953.
2. Fusion Inc., *Bulletin AD-503*, 1980.
3. Schwartz M. M., *Brazing*, ASM International, 1989.
4. Adamson A. W., *Physical Chemistry of Surfaces*, John Wiley & Sons, New York, 1976.
5. Milner D. R., 'Principles Related to Wetting and Spreading', *British Welding Journal*, Vol. 5, p. 90, 1958.

2

Parameters and elements of brazing

The intent of this chapter is to introduce the reader to a few basic ideas which shall be helpful in the understanding of the general characteristics of solids and liquids that influence the braze process.

2.1 WETTING:

Capillarity, a thermodynamic concept, deals with ideal geometric configurations under equilibrium conditions. In practical terms, this means that the concept has value only in situations where the interfaces of the phases present are sufficiently mobile to assume equilibrium shapes. Liquids generally fulfill this requirement, and most theoretical models of the braze process are based on the assumptions of capillarity.

The familiar problem of determining the configuration of a liquid droplet in contact with a solid substrate illustrates the central importance of the thermodynamic concept of surface free energy of condensed matter.[1] In simplest terms, the surface free energy of a solid or liquid is the excess free energy introduced by the presence of a surface or in terms of a simple bonding concept for condensed phases the 'unsatisfied' bonding forces at a surface. In general, the surface free energy of an equilibrated configuration of a liquid droplet on a solid surface will be a minimum, and the shape will be characterized by the parameters shown in Figure 2.1. In the absence of an appreciable gravitational influence, the shape of the liquid droplet is uniquely characterized by θ, the contact angle.

The relationship between the surface free energies for the various interfaces, liquid–vapor (LV), solid–vapor (SV), and solid–liquid (SL) and θ is given by the familiar equation:

$$\frac{\delta F}{\delta A} p, T, N = \gamma = \text{free surface energy, usually dynes cm}^{-2} \qquad [2.1]$$

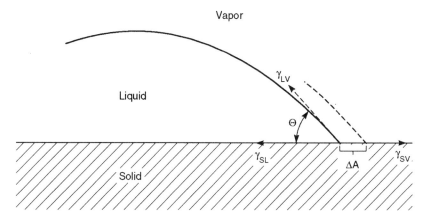

Fig. 2.1 Sessile drop showing the vectors of the surface energies for the system.

where F is the free energy of the substance and A is its surface (or interfacial) area.

The thermodynamic condition for spreading to occur is that, for an incremental increase in area, the entire system will have:

$$dF < 0 \qquad [2.2]$$

Conversely, for nonspreading:

$$dF > 0 \qquad [2.3]$$

If it is assumed that in the course of spreading of a liquid, L, in equilibrium with its own vapor, V, on a surface, S, in equilibrium with its own vapor, the following area relation exists:

$$dA_{LV} = dA_{SL} = -dA_{SV} \qquad [2.4]$$

then

$$\frac{\delta F}{\delta A} p, T = \gamma \cdot LV + \gamma \cdot SL - \gamma \cdot SV \qquad [2.5]$$

where $\gamma.LV$ is the liquid–vapor surface energy, $\gamma.SL$ the solid–liquid surface energy, and $\gamma.SV$ is the solid–vapor surface energy.

Let $-dF/dA$ be designated as the final spreading coefficient, S_{LS}; then, at constant temperature and pressure:

$$S_{LS} = \gamma \cdot SV - (\gamma \cdot LV + \gamma \cdot SL) \qquad [2.6]$$

For a measure of the attraction between different materials, the work of adhesion can be obtained as:

$$W_{ad} = (\gamma \cdot LV + \gamma \cdot SV) - \gamma \cdot LS \qquad [2.7]$$

The work of adhesion represents the decrease in energy in bridging together a unit area of liquid surface and a unit area of solid surface to form a unit of interface.[2]

Recent work in Japan[3] and the US[4] has shown that the effects of the reaction on the wetting of silicon nitride (Si_3N_4) and titanium-added active filler metals vary with reaction layer thickness and morphology as well as with the composition of the reaction layer. Several results of thermodynamic studies on the formation of the compound by the interfacial reaction are reported.[3,5] And moreover, to get a quantitative understanding of the formation of the reaction layer, kinetic studies were also performed.[3]

Although the chemical reaction at the interface is expected to cause the bonds to be strong, there are many points of disagreement on the effect of **wetting** and reaction on the bond strength. In order to evaluate the bond strength from the experimental results of wetting and reaction, work of adhesion (WAD),[6,7] has been used.

Experimentally, it is observed that liquids placed on solid surfaces usually do not completely wet but rather remain as a drop having a definite contact angle between the liquid and solid phases.[8] This condition is illustrated in Figure 2.1. The Young and Dupré equation (equation 2.5) permits the determination of change in surface free energy, ΔF^8, accompanying a small change in solid surface covered, ΔA. Thus:

$$\Delta F^8 = \Delta A(\gamma \cdot SL - \gamma \cdot SV) + \Delta A\gamma \cdot LV \cos(\sigma - \Delta\sigma) \qquad [2.8]$$

At equilibrium,

$$\lim_{A \to 0} \frac{\Delta F^8}{\Delta A} = 0$$

and

$$\gamma \cdot SL - \gamma \cdot SV + \gamma \cdot LV \cos\sigma = 0 \qquad [2.9]$$

or

$$\gamma \cdot SL = \gamma \cdot SV - \gamma \cdot LV \cos\sigma \qquad [2.10]$$

In [2.9] and [2.10] it can be seen that σ is greater than 90° when $\gamma.SL$ is larger than $\gamma.SV$, as shown on the left in Figure 2.2, and the liquid drop tends to spheroidize.

The contact angle, σ, is less than 90° when the reverse is true, as shown on the right in Figure 2.2, and the liquid drop flattens out and wets the solid. If the balance is such that σ is zero (0) and greater wetting is

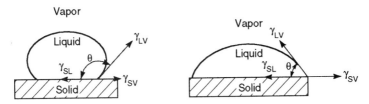

Fig. 2.2 Sessile drops and interfacial energies. The contact angle is greater than 90° on left; no wetting occurs. The contact angle is less than 90° on right; wetting occurs.

desired, σ should be as small as possible so that $\cos\sigma$ approaches unity and the liquid spreads over the solid surface.

These considerations show the importance of surface energies in brazing. If brazing filler metal is to form a joint, it must wet the solid material. The surface/energy balance must be such that the contact angle is less than 90° ($\cos\sigma > 0$), $\gamma.SV$ must be greater than $\gamma.SL$.

In a series of contact angle measurements of several liquid metals on beryllium at various test temperatures in argon and vacuum, although during the testing period minor fluctuations in angle were noted, the overall tendency was for the contact angles to decrease with increasing time at test temperature. The fluctuations may be attributed to alloying effects because, as alloying progresses, composition and temperature variations change the interfacial energies.

These data provide a means of comparison of the wettability of beryllium by the various liquid metals. If one assumes wettability for contact angles less than 90° and nonwetting for angles greater than 90°, then these data permit qualitative separation of the systems into two classes. Gold; Pd–2.1% wt Be; and Ti–6% wt Be, were found to wet beryllium at both temperatures. But even the latter two brazing filler metals, as well as silver and gold, did not wet when tested in argon atmospheres at 1010°C. It was found, however, that an increase of 50°C resulted in wetting. Aluminum exhibited a nonwetting behavior at both temperatures and in both atmospheres. Copper was nonwetting at both temperatures in vacuum, but underwent extensive alloying with the solid beryllium at these temperatures in argon. From these data it appears that the binary alloys have the greatest tendency to wet beryllium.[9]

Consider what effects the changes in the surface or interfacial energies have on the contact angle. If the initial contact angle is less than 90°C, the following changes will decrease the contact angle:

(a) increase $\gamma.SV$
(b) decrease $\gamma.SL$
(c) decrease $\gamma.LV$.

As mentioned previously, the presence of adsorbed molecules on a metal surface markedly reduces the surface energy $\gamma.SV$. This increases the contact angle and retards wetting. The importance of clean surfaces and their maintenance by brazing in vacuum is apparent. The presence of small amounts of impurities can affect the liquid/vapor surface energy.

It can be concluded that wetting is the ability of the molten brazing filler metal to adhere to the surface of a metal in the solid state and, when cooled below its solidus temperature, to make a strong bond with that metal. Wetting is a function not only of the brazing filler metal but also of the nature of the metal or metals to be joined. There is considerable evidence that in order to wet well a molten metal must be capable of dissolving, or alloying with, some of the metal on which it flows.

Wetting is only one important facet of the brazing process. A very important factor affecting wetting is the cleanliness of the surface to be wetted. Oxide layers inhibit wetting and spreading, as do grease, dirt, and other contaminants that prevent good contact between the brazing filler metal and the base metal. One of the functions of a **flux** is to remove the oxide layer on the joint area and expose clean base metal.

Good wetting and spreading of the liquid filler metal on the base metal are necessary in brazing because the mechanics of the process demand that the filler metal be brought smoothly, rapidly, and continuously to the joint opening. If the conditions within the capillary space of the joint do not promote good wetting, the filler metal will not be drawn into the space by capillary attraction.

Kim, *et al*[2] investigated the wettability and reactivity of pressureless sintered Si_3N_4 by powered Cu–Ti filler metal using sessile drop tests conducted in a vacuum where Si_3N_4 was bonded to itself and joint strength was evaluated by compressive shear testing. The wettability of Cu–Ti filler metals on Si_3N_4 was improved greatly by addition of titanium up to 50% wt. However, the reaction layer thickness was increased up to 10% wt and thereafter decreased up to 50% wt.

Wettability improved with the increment of titanium concentration and no linear proportionality was found between the reaction and the wettability. As titanium concentration increases, a continuous thin layer tends to be formed at the interface and improves the wettability remarkably.

The thickness of the reaction layer was not in proportion to titanium concentration. For a Cu–Ti filler metal with low titanium concentration (less than 5% wt), the initial reaction layer tends to be formed discontinuously and the reaction layer thickness increases with titanium concentration. On the contrary, in the filler metal with high titanium concentration (15% wt) the reaction layer thickness decreased as titanium concentration increased, because a continuous thin reaction layer tends to form.[10]

The shear strength of the Si_3N_4–Si_3N_4 joint using Cu–Ti filler metal is influenced by reaction layer thickness and morphology. Strength increased with the increment of titanium concentration, and it is also accompanied by improved wettability because the tendency of the continuous thin reaction layer to form increased, Figure 2.3.

With the development of new steels (21–6–9) Keller *et. al*[11] undertook an extensive study of the wettability of commercial braze filler metals on Type 304, the most common of the austenitic stainless steels, Type 316 which is often substituted for 304 when increased corrosion resistance is desired and 21–6–9, a high-manganese stainless steel, which is an attractive alternative to 304 or 316 for certain applications. Since it has a very stable austenitic structure, it is attractive for cryogenic applications.

Previous wettability studies have been reported [12,13] and because high wettability means that the thermocapillary attraction that fills the braze joint is strong, wettability is an important component of braze performance. The **wettability index** (WI) developed by Feduska[13] is used as the measure of wettability. The WI is defined as the area covered by

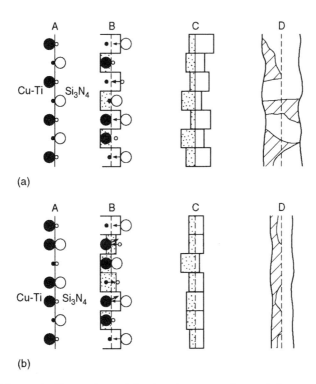

Fig. 2.3 Schematic illustration of the forming procedure of the reaction layer between Cu–Ti alloy and Si_3N_4 with (a) lower and (b) higher titanium. A. physical contact; B. reaction layer forming; reaction layer growth; D. structure after cooling. • Ti; ● Cu; ○ Si; ○ N; •→○ TiN.

the braze metal filler times the cosine of the contact angle between the braze and the base metal. Therefore, the higher the WI, the better the braze filler metal wets the base metal. It should be emphasized that the WI, as defined, is a relative measure and depends on the volume of filler metal used, Table 2.1.

The wettability indexes (WI) for each braze filler metal at each temperature on 316L stainless steel, 304L stainless steel and 21–6–9 stainless steel are provided in Table 2.2. WI greater than 0.05 are indicative of good performance during brazing, and WI greater than 0.10 are indicative of excellent performance during brazing.[13]

Comparison of WI of the braze filler metals on the three stainless steels revealed some trends:

1. Filler metals generally wet 316 stainless steel better than 304 stainless steel and generally wet 316 stainless better than 21–6–9.
2. The degree of wetting of most braze filler metals on 21–6–9 was equal to or better than on 304 stainless steel.[12]
3. The improved wettability of braze filler metals on 316 stainless steel is believed to be due to the presence of molybdenum in the surface oxide while the enhanced wetting on 21–6–9 versus 304 stainless steel is believed to be the higher manganese content of the surface oxide.

2.2 CAPILLARY FLOW

Capillary attraction is the physical force that governs the action of a liquid against solid surfaces in small, confined areas. As an example, dip one end of a tube with a centerhole no larger than a hair into a liquid. The liquid will rise in the tube through capillary attraction. (Appropriately, *capillus* is Latin for 'hair'.)

Capillary flow is the dominant physical principle that ensures good brazements provided that both faying surfaces to be joined are wetted by the molten brazing filler metal. The joint must also be properly spaced to permit efficient capillary action and resulting coalescence. More specifically, capillarity is a result of surface tension between base metal(s), brazing filler metal, flux or atmosphere, and the contact angle between base metal and filler metal. In actual practice, brazing filler metal flow characteristics are also influenced by dynamic considerations involving fluidity, viscosity, vapor pressure, gravity and metallurgical reactions between brazing filler metal and base metal.

2.2.1 Factors affecting capillary flow

As a matter of fact, present-day brazing practices have evolved as the result of an empirical approach to the phenomena of wetting and

Table 2.1 Filler metals used in various braze fillers

No.	Common name	Composition	Liq(°C)	Sol(°C)
1	Silver	99.99Ag	961	961
2	Cusil	72Ag–28Cu	780	780
3	Palcusil 5	68Ag–27Cu–5Pd	810	807
4	Palcusil 10	58Ag–32Cu–10Pd	852	824
5	Palcusil 15	65Ag–20Cu–15Pd	900	850
6	Palcusil 25	54Ag–21Cu–25Pd	950	900
7	Gapsil 9	82Ag–9Ga–9Pd	880	845
8	Nicusil 3	71.5Ag–28.1Cu–0.75Ni	795	780
9	Niculsil 8	56Ag–42Cu–2Ni	893	771
10	T50	62.5Ag–32.5Cu–5Ni	866	780
11	T51	75Ag–24.5Cu–0.5Ni	802	780
12	T52	77Ag–21Cu–2Ni	830	780
13	Cusiltin 5	68Ag–27Cu–5Sn	760	743
14	Cusiltin 10	60Ag–30Cu–10Sn	718	602
15	Braze 630	63Ag–28Cu–6Sn–3Ni	800	690
16	Braze 580	57Ag–33Cu–7Sn–3Mn	730	605
17	Braze 655	65Ag–28Cu–5Mn–2Ni	850	750
18	Silcoro 60	60Au–20Ag–20Cu	845	835
19	Nioro	82Au–18Ni	950	950
20	Palnioro 7	70Au–22Ni–8Pd	1037	1005
21	Incuro 60	60Au–37Cu–3In	900	860
22	Silcoro 75	75Au–20Cu–5Ag	895	885
23	Nicoro 80	81.5Au–16.5Cu–2Ni	925	910
24	Palcusil 20	52Au–28Cu–20Pd	925	875
25	Gold	99.99Au	1064	1064
26	Palniro 4	30Au–36Ni–34Pd	1169	1135
27	Palniro 1	50Au–25Ni–25Pd	1121	1102
28	Ticusil	68.8Ag–26.7Cu–4.5Ti	850	830
29	Palnicusil	48Ag–18.9Cu–10Ni–22.5Pd	1179	910
30	Palco	65Pd–35Co	1235	1230
31	Incusil 15	62Ag–24Cu–15In	705	630
32	Incusil 10	63Ag–27Cu–10In	730	685
33	BAg-8a	71.8Ag–28Cu–0.2Li	760	760
34	BAg-19	92.5Ag–7.3Cu–0.2Li	890	760
35	Braze 071	85Cu–7Ag–8Sn	986	665
36	Braze 852	85Ag–15Mn	970	960
37	Nioroni	73.8Au–26.2Ni	1010	980
38	Nicuman 23	67.5Cu–23.5Mn–9Ni	955	925
39	Palsil 10	90Ag–10Pd	1065	1002
40	Palni	60Pd–40Ni	1238	1238

Table 2.2 Wettability index for various fillers and temperatures on three grades of stainless steel

Alloy name	Temp (°C)	WI 316	WI 21-6-9	WI 304
Silver	975			0.000
	1000	0.008	0.013	0.015
	1050	0.041	0.007	0.023
	1100	0.053	0.016	0.020
	1150			0.000
Cusil	800			0.000
	850			0.000
	900	0.022		0.003
	925	0.032		
	950	0.037		
Palcusil 5	800	0.027		0.000
	850	0.047	0.014	0.011
	900	0.080	0.020	0.035
	950	0.035	0.061	0.015
Palcusil 10	850	0.057	0.020	0.025
	900	0.107	0.050	0.062
	950	0.107	0.101	0.068
Palcusil 15	900	0.152	0.092	0.119
	950	0.754	0.170	0.212
	1000	0.104	0.263	0.096
Palcusil 25	950	0.225	0.107	0.226
	975			
	1000			
Gapasil 9	900	0.023	0.283	
	950	0.068		
	1000	0.269		
Cusiltin 10	750		0.025	0.000
	800		0.017	0.023
	825	0.021		
	850	0.034	0.051	0.050
	875	0.043		
Braze 630	800	0.014	0.037	0.023
	850	0.046	0.065	0.024
	900	0.064		
Braze 580	750		0.060	0.060
	800	0.020	0.056	0.089
	850	0.051	0.120	0.102
	875	0.073		
Braze 655	825	0.078		
	850	0.080	0.037	0.074
	875	0.110		
	900	0.116	0.137	0.124
Silcoro 60	850	0.039	0.005	0.004
	900	0.051	0.007	0.011
	925	0.073	0.025	0.016
	950			
	1000			
Nioro	950	0.049	0.055	0.037
	975	0.061	0.000	0.000
	1000	0.000		
Palniro 7	1000	0.053	0.060	0.065
	1025			
	1050			
Palnicusil	950	0.025		0.038
	975	0.213		
	1000			
	1025	0.362		
	1050			0.000
	1075	0.336		
	1100			
Palco	1250	0.000		
	1275	0.073		
	1300	0.159		
Incusil 15	750	0.000	0.000	0.000
	800	0.000	0.000	0.000
	850	0.000	0.000	0.000
Incusil 10	750	0.000	0.000	0.000
	800	0.000	0.000	0.000
	850	0.000	0.000	0.000
BAg-8a	800	0.008	0.000	0.000
	850	0.005	0.000	0.000
	900	0.008		
	950	0.032		
BAg-19	950	0.045		
	1000	0.016		
	1050	0.016		
Braze 071	1000	0.066	0.106	0.101
	1050	0.000	0.125	0.101
Braze 852	1000	0.039	0.037	0.038

Table 2.2 (Continued)

Alloy name	Temp. (°C)	WI 316	WI 21–6–9	WI 304	Alloy name	Temp. (°C)	WI 316	WI 21–6–9	WI 304	Alloy name	Temp. (°C)	WI 316	WI 21–6–9	WI 304
Nicusil 3	800	0.006	0.000	0.012		1075	0.073				1050	0.029	0.038	0.042
	850	0.027	0.017	0.008	Incuro 60	900	0.010				1100	0.043	0.037	0.020
	900	0.041	0.039	0.021		950	0.025			Nioroni	1000	0.068		
	925	0.051				1000	0.091				1025	0.070		
	950	0.064			Silcoro 75	900	0.006				1100	0.116		
Nicusil 8	850	0.024		0.000		950	0.057			Nicuman 23	950	0.026		
	900	0.052		0.033		1000	0.170				975	0.099		
	950	0.085		0.064	Nicoro 80	950	0.041	0.019	0.016		1000	0.091		
	1000			0.299		1000	0.163	0.126	0.070		1025	0.091		
T–50	850	0.024		0.000		1050	0.413	0.190	0.084	Palsil 10	1050	0.113		
	900	0.038		0.026	Palcusil 20	875	0.061				1075	0.113		
	950	0.062		0.057		900	0.110				1100	0.126		
	1000			0.000		925	0.122			Palni	1225	0.000		
T–51	800	0.007		0.000	Gold	1070	0.087		0.088		1250	0.068		
	850	0.029		0.000		1075	0.358		0.238		1275	0.078		
	900	0.039		0.000		1100			0.355		1300	0.107		
	950			0.029	Palniro 4	1175	0.036			Palmansil 5	950			0.057
T–52	850	0.027		0.082		1200	0.061				1000			0.000
	900	0.045		0.001		1225	0.078				1050			0.000
	950	0.063		0.014	Palniro 1	1125	0.041				1100			0.000
	1000			0.045		1150	0.063							
Cusiltin 5	850	0.015		0.090		1175	0.075							
	875	0.043		0.000	Ticusil	875	0.032							
	900	0.047		0.013		900	0.057							
	950			0.033		950	0.083							
	1000			0.000										

spreading, which are of prime importance in the formation of brazed joints. Classical, physical, and chemical principles lead to equations governing the shape of liquid surfaces and the rate of filling of a capillary gap in noninteracting systems. However, the extension of theory to practical systems necessitates the consideration of a number of complicating factors which often arise in everyday practice.

A few of these factors include the condition of solid surface as to the presence of oxide films and their effects on wetting and spreading, surface roughness, alloying between the brazing filler metal and base metal and the extent to which this affects the thermodynamic properties of the liquid and solid surfaces, and the condition and properties of the brazing atmosphere. The factors that control the rate at which wetting, spreading and capillary flow occur are of great practical, as well as theoretical, interest. Studies have indicated profound influences of various kinds of surface activation which cannot be explained in terms of surface energies or alterations in equilibrium contact angle.[14,15] Some of the most spectacular of these effects have been observed in systems in which finite contact angle is thermodynamically unstable, because the solid/vapor surface energy exceeds the sum of the liquid/solid surface energies, that is, a system in which thermodynamics would predict complete spreading. In actual fact, spreading may or may not occur in this type of system and the rate of spreading can be markedly dependent on surface chemistry, although the fundamental mechanisms of this dependence are not all clear. Equations [2.1] through [2.10] refer.

In fact, capillary attraction is the result of the same forces that make for good wetting; the adhesive attraction of closely spaced parallel solid surfaces for a liquid are greater than the cohesive forces of the liquid. The imbalance of these forces causes the liquid (including molten metal) to flow between the closely spaced surfaces, even against the force of gravity. Clearly then, the joint must provide close parallel surfaces free of contamination.

The distance that a liquid will flow into a capillary space increases as the separation, or clearance, of the surface is reduced; the rate of flow of the liquid decreases as the clearance is reduced.

The wetting characteristics of the brazing filler metal have no effect on the reach or the rate of capillary action but, rather, determine whether capillary action will be generated at all. In effect, the wetting characteristics set up a go/no-go situation.

It all boils down to the fact that, for successful joining of components by brazing, the brazing filler metal selected must have a melting point above 450°C and must also wet the base metal without melting it. Then the joint must be designed so that the mating surfaces of the components are parallel and close enough together to cause capillary attraction.

2.3 ELEMENTS OF THE BRAZING PROCESS

In designing the braze joint, the engineer is concerned with reliability and cost. Joint strength, fatigue resistance, corrosion susceptibility, and high temperature stability are some of the other concerns which determine the selection of joint design, braze filler materials and processing parameters. Many parameters and elements influence and comprise the braze process and the properties of the joint produced.

The examination of the following elements of the brazing process are required:

- Temperature and time
- Surface preparation
- Joint design and clearance
- Rate and source of heating
- Filler-metal flow
- Base-metal characteristics
- Filler-metal characteristics.

2.3.1 Temperature and time

The temperature of the brazing filler metal naturally has an important effect on the wetting action. The selection of an optimum braze temperature requires an understanding of the influence of temperature on both the wetting and flow of the filler metal, because the wetting and alloying action improves as the temperature increases. Of course, the temperature must be above the melting point of the brazing filler metal and below the melting point of the parent metal. Within this range a temperature is generally selected that will give the best satisfaction from an overall standpoint.

Usually the lowest brazing temperatures are preferred to:

(a) economize on heat energy required;
(b) minimize heat effect on the base metal (annealing, grain growth, or warpage, for example);
(c) minimize base metal/filler metal interactions;
(d) increase the life of fixtures, jigs, or other tools.

Higher brazing temperatures may be desirable so as to:

(a) use a higher-melting but more economical brazing filler metal;
(b) combine annealing, stress relief, or heat treatment of the base metal with brazing;
(c) permit subsequent processing at elevated temperatures;
(d) promote base-metal interactions in order to modify the brazing filler metal (this technique is usually used to increase the remelt temperature of the joint);

(e) effectively remove surface contaminants and oxides with vacuum brazing;

(f) avoid stress cracking.[16]

Braze filler metals can be fairly complex alloys and their melting can take place over a range of temperatures. The implications of such behavior can be illustrated in terms of the phase diagram for the silver–copper system shown in Figure 2.4. Except for the eutectic composition (72% Ag–28% Cu), melting of the braze filler metals in this binary system occur over a range of temperatures as illustrated for the 50–50 composition.

Melting which begins at the solidus temperature of 780°C is not complete until a temperature in excess of 850°C is reached. Within that melting range there is a 'mushy' mixture of liquid and solid which has wetting and flow properties distinct from those of the filler metal above the liquidus temperature. Flow in the partially melted filler is much reduced and the wetting and spreading tendency of the liquid phase in the mixture leads to the phenomenon of liquidation or the tendency of the lower melting constituents of a filler metal to separate from the higher melting constituents. Liquidation results in melting point shifts and flow problems with the remaining higher melting constituents. The rate of heating through the melting temperature range and the brazing time are

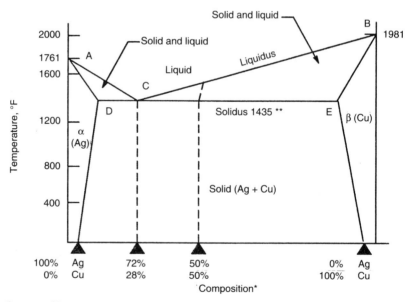

* Eutectic composition
** 780°C

°F = 9/5° C + 32

Fig. 2.4 Silver–Copper composition diagram.

important parameters which control parameters in avoiding these types of problems.

Alloying can occur between the liquid filler metal and the base materials during the braze process. This type of interaction occurs to some extent in all braze processing, and must be controlled in some cases where the consequences degrade the process and joint properties. In fact a number of metallurgical processes occur during the braze process that affect the solidification characteristics, grain size, and physical properties of the brazement. Both the filler metal and the base material compositions can be significantly altered depending upon the extent of liquid and solid solubility and the chemical affinity of the elements present for each other. Braze filler metal flow and wetting can be markedly influenced depending on the extent to which the melting points of filler metal and base material are affected, and new phases are formed. Significant increases and decreases in the volume of filler metal, filler dewetting, and joint embrittlement can occur as a result of this alloying process. Under certain circumstances, these alloying effects are not harmful and can be used to improve the properties of the brazement by increasing the solidus temperature and/or toughness of the filler metal.

The time at the brazing temperature also affects the wetting action, particularly with respect to the distance of creep of the brazing filler metal. If the filler metal has a tendency to creep, the distance generally increases with time. The alloying action between filler metal and parent metal is, of course, a function of both temperature and time. In general, for production work, both temperature and time are kept at a minimum consistent with good quality.

2.3.2 Surface preparation

Oxide layers and surface contaminants can impede the wetting and flow of liquid braze filler metals and must be removed for successful brazing.

When faying surfaces of parts to be brazed are prepared by blasting techniques, several factors should be understood and considered. The purpose of blasting parts to be brazed is to remove any oxide film and to roughen the mating surfaces so that capillary attraction of the brazing filler metal is increased. The blasting material must be clean and must be of a type that does not leave on the surfaces to be joined any deposit that restricts filler metal flow or impairs brazing. The particles of the blasting material should be fragmented rather than spherical so that the blasted parts are lightly roughened rather than peened. The operation should be done so that delicate parts are not distorted or otherwise harmed. Vapor blasting and similar wet blasting methods require care because of possible surface contamination.

Subsequent chapters on cleaning, fluxing and the various base metals deal with the removal of unwanted surface layers. For successful processing, it is important to understand the chemical nature of both the layer being removed and the surface of the base material that will come into contact with the filler metal during brazing. Base material surface finish can be influenced in pre-braze processing and should be considered when designing the brazement and selecting the braze process.[17]

A clean, oxide-free surface is imperative to ensure uniform quality and sound brazed joints. A sound joint may be obtained more readily if all grease, oil, dirt and oxides have been carefully removed from the base and filler metals before brazing, because only then can uniform capillary attraction be obtained. It is recommended that brazing be done as soon as possible after the material has been cleaned. The length of time the cleaning remains effective depends on the metals involved, atmospheric conditions, storage and handling practices, and other factors. Cleaning is commonly divided into two major categories: chemical and mechanical.

Chemical cleaning is the most effective means of removing all traces of oil or grease. Trichloroethylene and trisodium phosphate have been the usual cleaning agents employed. Various types of oxides and scale that cannot be eliminated by these cleaners are removed by other chemical means.

The selection of the chemical cleaning agent depends on the nature of the contaminant, the base metal, the surface condition, and the joint design.[18] Regardless of the cleaning agent or the method used, it is important that all residue or surface film be removed from the cleaned parts by adequate rinsing to prevent the formation of other equally undesirable films on the faying surfaces.

Objectionable surface conditions may be removed by **mechanical means** such as grinding, filing, wire brushing, or any form of machining, provided that joint clearances are not disturbed. In grinding of the surfaces of parts to be brazed, care also should be taken to ensure that the coolant is clean and free from impurities so that the finished surfaces do not have these impurities ground into them.

Mechanical cleaning may be adequate, in which case the design must permit this during manufacture. In some cases where chemical cleaning is required, it may be followed by protective electroplating, necessitating access to the faying surface by the liquids involved.

Another technique in surface protection is the use of solid and liquid **brazing fluxes**. At temperatures up to about 1000°C, fluxes often provide the easiest method of maintaining or producing surface cleanness, and in such cases the design must not only permit easy ingress of the flux but also allow the filler metal to wash it through the joint. Above 1000°C the flux residues can be difficult to remove, and surface cleaning by, for example, a furnace atmosphere is desirable, but the design must permit the gas to penetrate the joint.

Apart from cleanness and freedom from oxides, **surface roughness** is important in determining ease and evenness of flow of the brazing filler metal. This will vary with different manufacturing methods and may influence the engineer's choice, or it may require access to a surface treatment/roughening process. Generally a liquid which wets a smooth surface will wet a rough one even more. A rough surface will modify filler metal flow from laminar to turbulent, prolonging flow time and increasing the possibility of alloying and other interactions. Surfaces often are not truly planar, and in some instances surface roughening will improve the uniformity of the joint clearance.

Conversely, the designer and engineer may require that brazing filler metal should not flow onto some surfaces. **Stopoff materials** will often avoid this, but the design must permit easy application of the stopoff without danger of contaminating the surfaces to be joined. Self-fluxing filler metals, in a suitably protective environment such as vacuum, may provide the essential surface wetting.

In **vacuum furnace brazing**, the wetting action is a little different from that of water on glass, because we are dealing with metals at elevated temperatures and in most situations the attraction between the brazing filler metal and the parent metal is sufficiently great to cause alloying between the two. The condition of the surface of the parent metal, however, greatly influences the behavior of the brazing filler metal from the standpoint of wetting and the tendency to ball up. These tendencies depend on the relative surface tension and whether there is sufficient attraction by the surface of the parent metal to draw the brazing filler metal out in a thin film or whether the surface tension of the brazing filler metal is sufficient to draw it up into balls on the surface of the parent metal. (See Chapter 5 for discussion of fluxes, atmospheres, and vacuum.)

2.3.3 Joint design and clearance

A brazed joint is not a homogeneous body. Rather, it is a heterogeneous assembly that is composed of different materials with different physical and chemical properties. In the simplest case, it consists of the base metal parts to be joined and the added brazing filler metal. Diffusion processes, however, can change the composition and therefore the chemical and physical properties of the boundary zone formed at the interface between base metal and filler metal. Thus, in addition to the two different materials present in the simplest example given above, further dissimilar materials must be considered.

Why should small clearances be used? The smaller the clearance, the easier it will be for capillarity to distribute the brazing filler metal throughout the joint area and the less will be the likelihood that voids or shrinkage cavities will form as the brazing filler metal solidifies. Small clearances and correspondingly thin filler metal films make sound joints, and sound joints are strong joints. The soundest joints are those in which 100% of the joint area is wetted and filled by the brazing filler metal. They are at least as high in tensile strength as the filler metal itself, and often higher. If **brazing joint clearances** ranging from 0.03 to 0.08 mm are designed, they are designed for the best capillary action and greatest joint strength.

Brazing processes depend on the flow of braze filler metal between closely mated parts to achieve joint strength. In planning joint clearances, consider that brazing occurs at elevated temperatures and that joint clearance must be that which it obtains at the brazing temperature.

Thermal expansion of dissimilar metals is fundamentally important and complex. A 51 mm o.d. copper tube will expand 0.56 mm from room temperature to 649°C, while the carbon steel hole to receive it will expand only 0.41 mm. To calculate clearance for the brazing filler metal at its flow point, add the 0.15 mm difference to the desired clearance. On large dissimilar joints, a high expansion material inside one of low expansion may not flow enough filler metal between the parts to withstand stress of contraction. If possible, provide a flange surface and use a sluggish filler metal such as the silver–copper filler metal (BAg–3) so the joint will be in shear as the parts contract. Place a washer between the flange surfaces with sufficient filler metal to make a fillet. The excess filler metal will absorb some stress.

Localized heating can create a similar situation. The cold metal surrounding a hole may restrain the hole from expanding, while the tube inside is free to expand. The clearance decreases, and only a thin layer of filler metal can flow between joint surfaces. Upon contracting, the tube may rupture this thin layer. Heating slowly should permit the hole to expand. Additionally one could increase clearance or provide flange surfaces.

Low expansion inside a high-expansion material is the reverse situation. Starting with normal clearances would give too much clearance at the brazing temperature. Tight or interference fits give correct fit at brazing temperature. As the joint cools, the filler metal comes under compression, which is good for joint strength. The compression may slightly distort the weaker member of the assembly so that it will show some change from its pre-braze dimension.

Before the detailed design for a part to be brazed is made, the first decision is how and where the components are to be joined. Since brazing relies on capillary attraction, the design of a joint must provide an unobstructed and unbroken capillary path to enable flux, if used, to escape from it as well as allow the brazing filler metal to get into the joint. Where filler metal is added to a joint by hand, such as by feeding in a rod or wire, the joint entry must be visible and accessible. If preplaced rings or shims are used, the joint must be designed so that the **preform** can be placed in position easily and will remain in place until molten. (See Chapter 4).

Other factors influencing joint design

Some of the more important factors influencing joint design are:

- the required strength and corrosion resistance;
- the necessary electrical and thermal conductivity;
- the materials to be joined;
- the mode of application of brazing filler metal;
- post-joining inspection needs.

The actual design itself should follow from a consideration of all the above factors.

Consideration also should be given to the ductility of the base metal, the stress conditions of the joint, and the relative movements of the two surfaces during joining, which may introduce problems and inaccuracies requiring careful consideration by the designer. The lack of concentricity of circular components may cause gaps to vary from excessive to overtight. This can be avoided by designs and techniques such as knurling, splining, use of shims, and machining. (For more information, see Chapter 4.)

Viscosity, surface tension, and specific gravity of the brazing filler metal are not the only factors which determine the gap-filling capability of a given filler metal. Many other considerations are involved, such as the tendency of the filler metal and parent materials to alloy with one another. Joint strength increases as joint gap decreases, down to a minimum. For stressed applications, optimum joint clearance may have to be designed inside the gap-filling range of the brazing filler metal. Other factors influencing optimum joint gap with a specific brazing filler metal are joint length, brazing temperature, and base metal reactions.

Trials are necessary to take account of the sizes of components, temperature changes, any effect of the flux or filler metal on the parent metal and the method of holding the parts together. In general, the recommended gaps are given in Table 2.3 for joints in copper-based materials, ferrous (and similar) metals and aluminum alloys, but trials are necessary to determine gaps when dissimilar metals are involved.

Typical joint forms are shown in Figure 2.5. Joint design should take account of preplacing of the brazing filler metal if required and this entails care in setting the dimensions of preformed filler metal washers, rings or special shapes.[19]

The strength of the joint is determined largely by the tensile strength of the filler metal and size of the overlap, assuming that the brazing metal is

Parameters, elements of brazing

Table 2.3 Recommended brazing gaps for different parent metals (mm)

Type of brazing filler metal	Copper and copper alloys	Iron and steels, nickel alloys	Aluminium
Short melting range	0.05–0.15	0.04–0.15	0.15–0.25
Long melting range	0.05–0.25	0.04–0.15	0.15–0.25
Fluxless braze filler metal	0.04–0.20	Not suitable	Not suitable

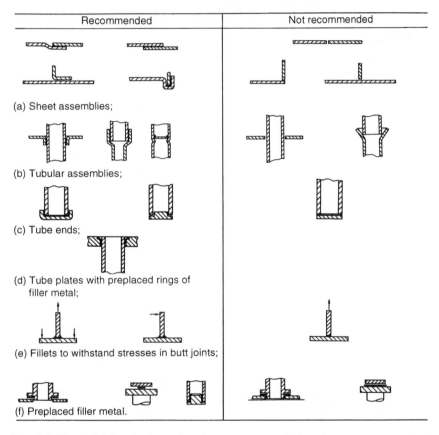

Recommended | Not recommended

(a) Sheet assemblies;

(b) Tubular assemblies;

(c) Tube ends;

(d) Tube plates with preplaced rings of filler metal;

(e) Fillets to withstand stresses in butt joints;

(f) Preplaced filler metal.

Fig. 2.5 Typical joint designs showing recommended and not recommended forms.[19]

free from defects, particularly voids of any kind. Maximum joint strength is developed, in general, when the overlap is four times the thickness of the thinnest component of the joint, but this varies with the parent metal; for example, on copper it can be as low as 2–2.5 times. Larger overlaps waste material and may result in incomplete filling of the gap with resultant voids or crevices, but they can help to overcome misalignment. In general, overlaps should not exceed 20 mm but when strength considerations require a greater length it may be possible to ensure that it can be filled by using several preforms or coiled rings of filler metal wire. Capillary action permits filler metal to flow into joints in any position, even against gravity, but for manual operations the joint must be readily accessible. When both sides are visible the molten filler metal can be seen to flow through the gap. Preplacement of the correct amount of filler metal is desirable, based upon the calculated volume required and checked by an initial trial.

A **brazeability test** that has been developed by engineers at a Japanese firm[20] for aluminum brazing is shown in Figure 2.6. Used in heat exchangers for automobiles and similar applications, aluminum alloys have been brazed by a growing variety of processes: dip, furnace, dry-air, vacuum, carrier gas, inert gas atmosphere, and the Nocolok process (an Alcan-patented process that uses a small amount of flux). Nocolok with the noncorrosive flux brazing process has established itself as an accepted

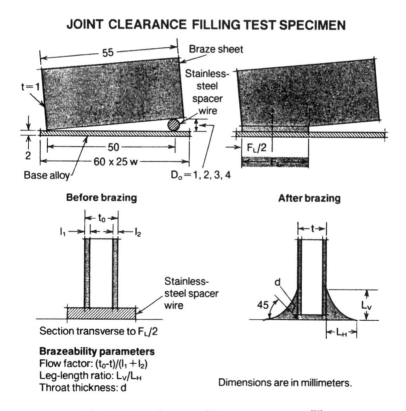

Brazeability parameters
Flow factor: $(t_0-t)/(l_1+l_2)$
Leg-length ratio: L_V/L_H
Throat thickness: d

Dimensions are in millimeters.

Fig. 2.6 Joint clearance filling test specimen.[20]

process for brazing aluminum heat exchangers of all types. Although brazing with Nocolok can be successfully achieved using flame and induction heating the process is at its most effective using a continuous furnace. This has led to the need for a brazeability test to evaluate the ability of a given process to produce a good brazed joint.

Engineers ran joint clearance/filling braze tests on alloy 3003 clad with filler metal braze sheet using the full range of braze processes. The test proved to be an economical and useful way to evaluate brazeability.

Its advantages:

- The specimen is of simple shape, easy to set up;
- Length of the filled joint clearance is immediately measurable, without sectioning the specimen;
- Filler-metal flowability can be measured by visual inspection brazing;
- Flow factor, throat thickness, and fillet/leg-length ratio can be obtained by inspection of the cross-section of a specimen cut vertically in the transverse direction of the fillet.

As a fallout from these tests it was found that flux brazing produced superior results to fluxless brazing relative to reproducibility, joint filling, and flowability.

The volume of brazing filler metal is calculated, as a first approximation, by multiplying the width of the gap by the total length of the gap at the brazing temperature, to which is then added 25% to allow for fillets and run-through. In general, this is sufficiently accurate for assemblies made up of similar materials or of metals of approximately the same coefficients of expansion, but when dissimilar metals are involved an adjustment must be made.

A simplified example is given in Figure 2.7, in which a steel rod is to be brazed into a solid plate of copper, and account must also be taken, of course, of the manufacturing tolerances on the components. This also shows a typically dimensioned item, with a preform calculator (Figure 2.7c) which provides a ready means of determining the wire diameter for a plain cylindrical joint with a diameter equal to the preformed ring. From the joint (8 mm) and the gap at brazing temperature (0.12 mm) the wire diameter can be read off by projecting lines (dashed in Figure 2.7c) at right angles to the axes to their meeting point within two of the curved lines, giving a wire diameter of 1.2 mm. The length of wire in each ring is, of course, πD, i.e. approximately 62.8 mm.

Joint dimensions as determined to meet the strength requirements of the joint must take account of the direction of applied stresses, as indicated in Figure 2.5e, which shows the value of the fillets.

It is important to remember that as an assembly expands during heating the joint gap may either widen or close by the time the brazing filler metal starts to melt and move.

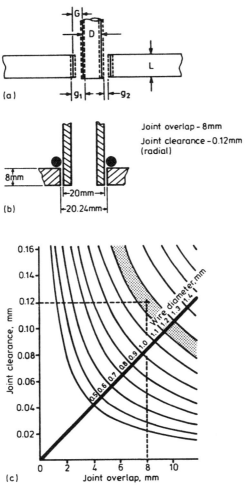

Fig. 2.7 Calculation of joint dimensions: (a) Showing a steel rod of diameter D to be brazed into a copper plate having a thickness L. The nominal gap width G may vary because of manufacturing tolerances on the hole and the rod, the actual gap being from a maximum g1 to a minimum g2, with an additional variation because of the change in temperature from normal to the melting point of the filler metal; (b) Typical dimensions of the tube to be brazed into the plate; (c) Calculator for determining the wire diameter for a range of joint gaps and joint overlap.

It is desirable to design the joint so that the solidifying filler metal is exposed to compressive rather than tensile stresses. This is much more important in brazing than in soldering, because brazing temperatures are higher, increasing the total expansion. With cylindrical joints, the component with the larger coefficient of expansion should, whenever possible, be on the outside (Figure 2.8). It is equally important to make sure that at the joining temperature the gap does not become impossibly wide. If the components in the assembly have to be reversed, design modifications can reduce stresses (Figure 2.8), but a sufficient gap must be provided to ensure that closure at brazing temperature does not provide insufficient clearance for filler metal flow. Finally, it is important to ensure that there is sufficient filler metal to absorb room temperature tensile stresses.

Influence of joint clearance

Joint clearance is probably one of the most significant factors in vacuum brazing operations. Naturally, it receives special consideration when joints are designed at room temperature. Actually, joint clearance is not the same at all phases of brazing. It will have one value before brazing, another value at brazing temperature, and still another value after brazing, especially if there has been diffusion of the brazing filler metal into the base metal. To avoid confusion, it has become general practice to specify joint clearance as the value at room temperature before brazing.

Joint clearance has a major effect on the **mechanical performance** of a brazed joint. This applies to all types of loading, such as static, fatigue, and impact, and to all joint designs. Several effects of joint clearance on mechanical performance are:

1. the purely mechanical effect of restraint to plastic flow of the filler metal by a higher strength base metal;
2. the possibility of slag entrapment;

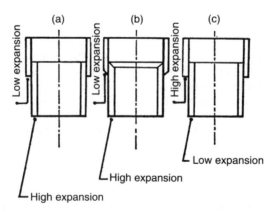

Fig. 2.8 Joint designs for cylindrical joints. (a) Avoid at all times. (b) Will reduce tensional stress if avoidable. (c) Ideally the joint cools in compression.

3. the possibility of voids;
4. the relationship between joint clearance and capillary force which accounts for filler metal distribution;
5. the amount of filler metal that must be diffused with the base metal when diffusion brazing.

If the brazed joint is free of defects (no flux inclusions, voids, unbrazed areas, pores, or porosity), its strength in shear depends upon the joint thickness, as illustrated in Figure 2.9. This figure indicates the change in **joint shear strength** with joint clearance.

Some specific clearance versus strength data for silver brazed butt joints in steel are shown in Figures 2.10 and 2.11. Figure 2.10 shows the optimum shear values obtained with joints in 12.7 mm round drill rod using pure silver. The rods were butt brazed by induction heating in a dry 10% hydrogen – 90% nitrogen atmosphere. Figure 2.11 relates tensile strength to joint thickness for butt brazed joints of the same size. Note how the strength decreased at extremely small clearances. (The data in Figures 2.10 and 2.11 were obtained with nonstandard test specimens.)

It should be noted that the type of fluxing will have an important bearing on the joint clearance to be used to accomplish a given brazement.

A mineral flux must melt at a temperature below the melting range of the brazing filler metal, and it must flow into the joint ahead of the filler metal. When the joint clearance is too small, the mineral flux may be held in the joint and not be displaced by the molten filler metal. This will produce joint defects. When the clearance is too large, the molten filler metal will flow around pockets of flux, causing excessive flux inclusions.

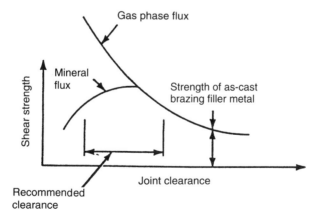

Fig. 2.9 Relationship between joint clearance and shear strength for two fluxing methods.

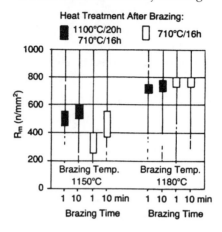

Fig. 2.10 Tensile strength of brazed joints brazed under defined load.[26]

The recommended joint clearances given in Table 2.4 are based on joints having members of similar metals and equal mass. When dissimilar metals and/or metals of widely differing masses are joined by brazing, special problems arise which necessitate more specialized selection among the various brazing filler metals, and joint clearance suitable for the job at hand must be carefully determined.

Although there are many kinds of brazed joints, selection of joint type is not as complicated as it may seem, because butt and lap joints are the two fundamental types. All others, such as the scarf joint, are modifications of these two. The scarf joint is identical with the butt joint at one extreme of the scarf angle and approaches the lap joint at the other extreme of the scarf angle. (See Chapter 6.)

2.3.4 Heat source and rate of heating

The heating methods available often place a constraint on the designer and engineer in selecting the best type of capillary joint. In principle there are many methods of heating available for brazing (see Table 2.5). Effective capillary joining requires efficient transfer of heat from the heat source into the joint. The heat capacity and thermal conductivity of the assembly must be considered.

The size and value of individual assemblies, the numbers required, and the rate of production necessary will influence selection of heating method. Many other factors must be considered before the choice is made. The rate of heating, differential thermal gradients, and cooling rates, both external and internal, will vary tremendously with different

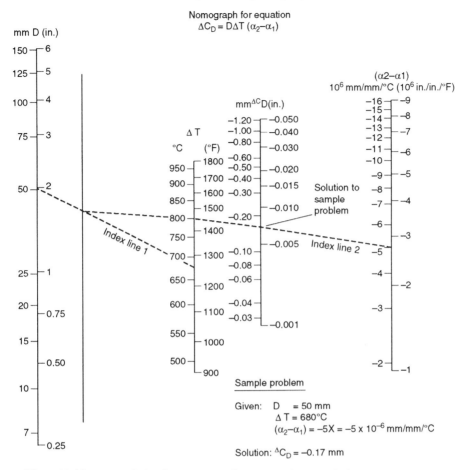

Fig. 2.11 Nomograph for determining changes in diametral clearance caused by heating of dissimilar metal joints.[27,28]

methods of heating, and the effects of these on dimensional stability, distortion, and metallurgical structure must be considered.

Getting heat to the joint can be accomplished in many ways and is best categorized by the actual method of heating.

2.4 FILLER METAL FLOW AND JOINT PROPERTIES

The base metal has a prime effect on joint formation and its ultimate strength. A high-strength base metal produces joints of greater strength than those made with softer base metals (other factors being equal). When hardenable metals are brazed, joint strength becomes less predictable.

Table 2.4 Recommended brazing gaps for different filler metals

Brazing filler metal system	Joint clearance (mm)
Al-Si alloys (a)	0.15 to 0.61
Mg alloys	0.10 to 0.25
Cu	0.00 to 0.05
Cu-P	0.03 to 0.13
Cu-Zn	0.05 to 0.13
Ag alloys	0.05 to 0.13
Au alloys	0.03 to 0.13
Ni-P alloys	0.00 to 0.03
Ni-Cr alloys (b)	0.03 to 0.61
Pd alloys	0.03 to 0.10

Notes (a) If joint length is less than 6 mm, gap is 0.12 to 0.75 mm. If joint length exceeds 6 mm, gap is 0.25 to 0.60 mm. (b) Many different nickel brazing filler metals are available, and joint gap requirements may vary greatly from one filler metal to another.

Table 2.5 Heating methods available for brazing and their advantages and disadvantages

Method	Capital cost	Running cost	Basic output	Flux required	Versatility	Operator skill required
Torch (flame)	L/M	M/H	L	Yes	H	Yes
Electrical resistance	M	M	M/H	Yes	L	No
Induction	M/H	M	M/H	Y/N	M	No
Furnace (atmosphere)	M/H	M/H	H	Y/N	M	No
Furnace (vacuum)	H	L	H	No	M	No
Dip (flux bath)	L/M	M/H	L/M	Yes	L	Yes
Infrared	M	L	M	Y/N	L	No

H = high; M = medium; L = low.

This is because there are more complex metallurgical reactions involved between hardenable base metals and the brazing filler metals. These reactions can cause changes in the base metal hardenability and can create residual stresses[21]

2.4.1 Metallurgical reactions

Some metals and alloys exhibit metallurgical phenomena that influence the behavior of brazed joints and base metal properties and in some cases necessitate special procedures. These phenomena may be classified as:

- base metal effects, which include carbide precipitation;
- hydrogen embrittlement;

- heat-affected zone;
- oxide stability;
- sulfur embrittlement;
- filler metal effects, such as vapor pressure;
- base metal/filler metal interactions, which include alloying, phosphorus embrittlement, and stress cracking.

Other factors that cause interactions between base metals and filler metals include post-brazing thermal treatments, corrosion resistance, and dissimilar metal combinations.

The extent of interaction varies greatly depending on compositions (base metal and brazing filler metal) and the duration and extent of the thermal cycles in the processing. There is always some interaction, except when mutual insolubility permits practically no metallurgical interaction.

In addition to the base metal effects above and the normal mechanical requirements of the base metal in the brazement, the effect of the **brazing cycle** on the base metal and the final joint strength must be considered.

Cold-worked strengthened base metals will be annealed and the joint strength reduced when the brazing process temperature and time are in the annealing range of the base metal being processed. 'Hot cold worked' heat-resistant base metals can also be brazed; however, only the annealed physical properties will be available in the final brazement.

The brazing cycle by its very nature will usually anneal the cold-worked base metal unless the brazing temperature is very low and the time at heat is very short. It is not practical to cold-work the base metal after the brazing operation.

When a brazement must have strength after brazing that will be above the annealed properties of the base metal, a heat-treatable base metal should be selected. The base metal can be an oil quench type, an air quench type that can be brazed and hardened in the same or a separate operation, or a precipitation hardening type that can be brazed and solution-treated in a combined cycle. Parts already hardened may be brazed with a low temperature filler metal using short times at temperature to maintain the mechanical properties.

The strength of the base metal has a profound effect on the strength of the brazed joint; thus, this property must be clearly kept in mind when designing the joint for specific properties. Some base metals also are easier to braze than others – particularly by specific brazing processes.

Carbide precipitation

If stainless steels are heated to temperature from 425 to 815°C the carbon in the base metal combines preferentially with chromium to form chromium carbide, usually at the grain boundaries. This chromium depletion reduces the corrosion resistance of the stainless steel. This condition has been defined as 'sensitization' by some investigators. In certain corrosive environments, the mechanical properties may be impaired with little or no apparent surface attack. A short brazing cycle will keep the chromium carbide precipitation to a negligible level with normal types of stainless steels. When this is not possible, one of the special grades of stainless steel may have to be used if its corrosion resistance is to be preserved after brazing.

Precipitated carbides in stainless steels may be redissolved by heat treating at 1010 to 1120°C followed by rapid cooling. Another stabilizing treatment that disperses the unprecipitated chromium uniformly throughout the structure consists of heating to 870°C for 2 h followed by furnace cooling to 540°C and subsequent air cooling.

If the cooling from the brazing temperature is rapid, no appreciable amount of carbides will be precipitated. Where this cannot be done due to mass and it is necessary to braze stainless steels for corrosive service, one of the stabilized compositions, such as type 347 or 321, or an extra low carbon grade, such as 304L, should be used.

There are several ways to prevent or minimize the deleterious effects of carbide precipitation. First, because the reaction is time-dependent, carbide precipitation can be minimized by keeping the brazing thermal cycle as short as possible. With short cycle times, such as would result from torch or induction brazing of small parts, even the unstabilized grades can be brazed without severe losses in corrosion resistance.

The susceptibility to carbide precipitation also depends on carbon content. Thus type 304 would be less susceptible than type 302, and the extra low carbon grades, such as type 304L, are relatively insensitive to carbide precipitation.

For critical applications, type 347, the niobium-stabilized grade, is recommended. It has good high-temperature strength and can be brazed without danger of impaired corrosion resistance. Type 321 is also a stabilized grade, but it has slightly lower general corrosion resistance than type 347 and is more difficult to braze because titanium is used as the carbide-stabilizing element.

When high-melting-point brazing filler metals are used, precipitated carbides can be redissolved by heat treatment after brazing.

Alternatively, corrosion resistance can be restored by diffusing chromium back into the depleted area around the carbide precipitates. Two hours at 870°C is the recommended homogenizing heat treatment.

Hydrogen embrittlement

Hydrogen can also be a source of trouble. Because of its small atomic size, it is able to diffuse quite rapidly through many metals, and the rate of diffusion increases with temperature. When hydrogen diffuses into a metal that has not been completely deoxidized it may reduce the oxide of the metal if the temperature is high enough. Metallic sponge and water vapor are the end products of this reaction.

Once hydrogen has diffused into the metal, several things can happen. If oxygen is present, the hydrogen may combine with it to produce water vapor. The water-vapor molecule, unlike the hydrogen molecule, is too large to diffuse out of the metal, and the high vapor pressures that develop can literally tear the metal apart by starting many fissures and blisters, mainly at the grain boundaries. The ultimate result is **hydrogen embrittlement**. It commonly occurs in copper and copper-based alloys that have not been deoxidized. Pressures developed for tough pitch copper have been calculated to be as high as 620 MPa.

Electrolytic tough pitch copper, silver, and palladium, when they contain oxygen, are subject to hydrogen embrittlement if heated in the presence of hydrogen. If tough pitch copper is to be brazed without embrittlement, hydrogen must not be present in the heating atmosphere. A better practice is to use deoxidized copper or oxygen-free copper where brazing is to be performed. Oxygen-free copper, if improperly heated, may too be oxidized and become subject to hydrogen embrittlement. It is impractical to salvage hydrogen-embrittled copper.

In a recently completed study, Dirnfeld, Gabbay and associates[22] examined several commercial brazing filler metals containing zinc, cadmium, or phosphorus which were found to cause embrittlement by migration of (copper) oxide to the grain boundaries, causing void formation and rupture of grain boundaries. (Oxides do not migrate as such, but rather dissolve in the grains. The oxygen diffuses to the grain boundaries, where it recombines, forming less stressed particles.) The brazing was performed both by the conventional fluxed and fluxless methods without the presence of any source of hydrogen. But this still resulted in the same embrittlement. Therefore, it was concluded that the influence of flux is insignificant, since embrittlement also persisted in joints brazed without flux under argon.

To narrow the possibilities of embrittlement a 72% silver –28% copper eutectic, braze filler metal was used to fill several joints and there was no such embrittlement, even in the most drastic brazing conditions employed. In comparing this filler metal with the others, the only difference was in the composition. While BAg–8 contains silver as the only addition to copper, the other filler metals are a ternary or quaternary formulation, containing additions of zinc, cadmium, and phosphorus. The difference between silver and these other alloying additions is in their ability to reduce copper oxide; silver cannot act as a reducing agent. This indicates that embrittlement occurs as a result of the interaction of the other alloying elements with the copper base metal and not directly because of other factors such as flux, atmosphere, and the time/ temperature cycle. These parameters were identical for all the filler metals, and their influences were as expected, that is, the embrittlement was enhanced when more drastic conditions were employed.

Furthermore, embrittlement occurred only in tough pitch copper (containing oxygen as Cu_2O precipitates) but not in phosphorus-deoxidized copper, which is completely free of oxides. As a consequence, the coexistence of the additional alloying elements of zinc, cadmium and phosphorus together with oxygen in the substrate is the necessary prerequisite for embrittlement.

The results indicate that the responsible elements are those that are capable of reducing copper oxide, by a mechanism analogous to the hydrogen embrittlement of tough pitch copper. It seems quite certain that the thermodynamic activity of zinc, cadmium, and phosphorus is sufficiently high to cause the following reaction to take place:

$$CuO + (X) \rightarrow Cu + XO$$

where X stands for one of these elements in the brazing filler metal.

The mechanism suggested by this study is embrittlement induced by the coexistence of copper oxides in the base material together with certain alloying additions in the filler metal, such as cadmium, zinc or phosphorus, capable of reducing the copper oxides.

Steel is especially prone to another mechanism for hydrogen embrittlement. In this type, hydrogen diffuses into the steel as atomic hydrogen in the same manner as it diffuses into copper, but it tends to accumulate in small voids, such as those around nonmetallic inclusions and at grain boundaries. Water vapor is not formed, as in copper, but the hydrogen atoms combine to form hydrogen molecules, which are less mobile and remain trapped at the discontinuities and as a result increase the concentration of molecular hydrogen, increase the vapor pressure, and lower the ductility of steel when stressed.

However, steel and other ferrous alloys may be salvaged by allowing the hydrogen to diffuse out by baking at slightly elevated temperatures (95 to 205°C), or by permitting the steel to stand for long periods of time until the ductility is regained.

A third type of embrittlement can occur when hydrogen combines with the metal to form hydride. The hydride lowers the notch toughness and affects the strain rate of the metal. Titanium, zirconium, niobium and tantalum and their alloys, are subject to this form of hydrogen embrittlement. Ductility can be restored if proper thermal treatments are followed after brazing; however, an inert or vacuum atmosphere should be used for brazing to avoid any embrittlement. Most other metals and alloys whose oxides may be reduced by hydrogen contain an excess of deoxidizing elements and are not subject to hydrogen embrittlement.

Heat-affected zone

The heating of base metals may cause changes in their properties, particularly if the metals are heated above their annealing temperatures. Base metals whose mechanical properties were obtained by cold working (hard tempers) may soften or undergo an increase in grain size if the brazing temperature is above the recrystallization temperature. Where mechanical properties are obtained by thermal treatment, they may be altered by the brazing operation. Materials in the annealed condition will generally experience no appreciable change due to brazing.

The width of the zone through which these changes may occur will vary with the process used. If the heating is localized, as in torch or induction brazing, the effects are confined to a narrow zone. If the whole assembly is heated, as in furnace brazing, the entire assembly is affected. In general, the heat-affected zone produced during brazing is wider and less sharply defined than those resulting from other welding processes.

Oxide stability and formation

When clean metals are heated to brazing temperature, their surfaces may form metal oxides if the atmosphere around the part contains oxygen. Oxidized metal surfaces are usually difficult to wet with most brazing filler metals. Fluxes and special atmospheres are designed to prevent oxide formation, or will reduce at elevated temperature any oxidation that occurs during initial heating.[23]

Chromium, aluminum, titanium, silicon, magnesium, manganese, and beryllium all have oxides that are difficult to remove, and therefore these metals usually require special preparation.[23] Fluoride-bearing fluxes can reduce some oxides, and hydrogen gas of sufficient purity can reduce them above certain temperatures, and techniques such as vacuum brazing may have to be used. Ideally, oxide formation should be prevented by brazing in low dew point or vacuum atmospheres.

Sulfur embrittlement

Nickel and certain alloys containing appreciable amounts of nickel, if heated in the presence of sulfur or compounds containing sulfur, may become embrittled. This occurs when a low-melting nickel sulfide is formed preferentially at the grain boundaries: this sulfide, being brittle and weak, will crack if subsequently stressed and the damage that has occurred cannot be salvaged.

Nickel and nickel–copper alloys are most susceptible to this attack, whereas alloys containing chromium are less susceptible. It is important that alloys in which nickel is the major component be clean and free of sulfur-containing materials (such as oil, grease, paint and drawing lubricants) prior to heating and that heating be done in relatively sulfur-free atmospheres.

Vapor pressure

Every metal is in equilibrium with its vapor pressure; some amount of the metal is present in the gaseous state. For most metals, at normal temperatures, this vapor pressure is so small as to be considered nonexistent. For vacuum tube applications, some metals, such as zinc and cadmium, have relatively high vapor pressures and will give off undesirable gases at normal brazing temperatures and cannot be permitted as constituents of the brazing filler metal. Accordingly, special vacuum-tube-grade filler metals have become commercially available and special fluxes are used in some situations.

Base metal/filler metal interactions

There are always some interactions between the brazing filler metal and the base metal. Although some of this interaction aids in wetting the base metal, other detrimental effects may occur. Such effects include:

- formation of brittle intermetallic compounds that lower joint strength;
- diffusion of the filler metal into the base metal to produce color changes;
- creation of new alloys, with higher melting points than that of the original filler metal, that choke off the flow of the filler metal.

Steffens, Wielage and Biermann[23-26] have reported on work whereby modified brazing processes for nickel-based materials were used to reduce the formation of brittle phases in the braze joint and also speed the joining operation.

The three modified brazing processes developed by Steffens *et al.* included:

(a) Brazing under defined load;
(b) High speed brazing;
(c) Application of mechanical excitation brazing.

Additives (silicon, boride) used to reduce the melting point of nickel base materials brazed with nickel brazing filler metals cause brittle phases which exert a negative influence on the mechanical properties of the brazed joints. Diffusion annealing and subsequent aging merges the brittle phases in the braze joint whereas the aging causes hardening of the base material. After this, the mechanical properties of the joint are comparable to those of the base material. However, applying this heat treatment may cause the formation of coarse grains.

Brazing under defined load pressures differs from conventional brazing in that a previously-defined load is set up quickly after the brazing temperature is reached. Brazing temperatures of 1150 and 1180°C with temperature retention times of 1 and 10 min. respectively have proved to be the best parameter combination. This type of **pressure brazed joint** with a more homogeneous microstructure produces good strength properties (see Figure 2.10). Brazing time is of little significance; in contrast, the brazing temperature is of great importance to the strength properties. However, strength values which can be attained without diffusion annealing are comparable with those of conventionally brazed specimens using cost-intensive and time-consuming heat treatments.

The formation of brittle phases can be influenced or even avoided by combining a considerable reduction of the brazing time with a simultaneous increase in the brazing pressure. The only useful method to achieve the required high gradient of temperature is conductive heat treatment technique.

Another method to prevent the formation of brittle phases is by the **mechanical excitation** of the components during the brazing process. While brazing, a transducer directly connected to one specimen transfers high frequency energy into the brazing couple.

Using ultrasonic vibrations of approximately 30 kHz (amplitude of approximately 2 μm in the longitudinal direction of the specimens) the accumulation of brittle phases is prevented. The wetting of the base metal with filler metal is improved because the ultrasonic vibrations destroy any existing surface oxides. The superposition of mechanical excitation produces seams of a quality which are comparable with those joints heat treated (1100°C for 20 hours) after conventional brazing.

Alloying

One of the significant base metal/filler metal interactions which has been used in determining the behavior of brazed joints is alloying. The extent of interaction varies greatly depending on the compositions of the base metal and the brazing filler metal and on in-process thermal cycles. There is always some interaction, except where mutual insolubility permits practically none.

The term 'alloying' is a general term covering practically every aspect of interaction. Some of these aspects are as follows.

First, the molten brazing filler metal can dissolve the base metal. Second, constituents of the filler metal can diffuse into the base metal, either through the bulk of the grains or along the grain boundaries, or can penetrate the grain boundaries as a liquid. The results of such base metal dissolution or filler metal diffusion may be to raise or lower the liquidus or solidus temperature of the filler metal layer, depending on the composition and thermal cycle.

Examples include nickel, cupro-nickel, or Monel joined with pure copper brazing filler metal, where enough dissolution and diffusion occur so that the solidus of the copper filler metal is increased and flow is terminated. This also means that the remelt temperature of the filler metal layer is higher than its original solidus temperature.

In brazing of ferrous-based high-temperature alloys with brazing filler metals containing boron, grain boundary penetration of the base metal by a low-melting complex can cause joint degradation. This effect is particularly damaging if the base metal is thin, as in the case of brazed honeycomb sandwich panels. Producers and experienced users of high-temperature brazing filler metals should be consulted during the design of parts for which the use of these filler metals is anticipated.

Formation of intermetallic compounds as a result of interactions between constituents of the base and filler metals can occur, and these compounds are usually brittle. Whether or not such compounds form depends on base metal and filler metal compositions, time, and temperature, and just because intermetallic compounds do form, it does not necessarily follow that the joint is so embrittled as to lose engineering utility. This depends on the nature of the specific compound, its quantity, and its distribution.

Phosphorus embrittlement

Phosphorus combines with many metals to form brittle compounds known as phosphides. For this reason copper–phosphorus filler metals are not usually used with iron- or nickel-based alloys; however, two nickel-based brazing filler metals – 10 to 12% P, 0.10% max C, remainder Ni, and 13 to 15% Cr, 9.7 to 10.5% P, 0.08% max C, remainder Ni – have been used in some applications for brazing heat-resisting alloys. The first filler metal is extremely free flowing and exhibits a minimum amount of erosion with most nickel- and iron-based alloys and is good for use in exothermic atmospheres. The second is used for brazing of honeycomb structures, thin wall tube assemblies, and other structures that are used at high temperatures. Erosion can be controlled because of low solubility with iron- and nickel-based alloys and produces strong, leakproof joints with heat-resistant base metals at relatively low brazing temperatures. Furthermore, it is recommended for nuclear applications where boron cannot be used.

Stress cracking

There are many high-strength materials, such as stainless steels, nickel alloys, and copper–nickel alloys, which have a tendency to crack during brazing when in a highly stressed condition and in contact with molten brazing filler metal. Materials with high annealing temperatures – and particularly those that are age-hardenable – are susceptible to this phenomenon. Such cracking occurs almost instantaneously during the brazing operation and is usually readily visible because the molten filler metal follows the crack and completely fills it.

This process has been described as **stress-corrosion cracking** (SCC) where the molten filler metal is considered to be the corrosion medium. Cracking of stressed steel in a caustic solution or stressed brass in an ammonia solution are widely known examples of SCC. Sufficient stress to cause stress cracking can be produced by cold work prior to brazing or by an externally applied stress from mechanical or thermal sources during the brazing operation.

When stress is encountered, its cause can usually be determined from a critical analysis of the brazing procedure. The usual remedy is to remove the source of stress. Stress cracking has been eliminated by:

• Using annealed-temper rather than hard-temper material;
• Annealing cold-worked parts prior to brazing;
• Removing the source of externally-applied stress, such as improper fit of parts or jigs that exert stress on the parts;
• Redesigning parts or revising joint design;
• Heating at a lower rate;
• Heating the fluxed and assembled parts in a torch brazing application to a temperature high enough to effect stress relief, cooling to the brazing temperature, and then hand feeding the brazing filler metal.

The age-hardenable high-nickel alloys are very susceptible to SCC. These alloys should be brazed in the annealed or solution-treated condition with a relatively high-melting filler metal (preferably above 870°C) that has sufficient strength to withstand handling during the age-hardening treatment.

2.4.2 Post-brazing thermal treatments

A postbrazing thermal treatment to improve mechanical properties in brazed assemblies is frequently desired. In ferrous alloys, this treatment entails quenching from an elevated temperature followed by tempering at some lower temperature. In other alloys, such as beryllium copper, 17–7 PH, Inconel X, and some Monels, the treatment consists of heating to some intermediate temperature followed by a controlled rate of cooling.

When a thermal treatment is performed subsequent to brazing, it is important that the brazing filler metal selected has sufficient strength at the thermal treatment temperatures to withstand the necessary handling. It is also important for the base metal, filler metal, and post-brazing thermal treatment to be compatible relative to temperatures in heating and cooling. Post-brazing thermal treatments may generate residual stresses in brazed joints and may result in lowered joint strength.

2.4.3 Dissimilar metal combinations

There are many dissimilar metal combinations which may be brazed. In fact, brazing can often be used where metallurgical incompatibility precludes the use of other joining processes.

One of the most important factors to consider in brazing dissimilar metals is **thermal expansion rates**. If a metal having high thermal expansion surrounds a low-expansion metal, clearances that are satisfactory for promotion of capillary flow at room temperature will be excessive at brazing temperature. Conversely, if the low-expansion metal surrounds the high-expansion metal, no clearance may exist at brazing temperature. For example, in brazing of a molybdenum plug in a copper block, the parts must be press fitted at room temperature. But, if a copper plug is to be brazed in a molybdenum block, a properly centered loose fit at room temperature is required.[27 - 29]

Nomographs have been developed that are useful for learning the actual changes in clearance in 'ring and plug' joints between dissimilar metals (Figure 2.11). The equation may be used in cases where more accuracy is important or where one of the variables is off the nomograph scale. In more complex joint configurations, it is usually best to prepare preproduction samples to establish ideal clearances.

A technique often used in brazing of materials with different coefficients of expansion is **sandwich brazing**. A common application of this technique is the manufacture of carbide-tipped metal cutting tools. A relatively ductile metal is coated on each side with the brazing filler metal and the composite is used in the joint. This places a third material in the joint which will creep during cooling and reduce the stresses caused by differential contraction. In some variations of the technique, wire mesh is used in place of the foil.

Figures 2.12 through 2.17 depict proper joint designs and techniques to use in brazing carbide cutting tools.

There are other factors that must be considered for successful brazing of dissimilar metals. The brazing filler metal must be compatible with both the base metals. Wide differences in base metal melting points must be considered when choosing the brazing filler metal. Where corrosion or oxidation resistance is needed, the filler metal should have properties at least equal to the poorest of the two metals being brazed. In addition, under conditions of the application, galvanic couplers that may promote crevice corrosion should be avoided. Brazing filler metals that form

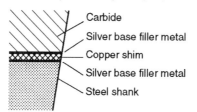

Fig. 2.12 Carbides are not readily wet by filler metals. Where possible, preplace shims rather than face-feed filler wire. Soft-core sandwich braze relaxes stresses set up by differential thermal expansions of carbide and base metal.

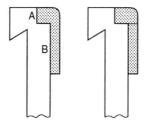

Fig. 2.13 Complex single-piece carbide cannot be brazed properly to two surfaces (see A and B). Use two-piece construction.

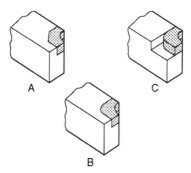

Fig. 2.14 Complex carbide shape cannot fit multiple shoulders (A). Design nonconflicting shoulders (B), or braze only bottom and end of tip (C).

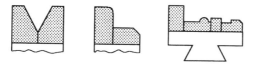

Fig. 2.15 Multiple inserts prevent strain cracks.

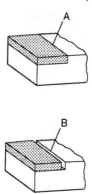

Fig. 2.16 Use stopoff paint (A) or relief gap (B) to prevent wetting more than one surface.

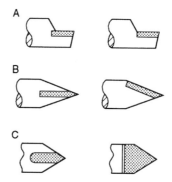

Fig. 2.17 Avoid enclosing carbide in steel slot. Change design to braze single surface.

low-melting phases with the base metals are not recommended unless, as in some special cases, techniques are used to form the final brazing filler metal *in situ*.

The metallurgical reactions that occur during brazing or subsequent thermal treatments between the brazing filler metal and the base metal are important. One example is brazing of aluminum to copper. The copper reacts with the aluminum to form a low-melting brittle compound. Such problems can be overcome by coating one of the base metals with a compatible metal. In the case of aluminum to copper, the copper can be coated with silver or a high-silver alloy and then the joints made with a standard aluminum brazing filler metal.

2.5 BRAZE FILLER METAL CHARACTERISTICS

It was not until recent years that the demands of the more sophisticated structures forced the craftsmen to appeal to metal producers to use their metallurgical knowledge to produce brazing filler metals that would meet more specific needs. Many types of filler metals, each with particular characteristics most suited for joining a particular combination of base metals, are now available to engineers, metallurgists, and technicians. All of these filler metals must, however, meet these criteria:

1. Ability to form brazed joints with mechanical and physical properties suitable for the intended service application.
2. Melting point or melting range compatible with the base metals being joined, and sufficient fluidity at brazing temperature to flow and distribute themselves into properly prepared joints by capillary action.
3. Composition of sufficient homogeneity and stability to minimize separation of constituents (liquation) during brazing.
4. Ability to wet surfaces of base metals and form a strong, sound bond.
5. Depending on requirements, ability to produce or avoid filler metal interactions with base metals.
6. Freedom from excessively volatile or objectionable constituents.

Aside from its ability to wet the base metal, the **melting characteristics** of the brazing filler metal are perhaps its most important characteristics which, to a large extent, determine selection for a particular application. Generally, a brazing filler metal is completely molten before it flows into a joint and is distributed by capillary attraction; therefore, its melting temperature is significant.

Pure metals transform from the solid state to the liquid state at one temperature when heated. Most brazing filler metals are alloys of two or more metals and, therefore, may not transform from one state to another at one single temperature. Typically, melting of an alloy progresses over a temperature range. While at temperatures within this **melting range**, the typical alloy goes through a change of state from solid to liquid or from liquid to solid – and both solid and liquid phases of the alloy are present together. In this semi-solid state, the filler metal is in a plastic or 'mushy' state.

It should be pointed out that some specific alloy compositions known as **eutectics** do melt at one temperature rather than over a range. Melting characteristics may be an important consideration when selecting the brazing filler metal for a specific application.

Melting and liquation

Most brazing filler metals are composed of three or more metals (see Appendix Table A.1). These filler metals, therefore, are much more complex than alloys of the simple binary copper–silver system, but the same conditions apply: some melt over a wide temperature range, some over a narrow range, and others at a single temperature. The wider the melting range, the more heterogeneous is the composition of the brazing filler metal at any temperature. In fact, the molten portion always has a composition distinctly different from that of the solid portion.

Because the solid and liquid alloy phases of a brazing filler metal generally differ, the composition of the melt will gradually change as the temperature increases from the solidus to the liquidus. If the portion that melts first is allowed to flow out, the remaining solid may not melt and so may remain behind as a residue or 'skull'. Filler metals with narrow melting ranges do not tend to separate, so they flow quite freely into joints with extremely narrow clearance. Filler metals with wide melting ranges need rapid heating or delayed application to the joint until the base metal reaches brazing temperature, to minimize separation, which is called **liquation**. Filler metals subject to liquation have a sluggish flow, require wide joint clearances, and form large fillets at joint extremities.

Brazing filler metals with wide melting ranges may be helpful, however, when clearances are large or nonuniform. If the filler metal has sluggish flow characteristics, the mushy consistency can cause a buildup that will help bridge the gap without liquating.

Transformation from the solid to the liquid phase does not occur uniformly. A relatively large percentage of some compositions will become fluid at a temperature slightly above the solidus and may become essentially fluid before the true liquidus is attained. Other compositions may remain essentially solid until heated well within the melting range. Recommended brazing temperatures take into account this difference in melting of each brazing filler metal. As a rule, brazing filler metals with narrow melting ranges are preferred when slow heating techniques, such as furnace brazing, are used and when the filler metal is preplaced. If the gap is wide, if the filler metal can be preheated and/or is hand fed, or if liquation can be prevented, either type of melting characteristic brazing filler metal can be used. A good general rule is to use a temperature 55 to 110°C above the liquidus temperature of the filler metal.

2.5.1 Diffusion and erosion

To be effective, a brazing filler metal must alloy with the surface of the base metal without:

(1) undesirable diffusion into the base metal;
(2) dilution with the base metal;
(3) base metal erosion;
(4) formation of brittle compounds.

Effects (1), (2) and (3) depend upon the mutual solubility between the brazing filler metal and the base metal, the amount of brazing filler metal present, and the temperature and time duration of the brazing cycle.

Some filler metals diffuse excessively, changing the base metal properties. To control diffusion, select a suitable filler metal, apply the minimum quantity of filler metal, and follow the appropriate brazing cycle. If the filler metal wets the base metal, capillary flow is enhanced. In long capillaries between the metal parts, mutual solubility can change the filler metal composition by alloying. This will usually raise its liquidus temperature and cause it to solidify before completely filling the joint.

Base metal erosion (3) occurs if the base metal and the brazing filler metal are mutually soluble. Sometimes such alloying produces brittle **intermetallic compounds** (4) that reduce the joint ductility.

Compositions of brazing filler metals are adjusted to control the above factors and to provide desirable characteristics, such as corrosion resistance in specific media, favorable brazing temperatures, or material economies. Thus, to overcome the limited alloying ability (wettability) of silver–copper alloys used to braze iron and steel, those filler metals contain zinc or cadmium, or both, to lower the liquidus and solidus temperatures. Tin is added in place of zinc or cadmium when constituents with high vapor pressure would be objectionable.

Similarly, silicon is used to lower the liquidus and solidus temperatures of aluminum and nickel-based brazing filler metals. Other brazing filler metals contain elements such as lithium, phosphorus, or boron, which reduce surface oxides on the base metal and form compounds with melting temperatures below the brazing temperature. These molten oxides then flow out of the joint, leaving a clean metal surface for brazing. Such filler metals are essentially self-fluxing.

Aggression of brazing filler metals, especially in brazing of superalloys, occurs by two general methods: the diffusion of filler metal elements into the base metal and the erosion of the base metal by the filler metal. There are several factors that affect the rate of attack. Primary among these are filler metal/base metal solubility, microconstituents in the filler metal, and diffusion mechanisms of small atoms such as carbon and boron.

Filler metal aggression reduces the effective thickness of the base metal, and is of greatest concern when thin material is brazed or when vibrational or impact loads are applied to the brazed structure.[30] For heavy sections, where brittle intermetallics form only a minor portion of the thickness, aggressive brazing filler metals may be beneficial. In such cases, diffusion may act as reinforcement to the brazed bond and provide additional strength.

Brazing filler metals containing boron, carbon, and silicon have been found to exhibit the greatest aggression. BNi–1 is the most aggressive and BNi–5 the least. Also, a high iron content in the base metal appears to decrease erosion but increase diffusion.[31 - 38]

There are three basic reactions which occur at the interface between brazing filler metals and austenitic base metals.[39] These three reactions are:

1. *Short-circuit paths.* Interstitial atoms, or atoms of relatively small atomic radii, in the brazing filler metal diffuse into the base metal along grain boundaries. Examples of such interstitial atoms are carbon, boron and, to a lesser degree, silicon.
2. *Irrigation effects.* These effects sometimes take place after rapid grain-boundary diffusion has occurred. The interstitial atoms literally migrate from the grain boundary into the lattice of the base metal, forming intermetallic compounds. Irrigation effects are most likely to occur during aging treatments and superheating.
3. *Incipient volume diffusion.* The elements of relatively large atomic radii, acting as substitutional atoms in the base metal, diffuse at a lower rate into the lattices of the base metal grains along the entire interface region. Generally, this migration in the austenitic alloys involves nickel, chromium, and silicon from the brazing filler metal and results in complex solid solutions, typically Fe–Cr–Ni–Si, at the interface regions. The solid solutions are sometimes unstable when cooled from the brazing temperature. They decompose to form a eutectoid mixture of solid-solution and intermetallic compounds.

In addition to the above mechanisms of diffusion, there is another reaction that can occur during the brazing operation: dilution of the brazing filler metal by base metal elements. The dilution can occur by dissolution of the base metal solid-solutions at the interface which have melting temperatures within the limits of the brazing temperatures employed.[40]

The degree to which the brazing filler metal penetrates and alloys with the base metal during brazing is referred to as diffusion. In applications requiring strong joints for high-temperature, high-stress service conditions (such as turbine rotor assemblies and jet engine components), it is generally good practice to specify a brazing filler metal that has high diffusion and solution properties with the base metal. Where the assembly is constructed of extremely thin base metals (as in honeycomb structures and some heat exchangers), good practice generally calls for filler metal with a low-diffusion characteristic relative to the base metal being used. Diffusion is a normal part of the metallurgical process that can contribute to good brazed joints when high-temperature metals are brazed with nickel-based filler metals.[41]

Solution occurs to some degree in every brazed joint but is harmful only when the amount is sufficient to cause erosion. Erosion (caused by excessive solution) is not affected directly by the manner in which the brazing filler metal is manufactured. The physical properties of the brazing filler metal that determine solution, erosion, strength, corrosion resistance, etc., are inherent chemical characteristics, which are not affected by melting practice or even by the purity of the filler metal.[42]

Probably the most important factor affecting the degree of erosion for a given filler metal/base metal combination is the brazing technique, which encompasses assembly and fixturing practices, alloy application, and the brazing cycle itself.

Several methods have been developed to minimize aggression by brazing filler metals. **Electroplating** the base metal with nickel has been found to be one solution in overcoming the problems arising from filler metal/base metal interface reactions. This barrier layer method has its greatest use in the joining of age-hardenable alloys containing aluminum and titanium. When not nickel plated, these alloys form stable and tenacious oxide films which reduce the wettability properties of the brazing filler metals.

2.5.2 Wetting and bonding

To be effective then, a brazing filler material must alloy with the surface of the base material without undesirable degrees of diffusion into the base material, dilution by the base material, base material erosion, or formation of brittle compounds at the interface. These effects are dependent upon the mutual solubility between the filler and the base materials, the amount of brazing filler material present, and the temperature and time profile of the brazing cycle.

Case Histories

A firm was brazing C110 copper plate, 101.6 mm × 25.4 mm thick × 152.4 long, to a silver–49% tungsten 3.175 mm sheet using BAg–8 silver filler metal foil. The furnace atmosphere was dry hydrogen and the parts were not brazing together and the filler metal was disappearing.

The initial item to be examined was the use of a hydrogen atmosphere at 816°C brazing temperature. Hydrogen gas is not suitably reactive to provide for good wetting of the filler metal on tungsten, thus, the silver–tungsten wettability would be marginal. To solve this part of the problem, flow on the silver tungsten can be obtained by coating the surface with a thin layer of thinned-out white flux. However, if no residual flux is desired in the brazed joint, wetting of the silver tungsten can be done by plating with electrolytic nickel or copper approximately 0.127 mm thick, or by raising the brazing temperature to 1010°C or above without the plating. The problem with increasing the brazing temperature is that the silver in the BAg–8 filler metal and in the silver tungsten will tend to make more of the 780°C eutectic silver–copper alloy, which would cause erosion of both base metals. Thus, raising the brazing temperature would not be recommended here.

Second, the copper is a grade of tough-pitch copper, which means that this grade has copper oxide at the grain boundaries. When tough-pitch copper is heated in a hydrogen atmosphere, or atmosphere containing hydrogen, the hydrogen penetrates the copper, reacts with the copper oxide, and turns into steam. The expanding steam makes the copper porous and no longer ductile. The porous copper soaks up the brazing filler metal, thus leaving no filler metal to produce a joint.

In order to solve the second part of the problem, copper should not be brazed in a hydrogen atmosphere, or hydrogen-containing atmosphere, at any time. Suitable atmospheres would be vacuum, argon or nitrogen.

In another instance a firm was brazing copper to 304L stainless steel with BNi–7 filler metal in a pure dry hydrogen atmosphere at 1010°C. One of the parts brazed perfectly while the other parts blistered and grew in length, width and thickness. The brazing filler metal essentially disappeared, leaving some dark residue where the filler metal was applied.

As indicated previously, blistering and a substantial increase in dimensions of copper parts typically occur when copper oxide is present in the grain boundaries, distributed throughout or in inclusions or layers in localized areas of the metals, and when the copper was processed in a hydrogen-containing atmosphere.

Copper containing copper oxide should not be brazed in a hydrogen-containing atmosphere. Copper grades known as electrolytic tough pitch, tough pitch and fire refined have cuprous oxide throughout the material and are subject to embrittlement when processed in a hydrogen-containing atmosphere. Grades of copper not subject to embrittlement when processed in a hydrogen-containing atmosphere are oxygen-free and phosphorus-deoxidized copper.

This problem occurs when the hydrogen diffuses into the copper and combines with the cuprous oxide to form steam. The larger water molecules are trapped and the steam pressure stretches the copper, increasing its dimension, making it porous, reducing copper ductility and in many cases, leaving the copper with essentially no ductility. When there are copper oxide stringer inclusions or oxide layers, blisters will result.

It was found that all of the parts were electroformed copper and that the part that brazed satisfactorily had previously been annealed in vacuum at 760°C. Electroformed parts that were not vacuum annealed prior to brazing in a hydrogen atmosphere showed the typical pressure of copper oxide in the plated copper.

Therefore, it was recommended that copper be brazed or annealed in an atmosphere of vacuum, dry argon or dry nitrogen. With 304L stainless being one of the components, the assembly can be brazed in the above atmospheres with BNi–7 at 1010°C. If results are not suitable in the argon or nitrogen atmosphere, the flow rate of atmosphere in the retort may be too low. However, an electrolytic nickel plating of 0.0127 to 0.0254 mm thickness on the stainless steel will assist in providing a good brazed joint.

A second example shows the precautions that need to be taken in the brazing of 300 stainless steel tubing to copper and brass with gas/air torches on a rotary index table or oxyacetylene hand-held torches with low-temperature silver brazing filler metals.

Stainless steels are readily torch brazed with a variety of filler metals, however, selection of the proper filler metal depends on the end use of the product.

Stainless steel does not absorb heat as readily as copper or brass and it takes longer to heat and can be easily overheated, or even melted locally by spot heating.

Torch brazing requires the use of either a standard white flux (AWS FB3–A) or the boron-modified black fluxes (FB3–C).

The green color observed in the flux residue after brazing is chromium oxide removed from the surface of the stainless steel. Chromium protects the surface from rusting and the strength of the braze joint is not in question, but if the product is subject to any kind of moisture, rust can form between the surface of the stainless and the filler metal. If the wrong brazing filler metal is used, progressive joint failure can result and it will fail.

The primary cure for this problem is to use a filler metal that contains nickel. Two good, low temperature silver-based brazing filler metals for this application are BAg–3 and BAg–24. Since BAg–24 is cadmium-free, it should be the preferred choice. The nickel in the brazing filler metal alloys with the base metal sufficiently to form a nickel-rich layer that is essentially immune to 'interface corrosion'. The flux is apparently the culprit since it can leech out the chromium from the stainless. Stainless steel brazed in a protective atmosphere furnace (no flux) does not have this problem.

Stress cracking can occur in stainless steels that are cold formed and not stress relieved. These stresses and rapid heating of the stainless can cause the steel to 'fire crack'. The cracking occurs before the filler metal melts and flows. When the filler metal does melt, it may flow into these cracks, giving the impression that the brazing filler metal caused the cracking.

A second type of cracking is affected by molten brazing filler metal. This is called liquid metal embrittlement or stress corrosion cracking. Stress must be present in the part for cracking to occur. In this case, molten filler metal weakens the surface layer of the stressed part and cracks form along the grain boundaries. The flux does not play a part in this failure mode. The lower-melting ingredients in the filler metal attack the stressed grain boundaries and cause the part to fail. Depending on the stress, the stainless can split wide open.

It should be noted that carbide precipitation generally does not appear as a problem in low-temperature torch silver brazing. The use of low-temperature silver brazing filler metals allows the brazing time to be kept so short that the amount of carbide precipitation is negligible. A change to a low-carbon grade of stainless is sometimes preferable if the problem shows up, such as when high-temperature silver torch brazing.

REFERENCES

1. Gibbs, J. W. *Collected Works*, Vol. 1, Yale University Press, New Haven, 1948.
2. Kim, D. H., Hwang, S. H., Chun, S. S., The wetting, reaction & bonding of silicon nitride by Cu–Ti alloys, *Jrnl of Matl Sci.* Vol. 26 (1991) 3223–3234.
3. Naka, M., Tanaka, T., Okamoto, I., *Quart. J., Jpn Weld. Soc.* Vol. 4 (1986) 597.
4. Kapoor, R. R., Eagar, T. W., *J. Amer. Ceram. Soc.* Vol. 72 (1989) 448.
5. Loehman, R. E., *Ceram. Bull.* Vol. 68 (1989) 891.
6. Ritter, J. E., Jr., Burton, M. S., *Trans. Metall. Soc.*, AIME, Vol. 239 (1967) 21.
7. Naka, M., Kubo, M., Okamoto, I., *J. Mater. Science*, Vol. 22 (1987) 4417–21.
8. Bondl, A., Spreading of liquid metals on solid surface chemistry of high energy substances, *Chem. Rev.* Vol. 52(2) (1953) 417–58.
9. Gilliland, R. G., Wetting of beryllium by various pure metals and alloys, *Weld. J.* Vol. 43(6), Jun. 1964, 248s–58s.
10. Nicholas, M. G., Ambrose, J. C., Real time observations of wetting and flow brazes, DVS 125, 19–23, *2nd International Conf. in Essen*, 19–20 Sept 1989.
11. Keller, D. L., McDonald, M. M., Heiple, C. R., *et. al*, Wettability of brazing filler, *Weld. J.* Vol. 69 (10), Oct. 1990, 31–34.
12. Bennett, W. S., Hillyer, R. F., Keller, D. L., *et. al*, Vacuum brazing studies on high manganese stainless steel, *Weld. J.* Vol. 53, 1974, 510s–16s.
13. Feduska, W., High-temperature brazing alloy – base metal wetting reactions, *Weld. J.* Vol. 38(3), Mar. 1959, 122s–30s.
14. Adams, C. M., Jr., Dynamics of wetting in brazing and soldering, *Technical Report WAL TR 650/1*, Army Matls. Res. Agency, Watertown Arsenal, Watertown, MA, July 1962.
15. Weiss, S., Adams, C. M., Jr., The promotion of wetting, *Weld. J. 46(2)*, Feb. 1967, 49s–57s.

16. Rabinkin, A., Fundamental aspects of the brazing process, *20th International AWS Brazing & Soldering Conf.*, Paper B4B, Wash. D.C., Apr. 5, 1989.
17. Milner, D. R., A survey of the scientific principles related to wetting and spreading, *Brit. Weld. J.* Vol. 5, 1958, 175–198.
18. Crabtree, G. E., Atmospheres and surface chemistry effects on copper brazing of Kovar, Joining Subgroup of the Interagency Mechanical Operation Groups, US Department of Energy, *22nd International AWS Brazing and Soldering Conf.*, Paper B5B, Detroit, MI, Apr. 1991.
19. Sheet 8, *Metal Construction*, Feb 1986, 101.
20. Wasase, H., *et. al*, Study of a method for evaluating the brazeability of aluminum sheet, *Weld. J.*, Vol. 68(10), Oct. 1989, 396s–403s.
21. *Metals Handbook*, 9th Ed., Vol. 6, Welding, Brazing and Soldering, American Society of Metals, Metals Park, Ohio, 1983, p. 956.
22. Hirnfeld, S. F., Gabbay, R., Ramon, J. J., *et. al*, Copper embrittlement by silver brazing alloys, *Matls. Characterization, Ref 2*, Vol. 26, 17–22, Jan 1991.
23. Steffens, H. D., Wielage, B., Biermann, K., Modified high temperature brazing techniques for nickel based materials, *Industrial Heating*, 17–18, May 1989.
24. Wielage, B., Doctoral Thesis, University of Dortmund, 1979.
25. Hartmann, K. -H, Doctoral Thesis, University of Dortmund, 1980.
26. Steffens, H. -D, Wielage, B., Kerns, H., *DVS-Berichte Band 92*, S. 11–18 Deutscher Verlad für Schweisstechnik GmbH, Dusseldorf, 1984.
27. *Welding Handbook*, Vol. 2, 8th Ed., pp. 380– 422, 1991.
28. *Brazing Handbook*, 4th Ed., 1991.
29. Schwartz, M. M., *Brazing*, ASM International, 439p, 1989.
30. Peterson, W. A., Brazing thin nickel sheet, *Weld. J.*, Vol. 42(4), April 1963, 190s–192s.
31. Barker, J. F., Mobley, P. R., Redden, T. K., A new brazing alloy for age-hardenable super alloys, *Weld. J.*, Vol. 41(9), Sept. 1962, 409s–410s.
32. Wysopal, R., Bangs, E. R., The importance of braze alloy application in high temperature brazing, *Second International AWS-WRC Brazing Conference and Colloquium, 52nd Annual AWS Meeting*, San Francisco, April 26–30, 1971.
33. Trimmer, R. M., Kuhn, A. T., The tensile strength of stainless steel wire and rod butt joints as a function of the brazing alloy, *Weld. J.*, Vol. 61(10), Oct. 1982, 327s–328s.
34. Lugscheider, E., Pelster, H., Nickel base filler metals of low precious metal content, *Weld. J.*, Vol. 63(10), Oct. 1984, 261s–266s.
35. Miller, F., A new look at the boron filler alloys, *The Tool and Manufacturing Engineer*, Vol. 51(4), Oct. 1963, 98–100.
36. Hoppin III, G. S., Brazing Rene' 41, *Metal Progress*, Nov. 1960, 75–79.
37. Lugscheider, E., Knotek, O., Klöhn, K., Development of nickel–chromium–silicon base filler metals, *Weld. J.*, Vol. 57(10), Oct. 1978, 319s–323s.
38. Sheward, G. E., Bell, G. R., Development and evaluation of Ni–Cr–P brazing filler metal, *Weld. J.*, Vol. 55(10), Oct. 1976, 285s–289s.
39. Feduska, W., The nature of high temperature brazing alloy – base metal interface reactions, *Weld. J.*, Vol. 37(2), Feb. 1958, 62s – 73s.
40. Feduska, W. The nature of the diffusion of brazing alloy elements into heat resisting alloys, *Weld. J.*, Vol. 40(2), Feb. 1961, pp 81s – 89s.
41. Pattee, H. E., *High Temperature Brazing*, WRC Interpretive Bull. 187, Sept. 1973.
42. Miller, F. M., Importance of purity in manufacturing brazing filler metals for high temperature service applications, *Weld. J.*, Vol. 40(8), Aug. 1961, 821–827

3

Brazing heating methods

Brazing processes are continuously designated according to the source or method of heating, not unlike welding processes. Manual processes are possible, but automated processes predominate. Some processes restrict heating to the joint proper, while others heat the entire braze assembly or brazement uniformly. The methods currently of most industrial significance, and described in detail in the following sections, are torch brazing, furnace brazing, induction brazing, resistance brazing, dip brazing and infrared brazing. Also several specialized brazing processes are worth noting, including laser brazing, electron-beam brazing, exothermic brazing and microwave brazing.

3.1 TORCH BRAZING

Manual torch brazing is the method most frequently used for repairs, one-of-a-kind brazing jobs, and short production runs as an alternative to fusion welding. Any joint that can be reached by a torch and brought to a brazing temperature (by the torch alone or in conjunction with auxiliary heating means) can be readily brazed by this technique.

Although any source of heat can be used for torch brazing, commercial torch brazing is accomplished with the same type of torch, controls, and gases used for fusion welding. Conversion to brazing merely requires changes in torch nozzles and goggle lenses.

The torch brazing technique is relatively simple and can be mastered by the mechanically adept in a short time. Those already experienced in torch welding and the brazing of other metals generally encounter no difficulty learning torch brazing.

Depending on the temperature and heat required, all commercial gas mixtures can be used to fuel the torch: oxyacetylene, oxyhydrogen, oxygen and natural gas, acetylene and air, hydrogen and air, propane, methane and natural gas and air. Oxyacetylene and oxygen and natural gas are the mixtures most often used commercially and are preferred in that order. The adjustment of the flame is very important. Generally, a slightly reducing flame is desirable.

Air-natural gas torches provide the lowest flame temperature as well as the least heat. Acetylene under pressure is used in the air-acetylene torch with air at atmospheric pressure.

Torches which employ oxygen with natural gas, or other cylinder gases (propane, butane) have higher flame temperatures. When properly applied as a neutral or slightly reducing flame, excellent results are obtainable with many brazing applications.

Oxyhydrogen torches are often used for brazing aluminum and nonferrous alloys. The lower temperature reduces the possibility of overheating the assembly during brazing. An excess of hydrogen provides the joint with additional cleaning and protection.

The torch tip orifice used for brazing is usually larger than the tip selected for gas welding. Gas pressure should be kept low so that the flame can be readily adjusted by means of the controls on the torch. When oxyacetylene is used, the flame can be adjusted by visual inspection.

The other gases vary little in appearance with changes in oxygen/gas ratio and can be adjusted only by means of flowmeters in the gas lines. Oxygen pressure is always half that of the accompanying gas.

In manual torch brazing, the brazing filler metal is usually face fed in the form of wire or rod, or preplaced. In the latter case, care must be exercised in the placement of the filler metal and the guidance of the torch to preclude premature melting of the filler metal. One way to prevent overheating is to use flux with a melting temperature not too far below that of the brazing filler metal. The proper brazing temperature is indicated when the flux becomes liquid. To ensure uniform heating throughout the joint, which is very important, it may be advisable to use a multiple-tip torch or more than one torch.

Torch heating for brazing is limited in use to filler metals supplied with flux or self-fluxing. The list includes aluminum–silicon[1], silver, copper– phosphorus, copper–zinc, and nickel. With the exception of the copper– phosphorus filler metals, they all require fluxes. For certain applications even the self-fluxing copper–phosphorus filler metals require added flux.

Overheating of the base metal and brazing filler metal should be avoided because rapid diffusion and 'drop through' of the metal may result. Natural gas is well suited for torch brazing because its relatively low flame temperature reduces the danger of overheating.

Brazing filler metal may be preplaced at the joint in the forms of rings, washers, strips, slugs, or powder, or it may be fed from hand-held filler metal, usually in the form of wire or rod. In any case, proper cleaning and fluxing are essential.

Torch brazing techniques differ from those used for oxyfuel gas welding. Operators experienced only in welding techniques may require instruction in brazing techniques. It is good practice, for example, to prevent the inner cone of the flame from coming in contact with the joint except during preheating, since melting of the base metal and dilution with the filler metal may increase its liquidus temperature and make the flow more sluggish. In addition, the flux may be overheated and thus lose its ability to promote capillary flow, and low melting constituents of the filler metal may evaporate.

3.1.1 Mass production

Manual torch brazing is particularly useful on assemblies involving sections of unequal mass. Machine operations can be set up, where the rate of production warrants, using one or more torches equipped with single or multiple flame tips. The machine may be designed to move either the work or the torches. See rotary system, Figure 3.1.

Torch brazing can be automated relatively easily with appropriate gas supplies, indexing fixtures, and cycle controls. Usually, such systems involve multiple-station rotary indexing tables.[2-4] The part is fed into a holding fixture at the first station, then indexed to one or more

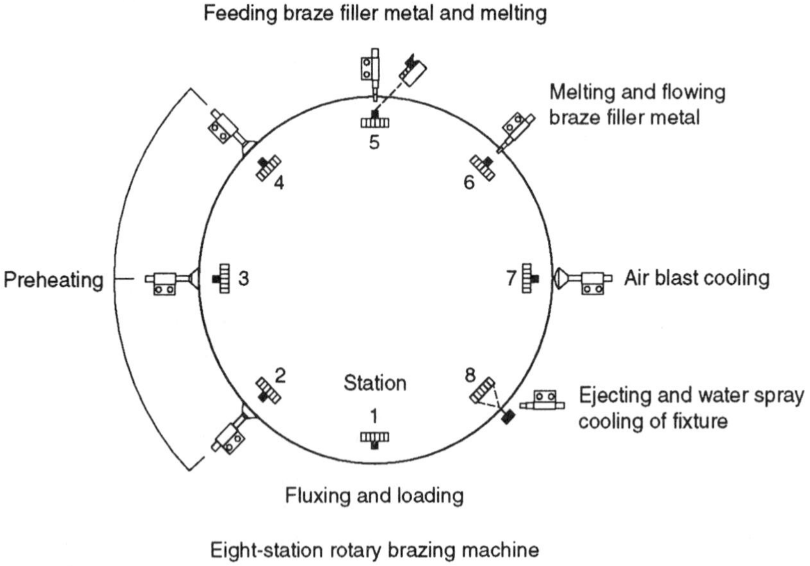

Eight-station rotary brazing machine

Fig. 3.1 Set-up for automatic torch brazing (at a rate of 230 per hour) of magnet armature assemblies used as striking members of a printing machine.[3,4]

preheating stations, depending on the heating time required. A brazing station is next, followed by a cooling station and an ejection station (see Figure 3.1).

In another variation, brazing filler metal in paste form, and perhaps including an appropriate flux, is automatically dispensed ahead of the brazing station.

3.2 FURNACE BRAZING

The popularity of furnace brazing derives from the comparatively low cost of equipment, the adaptability of the furnace, and the minimal jigging required. With many brazing assemblies, the weight of the parts alone is sufficient to hold them together. With other configurations, one or two rectangular blocks of metal are all the fixturing needed.

Furnace brazing is used extensively where the parts to be brazed can be assembled with the brazing filler metal preplaced near or in the joint.

Furnace brazing is used extensively when:

1. the parts to be brazed can be preassembled or jigged to hold them in the correct position;
2. the brazing filler metal can be placed in contact with the joint;
3. multiple brazed joints are to be formed simultaneously on a completed assembly;
4. many similar assemblies are to be joined;
5. complex parts must be heated uniformly to prevent the distortion that would result from local heating of the joint area.

Electric, gas, or oil heated furnaces with automatic temperature control capable of holding the temperature within ±6°C should be used for furnace brazing. Fluxes or specially controlled atmospheres that perform fluxing functions must be provided. Many commercial fluxes are available for both general and specific brazing operations. Satisfactory results are obtained if dry powdered flux is sprinkled along the joint. Flux paste is satisfactory in most cases, but in some cases it retards the flow of brazing filler metal. Flux pastes containing water can be dried by heating the assembly at 175 to 200°C for 5 to 15 minutes in drying ovens or circulating air furnaces.

Brazing time will depend somewhat on the thickness of the parts and the amount of fixturing necessary to position them. The brazing time should be restricted to that necessary for the filler metal to flow through the joint to avoid excessive interaction between the filler metal and base metal. Normally, one or two minutes at the brazing temperature is sufficient to make the braze. A longer time at the brazing temperature will be beneficial where the filler metal remelt temperature is to be increased and where diffusion will improve joint ductility and strength.

Times of 30 to 60 minutes at the brazing temperature are often used to increase the braze remelt temperature.

Furnace brazing is particularly applicable for high-production brazing in which continuous conveyor-type furnaces are used. For medium production work, batch-type furnaces are best. In any event, heating is usually produced by electrical resistance, although other types of fuel can be used in muffle-type furnaces.

The parts should be self-jigging or fixtured and assembled, with brazing filler metal preplaced near or in the joint. The preplaced filler metal may be in the form of wire, foil, powder, paste, slugs, or preformed shapes. Fluxing is used except when a reducing atmosphere, such as hydrogen, and either exothermic or endothermic combusted gas can be introduced into the furnace. In some instances, both flux and a reducing atmosphere may be necessary. Pure, dry, inert gases, such as argon and helium, are used to obtain special atmospheric properties.

When continuous-type furnaces are used,[5,6] several different temperature zones may be used to provide the proper preheating, brazing, and cooling temperature. The speed through a conveyor-type furnace must be controlled to provide the appropriate time at the brazing temperature. It is also necessary for the assembly to be properly supported so that it does not move while traveling on the belt. This may require special fixtures, but most often brazements are designed to be self-supporting.[7]

There are four basic types of furnaces used for brazing:

1. the batch-type with either air or controlled atmosphere, Figure 3.2;
2. continuous type with either air or controlled atmosphere;
3. retort type with controlled atmosphere;
4. vacuum.

A high-temperature, high-vacuum bottom-loading brazing furnace with control panel and charging carriage is shown in Figure 3.3. Most brazing furnaces have a temperature control of the potentiometer type connected to thermocouples and gas control valves or contactors. The majority of furnaces are heated by electrical resistance using silicon-carbide, nickel-chromium, or refractory metal (Mo, Ta, W) heating elements. When a gas or oil flame is used for heating, the flame must not impinge directly on the parts.

With controlled atmosphere furnaces, a continuous flow of the atmosphere gas is maintained in the work zone to avoid contamination from outgassing of the metal parts and dissociation of oxides. If the controlled atmosphere is flammable or toxic, adequate venting of the work area and protection against explosion are necessary.

Batch-type furnaces heat each workload separately. They may be top loading (pit type), side loading, or bottom loading. When a furnace is

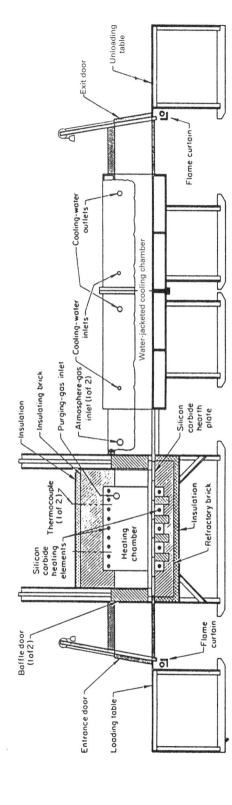

Fig. 3.2 Pusher batch-type brazing furnace with water-jacketed cooling chamber.[7]

Fig. 3.3 High vacuum bottom-loading type vacuum furnace.[8]

lowered over the work, it is called a bell furnace, Figure 3.4. Gas or oil fired batch type furnaces without retorts require that flux be used on the parts for brazing. Electrically heated batch type furnaces are often equipped for controlled atmosphere brazing, since the heating elements can usually be operated in the controlled atmosphere.

Continuous furnaces receive a steady flow of incoming assemblies. The heat source may be gas or oil flames, or electrical heating elements, Figure 3.5. The parts move through the furnace either singly or in trays or baskets. Conveyor types, Figures 3.6, 3.7, and 3.8, (mesh belts or roller hearth), shaker hearth, pusher or slot-type continuous furnaces are commonly used for high production brazing. Continuous furnaces usually contain a preheat or purging area which the parts enter first. In this area, the parts are slowly brought to a temperature below the brazing temperature. If brazing atmosphere gas is used in the brazing zone it also flows over and around the parts in the preheat zone, under positive pressure. The gas flow removes any entrapped air and starts the reduction of surface oxides. Atmosphere gas trails the parts into the cooling zone.

Retort-type furnaces are batch furnaces in which the assemblies are placed in a sealed retort for brazing. The air in the retort is purged by controlled atmosphere gas and the retort is placed in the furnace. After the parts have been brazed, the retort is removed from the furnace, cooled, and its controlled atmosphere is purged. The retort is opened, and the brazed assemblies are removed. A protective atmosphere is sometimes used within a high temperature furnace to reduce external scaling of the retort.

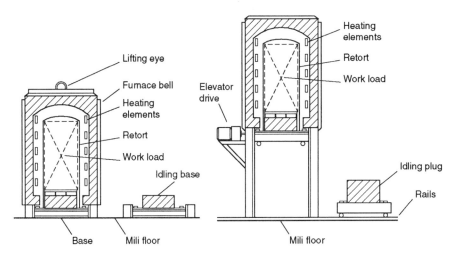

Fig. 3.4 Batch-type brazing furnaces, Bell on left and Elevator on right.[9]

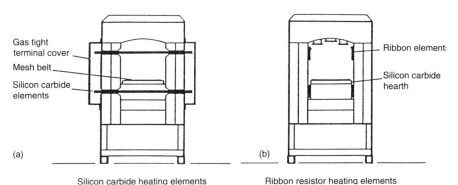

Fig. 3.5 (a) Cross-section of the high-heat zone in a continuous brazing furnace featuring SiC heating elements: (b) similar, but for a batch furnace featuring ribbon-type resistance elements and a SiC hearth.[10]

A large volume of furnace brazing is performed in-vacuum, which prevents oxidation and often eliminates the need for flux.[10] Vacuum brazing has found wide application in the aerospace and nuclear fields, where reactive metals are joined or where entrapped fluxes would be intolerable. If the vacuum is maintained by continuous pumping, it will remove volatile constituents liberated during brazing.

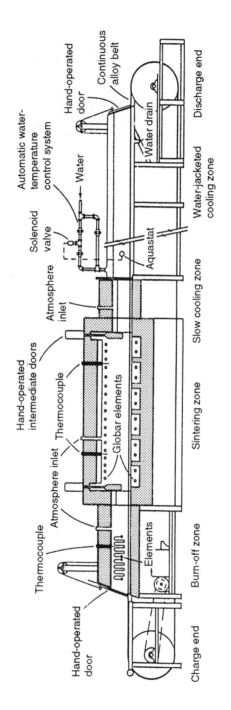

Fig. 3.6 Longitudinal section of a mesh-belt sintering furnace.[9]

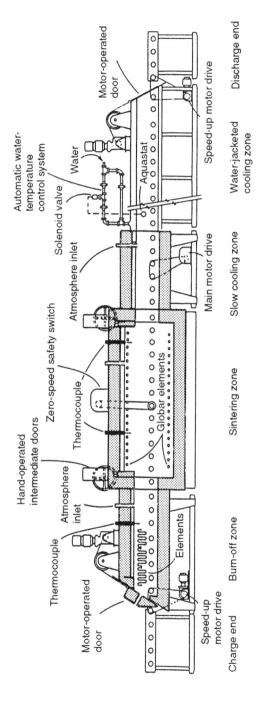

Fig. 3.7 Longitudinal section of a roller-hearth sintering furnace.[9]

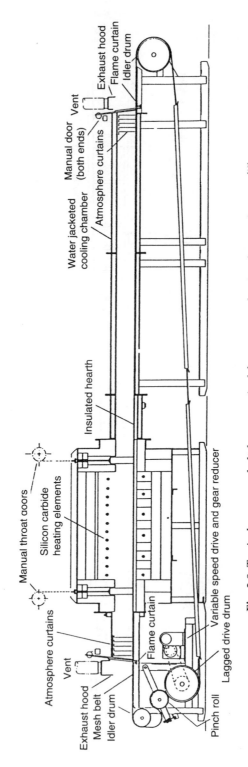

Fig. 3.8 Typical conveyor belt furnace suited for mass production furnace brazing.[10]

There are several base metals and brazing filler metals that can be harmed by brazing in a vacuum because their low-boiling-point or high-vapor-pressure constituents cause part of the metal to be lost.[11,12]

Vacuum is a relatively economical method of providing an accurately controlled brazing atmosphere and is an effective means of screening the work to be brazed from oxidizing gases and other impurities.

The vacuum pressures used for brazing generally range from 0.13 to 1.3 Pa Hg. This range corresponds to a gas that is several hundred times purer than the purest gas used for atmosphere brazing. Vacuum brazing does not allow as wide a choice of brazing filler metals as does atmosphere brazing.[13,14]

Vacuum furnaces are invariably heated by electricity, in any one of a number of forms. In a typical furnace the heating system is surrounded by radiation shields and mounted within a water-cooled steel shell of pressure-vessel proportions. Suitable pumping equipment is connected to the shell. The advantage of radiation shields over more conventional forms of thermal insulation is that maintaining the essential standard of cleanness within the furnace is easier with an all-metal system. Access to the furnace is gained through a door or removal panel in the chamber wall.

The radiation shields must perform the dual function of containing the heat and protecting the rubber vacuum seals that are incorporated in the door and which are fitted around all power leads or other controls that pass through the furnace wall. A less obvious consequence of effective insulation is that the cooling rate is severely retarded. Because of this, the overall time of a vacuum brazing operation is usually much longer than that of other batch brazing techniques and results in expensive equipment being tied up for long periods in processing comparatively small workloads. It should be remembered, however, that most of the metals on which vacuum brazing excels are also costly and that any process which permits material economies through fabrication and ensures the necessary joint properties is justified. Equipment developments that have reduced processing times include transfer mechanisms, multizone furnaces, and inert-gas quenching, Figure 3.9.

Vacuum brazing furnaces are of three types as follows.

Hot retort, or single pumped retort furnace

This is a sealed retort, usually of fairly thick metal. The retort with work loaded inside is scaled, evacuated, and heated from the outside by a furnace. Most brazing work requires vacuum pumping continuously throughout the heat cycle to remove gases being given off by the workload. The furnaces are gas-fired or electrical. The retort size and its maximum operating temperature are limited by the ability of the retort to withstand the collapsing force of atmospheric pressure at brazing

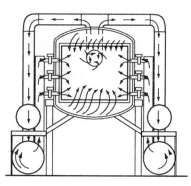

Fig. 3.9 Gas quench flow schematic utilizing positive pressure gas cooling.[11]

temperature. Top temperature for vacuum brazing furnaces of this type is about 1150°C.

Argon, nitrogen, or other gas is often introduced into the retort to accelerate cooling after brazing.

Double pumped or double wall hot retort vacuum furnace

The typical furnace of this type has an inner retort containing the work, within an outer wall or vacuum chamber. Also within the outer wall are the thermal insulation and electrical heating elements. A moderately reduced pressure, typically 133 to 13.3 Pa, is maintained within the outer wall, and a much lower pressure, below 1.3 Pa, within the inner retort. Again most brazing requires continuous vacuum pumping of the inner retort throughout the heat cycle to remove gases given off by the workload.[15]

In this type of furnace, the heating elements and the thermal insulation are not subjected to the high vacuum. Heating elements are typically of nickel–chromium alloy, graphite, stainless steel, or silicon carbide materials. Thermal insulation is usually silica or alumina brick, or castable or fiber materials.

Cold wall vacuum furnace

A typical cold wall vacuum furnace has a single vacuum chamber, with thermal insulation and electrical heating elements located inside the chamber. The vacuum chamber is usually water cooled. The maximum operating temperature is determined by the materials used for the thermal insulation (the heat shield) and the heating elements, which are subjected to the high vacuum as well as the operating temperature of the furnace.

Heating elements for cold wall furnaces are usually made of high-temperature, low-vapor-pressure materials, such as molybdenum, tungsten, graphite, or tantalum. Heat shields are typically made of multiple layers of molybdenum, tantalum, nickel, or stainless steel. Thermal insulation may be high purity alumina brick, graphite, or alumina fibers sheathed in stainless steel. The maximum operating temperature and vacuum obtainable with cold wall vacuum furnaces depends on the heating element material and the thermal insulation or heat shields. Temperatures up to 2200°C and pressures as low as 1.33×10^{-4} Pa are obtainable.

Configurations for all three types of furnaces include side loading (horizontal), bottom-loading, and top-loading (pit-type), Figures 3.10 and 3.11.

Vacuum pumps

Vacuum pumps for brazing furnaces may be oil sealed mechanical types for pressures from 13 to 1300 Pa. Brazing of base metals containing chromium, silicon, or other rather strong oxide formers usually requires pressures of 1.3 to 0.13 Pa, which are best obtained with a high-speed, dry Roots, or turbo-mechanical pump type. Vacuum pumps of this type are not capable of exhausting directly to atmosphere and require a roughing vacuum pump.

Brazing of base materials containing more than a few percent of aluminum, titanium, zirconium, which form very stable oxides, requires a

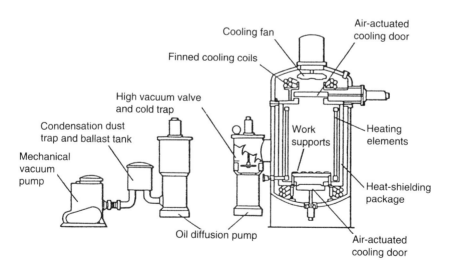

Fig. 3.10 Top-loading cold wall vacuum furnace.[13]

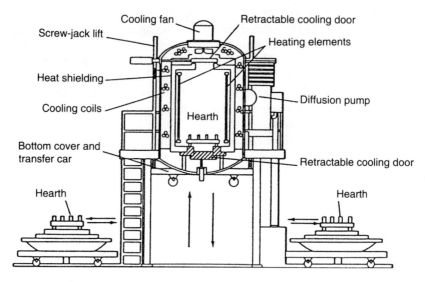

Fig. 3.11 Bottom-loading cold wall vacuum furnace.[9,13]

vacuum of 0.13 Pa or lower. Vacuum furnaces for such brazing methods usually require a diffusion pump that will obtain pressures of 1.3 to 0.0001 Pa.

3.3 INDUCTION BRAZING

The high-frequency induction heating method for brazing is clean and rapid, lending itself to close control of temperature and location, and requiring little operator skill.

The heat for induction brazing is created by a rapidly alternating electric current which is induced into the workpiece by an adjacent coil. The workpiece is placed in or near a coil carrying alternating current, which induces the heating current in the desired area of the coils, which are water cooled, and are designed specifically for each part. As a result heating efficiency relies on establishing the best coil design and power frequency for each application. Furthermore, the coils provide heat only to the joint area. The time and effort spent on coil design and establishment of proper power frequency depend on the complexity of the shapes of the component parts and on the materials involved. Magnetic materials are heated far more readily, and unbalanced heating can easily occur if, for example, copper and steel parts are being joined. Similarly, local overheating is likely to occur if the shape of the parts is such that a coil cannot follow the surface contours. As a result, less efficient heating occurs, whereas slower heating allows time for temperature gradients within the parts to be smoothed out.[16]

The capability to heat selectively is one of the main reasons that the induction method is used when the nature of the brazement demands localized brazing. This may be because of a heat treating consideration or because it is uneconomical to heat an entire assembly. Selective heating of the joint may also prove advantageous in preventing distortions due to overheating or annealing of cold-worked areas. Induction heating also provides very fast heating, which may be important in certain production operations.

Frequencies for induction brazing generally vary from 5 KHz to 500 kHz. The lower frequencies are obtained with solid-state generators and the higher frequencies with vacuum tube oscillators. Induction generators are manufactured in sizes from one kilowatt to several hundred kilowatts output. One generator may be used to energize several individual workstations in sequence, using a transfer switch, or assemblies in holding fixtures may be indexed or continuously processed through a conveyor-type coil for heating to brazing temperature.

The frequency of the power source determines the type of heat that will be induced in the part: high-frequency sources produce skin heating; lower-frequency sources, deeper heating. The brazing heat is usually developed within 10 to 60 s.

The brazing filler metal is preplaced. Careful design of the joint and the coil setup[17] are necessary to ensure that the surfaces of all members of the joint reach the brazing temperature at the same time. Flux is employed except when an atmosphere is specifically introduced to perform the same function, Figure 3.12.

Assemblies may be induction brazed in a controlled atmosphere by placing the components and coil in a non-metallic chamber, or by placing the chamber and work inside the coil. The chamber can be quartz Vycor or tempered glass.[18]

Experienced users of high-frequency equipment often show considerable ingenuity in devising coils for seemingly impossible applications. Any new user would, therefore, do well to seek the advice of the equipment manufacturer.

Induction brazing is well suited for mass production, and mechanized systems for moving the assemblies to and from the coils are very common. The rapid heating rates available with induction heating are a major advantage when brazing filler metals that tend to vaporize or segregate are used. The heating cycle for induction brazing is invariably automated, even when manual loading of assemblies is used.

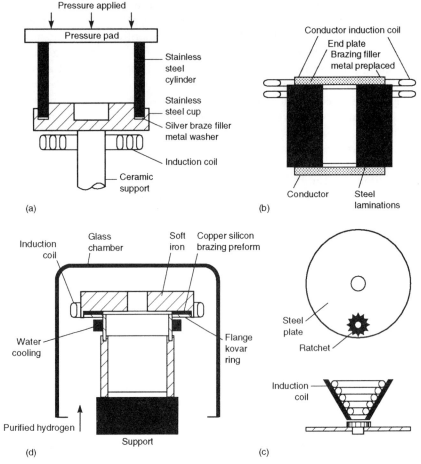

Fig. 3.12 Coil work arrangements for fluxless brazing in a bell jar: (a) stainless steel parts joined in purified hydrogen atmosphere; (b) brazing copper conductors to endplates of motor rotors; (c) steel ratchet wheels brazed to copper-plated disk; and (d) vacuum tube assembly.[16–18]

3.4 DIP BRAZING

Dip brazing is divided into two methods; chemical bath and molten metal bath dip brazing. The molten material is contained in a 'pot' furnace heated by oil, gas, or electricity. In some instances, electrical resistance units in the bath are used for heating. In the former process, the parts being joined are held together and immersed in a bath of molten bonding metal which flows into the joints when the parts reach a temperature approaching that of the bath. For brazing, the molten metal is covered by a layer of flux. This cleans the workpiece as it is introduced and also protects the brazing filler metal by preventing oxidation and the loss of volatile elements from the bath.

Molten metal bath method

This method is usually limited to the brazing of small assemblies, such as wire connections or metal strips. A crucible, usually made of graphite, is heated externally to the required temperature to maintain the brazing filler metal in fluid form. A cover of flux is maintained over the molten filler metal. The size of the molten bath (crucible) and the heating method must be such that the immersion of parts in the bath will not lower the bath temperature below brazing temperature. Parts should be clean and protected with flux prior to their introduction into the bath. The ends of the wires or parts must be held firmly together when they are removed from the bath until the brazing filler metal has fully solidified.

Jigging to maintain alignment is generally necessary. Because of the difficulties of heating and containing metals at high temperatures, alloys which require a brazing temperature above 1000°C are rarely used. The choice of brazing filler metal is therefore restricted to straight brasses and silver-based alloys.

Molten chemical (flux) bath method.

This brazing method requires either a metal or ceramic container for the flux and a method of heating the flux to the brazing temperature. Heat may be applied externally with a torch or internally with an electrical resistance heating unit. A third method involves electrical resistance heating of the flux itself; in that case, the flux must be initially melted by external heating. Suitable controls are provided to maintain the flux within the brazing temperature range. The size of the bath must be such that immersion of parts for brazing will not cool the flux below the brazing temperature.

In the molten flux method, the brazing filler metal is located in or near the joints and is heated to the required temperature by immersion in a bath of molten salt (Figure 3.13). Salt bath brazing has a greater scope than any other single brazing process; it can be used on as wide a range of parent metals as torch brazing but is not subject to the same maximum temperature limitations. It is, unfortunately, an inflexible process. The type of salt used for a particular application depends on the ease with which the parent metal surface oxides can be removed and on the temperature required for brazing.

Parts should be cleaned, assembled, and preferably held in jigs prior to immersion into the bath. Brazing filler metal is preplaced as rings, washers, slugs, paste or as a cladding on the base metal. Preheat may be

Fig. 3.13 Aluminum assemblies being immersed in a molten salt bath.[19,20,26,31]

necessary to assure dryness of parts and to prevent the freezing of flux on parts which may cause selective melting of flux and brazing filler metal. Preheat temperatures are usually close to the melting temperature of the flux. A certain amount of flux adheres to the assembly after brazing. Molten flux must be drained off while the parts are hot. Flux remaining on cold parts must be removed by water or by chemical means.

The molten flux method is used extensively for brazing aluminum and its alloys. The brazing filler metal is preplaced and the assembly immersed in the flux bath, which has been raised to brazing temperature. The flux bath provides excellent protection against reoxidation of the metal, which can occur quite easily with aluminum.

The molten flux brazing method generally causes less distortion than torch brazing does, because of the uniform heating. It may require relatively complex tooling, however, and is therefore best suited to medium- to high-production runs. This process is particularly well suited for small- to medium-size parts with multiple hidden joints

3.5 RESISTANCE BRAZING

The heat necessary for resistance brazing is obtained from the flow of an electric current through the electrodes and the joint to be brazed. The process is most applicable to relatively simple joints in metals that have high electrical conductivity. In the usual application of resistance brazing, the heating current, which is normally alternating current, is passed through the joint itself. The joint becomes part of an electrical circuit, and the brazing heat is generated by the resistance at the joint. Equipment is the same as that used for resistance welding, and the pressure needed for establishing electrical contact across the joint is ordinarily applied through the electrodes.

The brazing filler metal, in some convenient form, is preplaced or face-fed. Fluxing is done with due attention to the conductivity of the fluxes. (Most fluxes are insulators when dry.) Flux is employed except when an atmosphere is specifically introduced to perform the same function. The parts to be brazed are held between two electrodes, and proper pressure and current are applied. The pressure should be maintained until the joint has solidified. In some cases, both electrodes may be located on the same side of the joint with a suitable backing to maintain the required pressure.

Brazing filler metal is used in the form of preplaced wire, shims, washers, rings, powder, or paste. In a few instances, face feeding is possible. For copper and copper alloys, the copper–phosphorus filler metals are most satisfactory since they are self-fluxing. Silver-based filler metals may be used, but a flux or atmosphere is necessary.

The parts to be brazed must be clean. The parts, brazing filler metal, and flux are assembled and placed in the fixture and pressure applied. As current flows, the electrodes become heated, frequently to incandescence, and the flux and filler metal melt and flow. The current should be adjusted to obtain uniform rapid heating in the parts. Overheating risks oxidizing or melting the work, and the electrodes will deteriorate. Too little current lengthens the time of brazing. Experimenting with electrode compositions, geometry, and voltage will give the best combination of rapid heating with reasonable electrode life.

Quenching the parts from an elevated temperature will help flux removal. The assembly first must cool sufficiently to permit the braze to hold the parts together. When brazing insulated conductors it may be advisable to quench the parts rapidly while they are still in the electrodes to prevent over-heating of the adjacent insulation. Water-cooled clamps prevent damage to the insulation.

Resistance brazing is most applicable to joints which have a relatively simple configuration. It is difficult to obtain uniform current distribution, and therefore uniform heating, if the area to be brazed is large or discontinuous or is much longer in one dimension. Parts to be resistance brazed should be so designed that pressure may be applied to them without causing distortion at brazing temperature. Wherever possible, the parts should be designed to be self-nesting, which eliminates the need for dimensional features in the fixtures. Parts should also be free to move as the filler metal melts and flows in the joint.

The equipment consists of tongs or clamps with the electrodes attached at the end of each arm. The arms are current-carrying conductors attached by leads to a transformer.

Electrodes for resistance brazing are made of high resistance electrical conductors, such as carbon or graphite blocks, tungsten or molybdenum rods, or even steel in some instances. The heat for brazing is mainly generated in the electrodes and flows into the work by conduction. It is generally unsatisfactory to attempt to use the resistance of the workpieces alone as a source of heat.

The pressure applied by a spot welding machine, clamps, pliers, or other means must be sufficient to maintain good electrical contact and to hold the pieces firmly together as the filler metal melts. The pressure must be maintained during the time of current flow and after the current is shut off until the joint solidifies. The time of current flow will vary from about one second for small, delicate work to several minutes for larger work. This time is usually controlled manually by the operator, who determines when brazing has occurred by the temperature and the extent of filler metal flow. The process is generally used for low-volume production in joining electrical contacts and related electrical elements.

3.6 INFRARED BRAZING

The development of high-intensity quartz lamps and the availability of suitable reflectors have made infrared heat a commercially important generator of heat for brazing. Infrared heat is radiant heat obtained below the red rays in the light spectrum. Although with every black heat source there is some visible light, the principal heating is done by the invisible radiation.

Infrared brazing may be considered a form of furnace brazing with heat supplied by long-wave light radiation. Heating is by invisible radiation from high intensity quartz lamps capable of delivering up to 5000 watts of radiant energy. Heat input varies inversely as the square of the distance from the source, but the lamps are not usually shaped to follow the contour of the part to be heated. Concentrating reflectors focus the radiation on the parts.

For vacuum brazing or inert-gas protection, the assembly and the lamps are placed in a bell jar or retort that can be evacuated or filled with inert gas. The assembly is then heated to a controlled temperature, as indicated by thermocouples. The part is moved to cooling platens after brazing.

Lamps are often arranged in a toaster-like configuration, with parts travelling between two banks of lamps. Infrared brazing can concentrate large amounts of heat in small areas, which can be advantageous in certain applications.

Infrared brazing set-ups are generally not as fast as induction brazing, but the equipment is less expensive. Honeycomb panels have been successfully brazed using opposite banks of lamps, and spot brazing of smaller parts has been accomplished using parabolic mirrors.

3.7 BRAZE WELDING

The common application for the oxyfuel welding (OFW) torch is braze welding. Unlike brazing, filler metal does not feed into the joint by capillary action. In this process, the oxyfuel flame melts and flows filler metal into a joint without melting the base metal.

Braze welding is a joining process in which a filler metal is melted and deposited in a specific joint configuration, and in which metallurgical bonding is obtained by a wetting action often accompanied by some degree of diffusion with the base metals. Braze welding requires heating, but not melting (liquidus temperature) of the base metal having a melting (liquidus) temperature above 450°C.

Stringent fit-up is not critical because the filler metal is deposited in grooves and spaces. The filler metal flows into gaps wider than those used for brazing. Fabricators use braze welding as a low-temperature substitute for oxyfuel welding or as a low-cost substitute for brazing. Joint designs for braze welding are the same as for OFW, Figure 3.14.

Braze welding was originally developed for the repair of cracked or broken cast iron parts. Compared to OFW, it is less likely to cause weld cracking and to form hard microconstituents. The process has since been expanded to several other applications where low temperature, high strength joints with a narrow heat–affected zone is required. Braze welding also has been used to join steels, copper, nickel and nickel alloys.

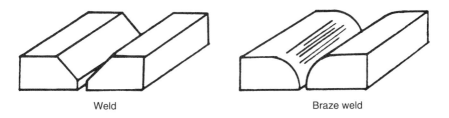

Weld   Braze weld

Fig. 3.14 Configuration of butt joint preparation.[19,20,27,31]

Foundries braze-weld to repair iron castings; machine shops use the process to correct machining errors; fabricating shop maintenance departments and tool rooms repair tooling and hand tools and press frames by braze welding. Compared to conventional fusion welding processes, braze welding requires less heat input, permits higher travel speeds, and causes less distortion; deposited filler metal is soft and ductile for good machinability; and residual stress is low.

To braze-weld steels, heat the work only to 732°C before depositing filler metal; cast irons need preheat to 538°C or less. The process joins brittle cast irons without extensive preheating.

Although most braze welding initially used an OFW torch, copper alloy brazing rod, and a suitable flux, present applications use carbon arc (CAW), gas tungsten arc (GTAW), gas metal arc (GMAW) or plasma arc torch welding (PAW) without flux in the manual, semi-automatic, or automatic modes to economically bond and deposit the filler metal in the braze-welded joints.[19-27]

Filler metal selection, proper wetting and compatibility with the base metals, and shielding from air are important considerations for effective use of the process with any suitable heating method.[19-27] Filler metals for oxyfuel braze welding contain approximately 60% Cu–40% Zn. These fall under AWS specification 5.27, *Specification for Copper and Copper Alloy Rods for Oxyfuel Gas Welding*, classifications RBCuZn–A, –C, and –D, Table 3.1.

A wide variety of parts can be braze-welded with the use of typical weld joint designs. Groove, fillet, and edge welds can be used to join simple and complex assemblies, while avoiding sharp corners that are easily overheated and may become points of stress concentrations, Figure 3.15.

To obtain good joint strength, an adequate bond area is required between the brazing filler metal and the base metal. Weld groove geometry should provide an adequate groove face area so that the joint will not fail along the interfaces. Selection of a proper joint design will

Table 3.1 Filler rods for braze welding of cast iron and steel[19, 20, 26, 27, 31]

| Classification | Common name | Composition, maximum weight percent | | | | | | | Weld tensile strength, $(\times 10^5 \, Pa)$ |
		Cu	Zn	Sn	Fe	Ni	P	Si	
RBCuZn–A	Naval brass	61	rem.	1.00	—	—	—	—	2.7
RBCuZn–C	Low-fuming brass	60	rem.	1.10	1.20	—	—	0.15	3.4
RBCuZn–D	Nickel brass	50	rem.	—	—	11.0	0.25	0.25	4.1

produce deposited metal filler metal strengths which may meet or exceed the minimum base metal tensile strengths. Because of the inert shielding gas, electrical arc methods have less included flux compounds and oxides at the faying surfaces. The result is higher joint strengths and improved corrosion resistance. Original surfaces are restored by overlayments and subsequent machining, Figures 3.15 and 3.16.[19, 20, 26, 27]

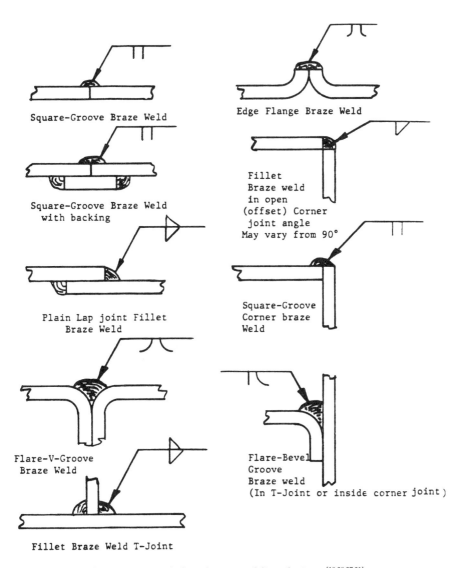

Square-Groove Braze Weld

Square-Groove Braze Weld with backing

Plain Lap joint Fillet Braze Weld

Flare-V-Groove Braze Weld

Fillet Braze Weld T-Joint

Edge Flange Braze Weld

Fillet Braze weld in open (offset) Corner joint angle May vary from 90°

Square-Groove Corner braze Weld

Flare-Bevel Groove Braze weld (In T-Joint or inside corner joint)

Fig. 3.15 Typical sheet braze-welding designs.[19,20,27,31]

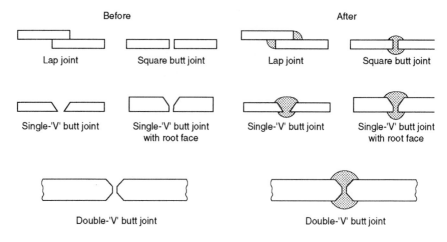

Fig. 3.16 Typical joint designs for braze-welding.[19,20,26,27,31]

3.8 MISCELLANEOUS PROCESSES

3.8.1 Blanket Brazing

In blanket brazing, a blanket is resistance heated and most of the heat is transferred to the parts by two methods – conduction and radiation. Savings in heat energy, time, cost of furnaces, and other factors added up to major economies in the manufacture of honeycomb panels by blanket brazing.

3.8.2 Exothermic Brazing

Exothermic brazing is a special process which heats a commercial filler metal by a solid-state exothermic chemical reaction. An exothermic chemical reaction generates heat released as the free energy of the reactants. Nature has provided countless numbers of such reactions; those solid-state or nearly solid-state metal/metal oxide reactions are suitable for use in exothermic brazing units.

Exothermic brazing uses simplified tooling and equipment. The reaction heat brings adjoining metal interfaces to a temperature at which preplaced brazing filler metal melts and wets the base metal interface surfaces. Several commercially available brazing filler metals have a suitable flow temperature. The process is limited only by the thickness of the base metal and the effect of brazing heat, or any previous heat treatment, on the metal properties.[26]

3.8.3 Electron and Laser Beam Brazing

In a limited number of applications, a laser has been used where a small localized area of heat was required, such as for brazing of small carbide tips on printer heads for electronic printers. The electron beam has generally been used for brazing by defocusing the beam to provide a wider area of heating. Because this is done in a vacuum, fluxes cannot be used and the brazing filler metal must be selected so that there is little or no vaporization during brazing.[26]

3.8.4 Microwave Brazing

One of the newest joining methods to be developed is the use of microwaves. This technique is especially being expanded to include the range of ceramics in high-temperature, corrosion-resistant, and high-performance applications. Applications such as these require complex ceramic parts with strong, durable joints.

The technique uses a single-mode microwave cavity whereby the iris, Figure 3.17, controls the percent of microwaves reflected in the cavity, and the plunger adjusts the frequency. Together, they focus microwaves on the joint. The major advantage to microwave joining[28] is not having to heat the entire ceramic part in order to make a joint. Reheating finished ceramic parts can subject them to thermal stresses, causing cracking, which weakens the part. With microwave joining, only the interface

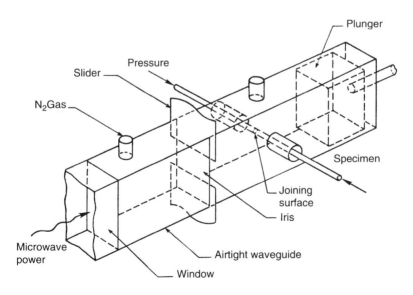

Fig. 3.17 Single-mode microwave cavity.[28,29]

between the pieces is heated. It is faster than conventional joining and large pieces can in principle, be joined at a lower cost.

Microwave joining does have a few disadvantages – commercial microwave equipment is only produced by a few companies, mostly for research and development.

One unit, for example, features 2.45 GHz single-mode or controlled multimode operation. Its cavity is designed to provide controlled mode patterns which can focus high energy microwave fields for heating a part only where it is needed, for instance, a joint. It can heat a joint or part up to 2200°C. Researchers[29] have joined 92% alumina ceramic parts in a 6 GHz single-mode microwave cavity at various temperatures and times. The bending strength of this joint gradually increased from the solidus temperature of 1400°C and reached a peak of 420 MPa at about 1750°C. The ceramics were directly butt-joined with no interface layer, loaded at 0.6 MPa, and heated for 3 min. with microwave energy. This value was equal to the strength of a monolithic ceramic part.

Others[30] have joined 85% mullite ($3Al_2O_3 \bullet 2SiO_2$) in a 2.45 GHz single-mode microwave cavity with a magnetron as its source of electromagnetic energy. Unfortunately, the magnetron does not have the fine control of the klystron and can only be turned on and off. These researchers made butt joints using 89 mm by 9.2 mm diameter rods, heating them to 1300°C while loaded at 5 to 10 MPa.

Inherent with microwave joining is thermal runaway. As a ceramic gets hotter, it absorbs energy more readily and therefore, heating has to be controlled.

Silicon nitride (Si_3N_4) has been microwave-joined in a nitrogen gas atmosphere at 1400°C. At this temperature, the Si_3N_4 joint strength is equal to the bulk material's strength of 340 MPa. Again, the joining time influences the joint strength. Researchers also found that a Si_3N_4 joint heated to 1720°C with a Si_3N_4 interface thickness of 0.6 mm reached its maximum strength of 390 MPa in 3 min. This value was higher than the original strength of the interface material and in excess of 70% of the base material. Joint strength tends to decrease at more than 3 min. by evaporation of elements or oxidation of the surface. The boundary line of the above is unobservable.[28, 29]

Case Histories

A company installed a bell furnace to copper braze stainless steel (304) heat exchangers. The parts would not braze and the parts came out of the furnace with a green coating. All purging of the furnace was done with a mechanical pump. The furnace was purged with gasified liquid nitrogen with –57°C dew point and gaseous hydrogen added at 760°C and shut off when the temperature dropped below 760°C.

In trouble-shooting the above situation the first item to be examined was the atmosphere since it is an essential component of brazing stainless steel. The furnace, since it was purged with a vacuum pump, should have been checked for leaks by evacuating (shutting the main valve to the vacuum pump and watching the pressure rise in the furnace). Some outgassing will occur, thus, a slow increase in vacuum pressure if the furnace is pumped down to 500 microns.

Secondly, since the very dry nitrogen and hydrogen gases will pick up oxygen and moisture from the smallest leaks in the transmission lines you must examine this area. If you shut off the main atmosphere lines on your control panel, and open the valves at the trailer and the nitrogen tank, to pressurize these lines, an all-inclusive test of the transmission lines can be conducted by shutting off the valves just past the regulator and observing the drop in pressure for a 15- or 30-min period.

Finally the next item to consider is the flow rate of atmosphere in the retort. For a very good atmosphere, one would normally use 0.009 m^3 min^{-1} of atmosphere per cubic meter of retort space to sweep out all of the outgassing from the parts, the retort work base and the binders, so that a low dew point can be maintained. While this is a rather large volume of gas, it has been found that under normal circumstances 0.007 and 0.0045 m^3 min^{-1} m^{-3} of retort can be used quite satisfactorily. As a minimum in handling stainless steel, 0.0035 m^3 min^{-1} m^{-3} of retort can be used if the system is very tight and the dew point of the incoming atmosphere is quite good; however, normally it's recommended to be at least 0.0045 m^3 min^{-1} m^{-3} of retort.

In a second situation a firm planned to install a continuous mesh belt furnace for the brazing of stainless steel. They were planning to use brazing filler metals of copper, nickel, and possibly silver. Their concern was:

- What atmosphere and dew point would be required?
- Would any particular base metals present difficulties with these atmospheres?

The continuous mesh belt furnace used for brazing stainless steel has two major design configurations. The most frequently used is the straight-through mesh belt furnace. The second design is best suited to hydrogen atmosphere, which is lighter than air, and dissociated ammonia, which is 75% hydrogen. However, other atmospheres can readily be used in this furnace.

To braze stainless steel in a straight-through mesh belt furnace, it is necessary to have highly specialized ceramic cloth curtains and gas curtains at each end of the furnace, to minimize the amount of oxygen and moisture diffusing back into the furnace. The atmospheres that are most commonly used when brazing stainless steels are: 1) a nitrogen/hydrogen mixed gas. The percent of hydrogen added to the nitrogen can vary from 20 to 50% in the normal installations. 2) Dissociated ammonia atmospheres, which are 75% hydrogen and 25% nitrogen.

The dewpoint of the atmosphere in the furnace in the hot zone is all important and will determine whether the stainless steel is bright and shiny or is discolored. To obtain satisfactory brazing, the stainless steel must be bright and clean so that the filler metal will wet the surfaces and flow into the capillary joint. Thus, the atmosphere in the hot zone of the furnace is the critical key to good quality brazed parts. The dewpoint of atmosphere going into the furnace must be as low as possible, to compensate for the increase of oxygen and moisture which are brought into the furnace by the parts, the belt and diffusion through the door openings. The atmosphere in the hot zone should be −40°C or lower, for suitable brazing quality.

In reference to base metals, any of the chromium irons and chromium-containing stainless steels that do not contain aluminum and titanium as alloying elements will be readily brazeable. However, depending on the dewpoint of the atmosphere, small amounts of aluminum and titanium in the base metal may or may not cause a problem in brazing. When the atmosphere is marginal and the titanium and/or aluminum are sufficiently high, then the braze may not wet and flow.

Examples of marginal materials would be 321 and 409 stainless steels. Both of these contain small amounts of titanium, but depending on the chemistry, surface condition and dewpoint, they may or may not braze satisfactorily. Base metals containing higher amounts of aluminum and titanium, can also be brazed in the continuous mesh belt furnace with these atmospheres, if they are plated with electrolytic nickel. The thickness of the electrolytic nickel will depend on the amount of titanium and aluminum in the base metal and the time and temperature of the brazing process. A good plating thickness would be 0.01 mm for the 321 or 409 and get satisfactory results. However, the color of the final part is the indicating footprint that shows whether or not the plating thickness is satisfactory, everything else being in the proper condition. Electroless nickel should not be used, as this melts and brings the titanium and aluminum to the surface much faster and does not provide a good barrier coating.

Brazing filler metals of many kinds can be used in the straight through continuous furnace when brazing stainless steel parts. Copper (BCu–1) is a satisfactory brazing filler metal if the service conditions will permit; however, if the application requires a stainless steel, a more corrosion resistant filler metal may be desirable. There are several types of nickel filler metals that are usable in the continuous mesh belt furnace, and one of these is BNi–7, which is a nickel–chromium–phosphorus filler metal. The importance with this filler metal is that it needs to be adequately diffused to obtain good physical properties, and it should be limited to very small fillets.

Most of the nickel filler metals are brazed at 1121°C to provide suitable diffusion. The BNi–5 filler metal would be suitable in a nitrogen/hydrogen atmosphere; however, it is not suitable in continuous mesh belt furnaces because the brazing temperature is above 1149°C, which is extremely hard on the furnaces. The boron-containing nickel filler metals are not suitable for nitrogen atmospheres, as the boron converts to nitride and takes the low melting phase out of the alloy, thus giving unsuitable braze results.

The silver filler metals are suitable for brazing at the higher temperatures, where the stainless is bright and clean, however, at the lower temperatures such as 760–816°C the atmosphere is not sufficiently reactive to give suitable braze results. Thus, electrolytic nickel plating, again, can be used to hide the chromium and provide a suitable surface for brazing at these temperatures.

As a note of caution, silver filler metals containing zinc and cadmium should not be used in an atmosphere furnace, since the atmosphere increases the vaporization and thus vaporizes the zinc and cadmium and removes them from the filler metals. This, of course, changes the melting point of the resulting filler metal, as well as dumping out the zinc and cadmium into the atmosphere. Since cadmium is listed as poisonous, filler metals without the zinc and cadmium should be used in these furnace atmospheres. Some of the filler metals that would be suitable are BAg–18, BAg–13a, BAg–19, and BAg–21, and in some conditions, BAg–8 and BAg–8a.

REFERENCES

1. Schwartz W. H., Amazing brazing, *Assembly Engineering*, Aug 1989, pp. 31–33.
2. Brazing system joins 22 different parts, *Welding Journal*, Oct. 1989, 68(10), p. 55.
3. Brazing and riveting combine to produce 1,200 parts per hour, *MAN*, Jan 1990, p. 26.

4. Automated brazing increases uptime and productivity, *MAN*, Feb 1991, p. 40.
5. Steffan W. S., Nitrogen-based atmosphere furnace brazing system provides strength & leak-proof performance of newly-designed fuel injection rail, *Indust. Heating*, Aug 1989, pp. 21–23.
6. Continuous roller hearth furnace installed for brazing automotive radiators, *IH*, June 1991, p. 9.
7. *Metals Handbook*, Vol 6, 8th Ed., 1971, p. 595.
8. *IH*, Nov 1988, p. 8.
9. *Metals Handbook*, Vol 7, 9th Ed., 1984, p. 313, pp. 355–358.
10. Seco-Warwick Corp., Sunbeam 2M Furnace Sep 1977 brochure.
11. Sakamoto A., Study of furnace atmosphere for vacuum-inert gas partial-pressure brazing, *Welding Journal*, 70(11), Nov. 1991, pp. 311s–20s.
12. New generation of vacuum heat-treatment furnaces feature high gas pressure quenching and oil quench vacuum furnace, *Metallurgia*, May 1989, p. 205.
13. Jones W. R., Vacuum – another atmosphere? *Heat Treating*, Oct 1986, pp. 39–41.
14. Hooven W. T., Nokes K. W., Make your vacuum brazing furnaces user-friendly, *Welding Journal*, 69(10), Oct. 1990, pp. 25–9.
15. Dressler S., Role of hot wall vacuum furnace in plasma-assisted surface treatment: Part II Operating influences, *IH*, Sep 1989, pp. 31–3.
16. Rupert W. D., Induction brazing guidelines: when, how to use this method, *Heat Treating*, Vol 21(10), Oct 1989, pp. 28–30.
17. Lasday S. B., Flexibility of induction heating increased with newly developed hand-held system, *IH*, Nov 1989, pp. 20–2.
18. Gerbossi P. F., Libsch J. F., Controlled-atmosphere brazing with induction heating, *Welding Journal*, 68(10), Oct. 1989, pp. 32–7.
19. *Welding Handbook*, Vol 2, 8th Ed., 1991, pp. 380–422.
20. *Brazing Handbook*, 4th Ed., 1991.
21. Riad S. M., El-Naggar A., Brazing of gray cast iron, *Welding Journal*, 60(10), Oct. 1981.
22. Metzger G., Lison R., Electron beam (braze) welding of dissimilar metals, *Welding Journal*, 55(8), Aug. 1976.
23. Grieshiem G., *Electrode for 0.002 shielded braze welding of vehicle body panels*, British Patent, published Sep 19, 1979.
24. *D9.1 Specification for the welding of sheet metal*.
25. Arco, Pulsed-arc machine MIG brazing improves Chrysler sheet metal welds, Arco Welding Products, *Productivity News*, Vol 2, No 1, 1981.
26. Schwartz M. M., *Brazing*, ASM International, 1989.
27. *ASM Braze Welding Handbook*, Vol 4.
28. Schwartz M. M., *Ceramic Joining*, ASM International, 1990.
29. Kempfer L., New waves in ceramic joining, *Materials Engineering*, Apr 1991, 31–8.
30. Schwartz M. M., *Structural Ceramic Handbook*, McGraw-Hill Book Co. Inc, New York, NY, Jan 1992.
31. Naka M., Tanaka T., Okamoto I., *Quart. J. Jpn Weld. Soc.* Vol 4 (1986) 597.

4

Braze filler metal and base material families

For satisfactory use in brazing, filler materials must possess certain basic characteristics. First, filler materials must have the ability to form brazed joints possessing suitable **mechanical and physical properties** for the intended application. This often means strength but may include ductility, toughness, electrical or thermal conductivity, temperature resistance, and stability.

An extremely important physical property for intended filler materials is that they have a **coefficient of thermal expansion** (CTE) that closely matches the substrates being joined or, where severe temperature gradients persist, bridges the difference in CTEs between the two joint element materials. This is so that thermal mismatch stresses across the joint do not cause failure by fracture.

Second, the **melting point or range** of an intended filler material must be compatible with the base materials being joined and have sufficient fluidity at the brazing temperature to flow and distribute into properly prepared joints by capillary action. 'Suitable' melting range means below the solidus of the base materials but as high as necessary to meet service operating temperature requirements.

Third, the composition of the intended filler must be sufficiently homogeneous and stable that separation of constituents, **liquation**, does not occur under the brazing conditions to be encountered. Obviously, the intended filler material composition must also be chemically compatible with the substrates to avoid adverse reactions during brazing or by subsequent sacrificial (i.e., galvanic) corrosion.

Fourth, intended fillers must have the **ability to wet** the surfaces of the base materials being joined to form a continuous, sound, strong bond.

Fifth, depending on requirements, intended fillers must have the ability to produce or avoid reactions with the base materials. Usually it is desirable to avoid such reactions, since **brittle intermetallics** may result, degrading joint properties. However, for so-called active-metal or reactive brazing, it is also necessary for the filler and the substrates to react chemically in a particular way (see ceramic/ceramic and metal/ceramic brazing).

4.1 FILLER METAL TYPES

One must consider several factors when selecting a brazing filler material, whether it is a metal or a ceramic. First **compatibility**:

- with the base material means properly matching chemical, mechanical, and physical properties;
- with the joint design means proper mechanical properties for the type and magnitude of loading (e.g., static or fatigue; tension, shear, or peel).

Second, the filler material must be suitable for the planned **service conditions** for the brazed assembly, including:

- service temperature;
- thermal cycling;
- life expectancy, stress loading;
- corrosive conditions;
- radiation stability;
- vacuum operation (i.e., outgassing).

4.1.1 Coatings

Another technique used to preserve the surface of the pre-brazed component and be free of oxides is to electrochemically **metallize** the joint surfaces prior to brazing. The process removes surface oxides and applies a thin adherent coating to give a protected surface for brazing. An operator can direct a handheld stylus over the joint; the stylus holds a conductive fluid that electrochemically coats the work. No heat, no distortion, no masking are involved.

The possibility of using the new coating technologies in brazing opens many new avenues for heretofore difficult-to-join materials. By depositing one or more layers of filler metals on the joint surfaces, traditionally difficult-to-join materials can be brazed together. Barrier coatings, coatings for dissolution/solidification, and reactive metal coatings (interlayers) are three different groups of coatings. To control brazing, the optimal amount of coating applied must be determined and the interfacial chemical reactions must be understood, with the characterization of the products that result in the braze filler metal.

The brazing of advanced engineering materials, and especially dissimilar materials, is coupled to the effective use of coatings. The most obvious application is to use the coating process to establish a highly wettable surface for the liquid filler metal. The coating may also be used as a barrier when the brazing filler metal is incompatible with the base material, to avoid diffusion of certain alloying elements and to prevent intermetallic formations.

Recent approaches [1-3] to coating technology in brazing recognize that coatings are no longer just an intermediate passive barrier, but they take an active part in the brazing process. Mass transport, either from the brazing filler metal into the base material or the dissolution of the coating into the braze metal, can promote physical changes, such as increasing the liquidus temperature of the braze metal, which affect significantly the brazing process. With proper selection of the coating material and brazing filler metal, very consistent brazing processes can be developed. The control of the chemical composition and thickness of the coating layers are far more critical than that required in traditional brazing filler metals.

Application

There are many ways to apply coatings on parts to be brazed. They are:

1. *Electrolytic*: electroplating or electroless deposition;
2. *Thermal*: dip coating, barrel coating, roller coating, flow melting, etc.;
3. *Surface modification*: cladding, thermal spraying, plasma spraying, etc.;
4. *Physico-chemical*: sputtering deposition, vapor deposition, ion implantation, etc.

Since both chemical homogeneity and thickness are important properties of a coating, the coating procedure must be well controlled. For components that require critical dimensional control, continuous **sputter-cleaning** and **vapor deposition** processes are preferred because of the uniform adhesion obtained at the interfaces.

Three different coating schemes are the primary methods used in brazing. They are:

1. barrier coatings;
2. reactive metal coatings;
3. coatings for dissolution-solidification.

Barrier coatings

It has been common practice, when no known suitable brazing filler metal is available for the materials to be joined, to use a coating that can be wetted by the liquid braze.[4] The coating also serves as a barrier between the base metal and the braze during processing. A barrier layer must be dense, ductile and free from defects such as voids. It must also show good adhesion and wetting toward both the base and filler metals. Copper–phosphorus filler metal had been used in the past to join ferrous alloys; however, intergranular penetration of phosphorus was observed to promote subsurface embrittlement of the joint. A thin undercoating of nickel reduced phosphorus diffusion and provided good wetting for the copper–phosphorus filler metal.[4]

Sometimes the barrier coating scheme may require a double coating. Material for the first layer is selected for its compatibility with the base metal and the second layer (overlayer or outer coating) is used to promote wetting with the liquid brazing filler metals, as illustrated in Figure 4.1.

The primary layer is generally a transition metal solution containing polyvalent elements such as titanium, manganese, zirconium, tantalum, or molybdenum, where the polyvalent states increase the bonding tendency. The second layer is usually a thin layer of noble metal which provides a surface that is oxide-free, or with a very thin oxide layer which readily decomposes or dissolves during brazing. Such oxygen-free surfaces have, in general, high surface energy and offer excellent wettability. Metals such as silver, gold and copper serve this function, requiring little, if any, fluxing agent. The double coating scheme will also allow for longer brazing times since the thickness of the other coating can be adjusted.

Reactive metal coatings

The use of reactive metals in brazing has been shown to promote the formation of a thin interlayer between the brazing filler metal and the base metal.[5–8] This layer promotes adhesion between the faying surfaces of similar and dissimilar materials. Hence, the resulting strength of the braze

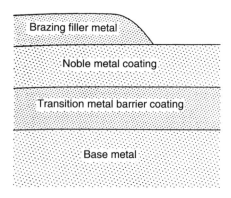

Fig. 4.1 Schematic diagram showing the noble metal and transition metal barrier coatings inserted between the base metal and the brazing filler metal.[4]

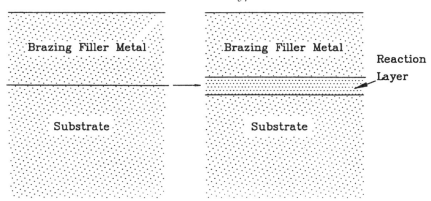

Fig. 4.2 Schematic drawing showing the formation of a reaction layer between the substrate and brazing filler metal.[4]

is dependent upon the nature of the product layer and its thickness. An example is the use of titanium in noble-metal-based brazing filler metals for the joining of metals to ceramic materials. Titanium in the liquid braze metal reacts with the substrate to form a thin reaction layer as shown in Figure 4.2. This product layer can be a complex oxide such as Cu_2Ti_4O and $Cu_3Ti_3O_2$, or an intermetallic compound, depending whether the base material is an oxide ceramic or a metal. Certain types of complex oxide, for example, (Cr, Mn, Mg)O. (Cr, Mn, Al_2O_3 spinel) also promote adhesion between metal and oxides. The characterization and understanding of the crystal structure and microstructure of the product layers must be carried out so that adequate process control can be achieved.[9,10]

Dissolution–solidification coatings

Some coatings are deposited for the purpose of dissolution and solidification. In contrast to the passive barrier coatings, these dissolution–solidification coatings are considered active because of their contribution to the braze metal composition and its thickness.

A good example is the **transient liquid phase bonding** (TLP).[11] TLP bonding is a diffusion brazing process that combines the features of both brazing and diffusion welding.[12–14] It uses as filler metal a thin interlayer or brazing filler metal of specific composition and melting temperature. At the bonding temperature, the interlayer may melt or a liquid may form by alloying between the interlayer metal and the base metal. The liquid, by capillary action, fills the joint clearance and contributes to the elimination of voids at the braze interface. While the joint members are held at the bonding temperature, diffusion of alloying elements occurs

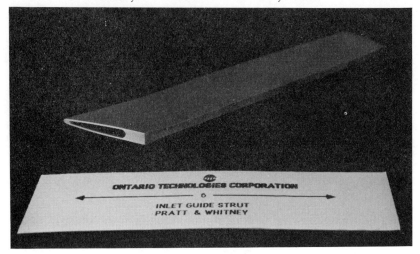

Fig. 4.3 Ti–6A1–4V inlet guide strut SPF/DB (super plastic forming/diffusion bonding).

between the liquid and base metal. Isothermal solidification of the joint results because of the solute composition change in the braze. Maintaining the joint at the bonding temperature after solidification will promote further homogenization of the chemical composition and microstructure. Solid-state diffusion of elements away from the interfacial region reduces the initially large chemical composition gradient, avoiding the formation of intermetallics at the braze. An element of high mobility, both in the liquid and solidified braze metal, will decrease the time for completion of the TLP process.

Conventional transient liquid phase bonding is mostly applied to binary alloy systems which show some intermediate, low-temperature reactions such as eutectic transformation.

Another example is **liquid interface diffusion** (LID) which is most appropriately used on titanium and titanium alloys shown in Figure 4.3[15] as well as **activated diffusion bonding** (ADB).

4.1.2 Filler metal selection

In choosing a braze filler metal the following information should be determined for specific brazing applications:

1. *The base metals being joined.*
2. *The method of heating to be used.* Brazing filler metals with narrow melting ranges of less than 25°C between solidus and liquidus can be used with any heating method, and the filler metal may be preplaced

in the joint area in the form of rings, washers, formed wires, shims, powder, or paste. Alternatively, such filler metals may be manually or automatically face-fed into the joint after the base metal is heated. Filler metals that tend to liquate are used with heating methods that bring the joint to brazing temperature quickly or allow introduction of the filler metal after the base metal reaches the brazing temperature.

3. *The brazing temperature required*. Low brazing temperatures are usually preferred to economize on heat energy; to minimize heat effects on the base metal (annealing, grain growth, warpage, etc.); to minimize filler metal/base metal interactions; to increase the life of fixtures and other tools. High brazing temperatures are preferred in order to take advantage of a higher melting, but more economical, brazing filler metal; to combine annealing, stress relief or heat treatment of the base metal with brazing; to permit subsequent processing at elevated temperatures; to promote filler metal/base metal interactions that increase the joint-remelt temperature;[16] or to promote removal of certain refractory oxides by vacuum or an atmosphere.

4. *Service requirements of the brazed assembly*. Compositions should be selected to suit operating requirements, su ch as service temperature (high or cryogenic), thermal cycling, life expectancy, stress loading, corrosive conditions, radiation stability, and vacuum operation.

4.1.3 Filler metal forms

Brazing filler metals are available as rod, ribbon, powder, paste, creams, wire, sheet, and preforms (stamped shapes, washers, rings, or shaped wires) shaped to fit a particular part. Depending on the joint design, heating method, and level of automation, the filler metal can be preplaced before the heating cycle starts or face-fed after the work is heated. High-production brazing, such as furnace or induction brazing, which typically involves a high level of automation, usually requires pre-placement of the filler metal, Table 4.1.

Rod and wire forms are usually used for manual face-feeding. Many special shapes and forms designed for specific applications are utilized as preplaced or preformed filler metals. Such pre-placing of the filler metal ensures that there is a uniform amount of filler metal in the correct position on each assembly. Pre-placed filler metal may be useful with manual brazing, but is usually a fundamental part of any mechanized brazing procedure. Where joint areas are large, filler metal may be located between the faying surfaces. Brazing rings are sometimes inserted into grooves machined into the work. Totally enclosed rings in such grooves may be necessary for long sleeve joints or for salt bath dip brazing where it is desirable to avoid melting of the ring before the work is heated to the brazing temperature.

Braze filler metal, base material families

Table 4.1 Available physical forms of brazing filler metals[15]

Brazing filler-metal groups	Available forms (a)													
	Ad	Cl	Cr	Fl	Fo	Pa	Pb	Po	Pr	Pt	Sh	Sk	Sp	Wi
Ni and Co alloys	x				x	(b)	x	x	x	(b)	x	(b)		(c)
Pd alloys					x	x		x	x		x		x	x
Cu and Au alloys					x	x	x	x	x		x		x	x
Ni alloys	x				x	(b)	x	x	x	(b)	x	(b)		(c)
Cu–Sn, Cu–Zn, Pd–Ag–Cu					x	x		x	x		x		x	x
Ni-P, Ni–Cr–P	x				x	x		x						
Cu–P, Cu–Ag–P					x			x					x	x
Ag–Cu–Zn, Ag–Cu–Zn–Cd					x	x		x	x		x		x	x
Al–Si		x						x			x			x
Mg–Al–Zn											x		x	x

(a) Ad = adhesive sheet; Cl = cladding; Cr = cream; Fl = flux paste; Fo = foil; Pa = paste (nonfluxing); Pb = plastic-bonded sheet; Po = powder; Pr = preform; Pt = paint; Sh = shim; Sk = stick; Sp = strip; Wi = wire. (b) A few alloys only. (c) Plastic bond.

Although the use of preforms or automatic filler-metal feed is virtually mandatory for mechanized brazing, there is still a wide choice in the form in which to apply the filler metal. The most suitable form for any particular application must be decided on the basis of the following factors:

- Joint design (size of assembly, depth of braze joint, cross-section of components, joint complexity);
- Heating method;
- Desired degree of automation;
- Desired appearance of completed assembly;
- Number of assemblies to be manufactured;
- Range of assembly types to be manufactured.

Competitive forces have pressured metalworking manufacturers to re-evaluate their manufacturing processes. As a result, demands on brazing operations have increased to improve efficiency, produce higher-quality products and reduce costs.

There are a number of factors which can affect the performance of brazing operations. Of particular importance is the proper selection of the correct form of braze filler metal.

There are three general categories of brazing filler metals in production applications. They are:

1. *Non-fabricated wire and strip.* Available on spools which contain a specified amount of filler metal at a standard size.

2. *Fabricated wire and strip forms.* Commonly known as **preforms**, this filler metal can be fabricated as rings, washers, discs, shims or other engineered shapes.[11]

3. *Paste alloys.* A combination of atomized brazing filler metal powder, a neutral binder and, depending on the heating method to be used, a flux. The binder is used to keep the components of the paste in suspension and facilitate dispensing. Each of these filler metal forms has its specific advantages and disadvantages.

Preforms

Of all the filler metal forms, **preforms** offer the most precise control over the amount of filler metal placed in a joint; such control can reduce excess alloy consumption. Example: A wire preform 1.6 mm diameter in the form of a ring with a 25.4 mm I.D. contains more material than the joint needs, as evidenced by a large fillet in the joint area. By reducing the wire diameter to 1.4 mm and retaining the 25.4 mm inner ring diameter, the amount of material used is reduced by 13%.

Where workers are applying excessive amounts of filler material, cost of preforms will likely be less than hand application of bulk filler metal. Higher production rate, improved joint quality, and reduced cost of postbraze clean-up contribute to further cost reductions.

Preforms can be buried in the joint to improve inspectability. A preform placed in the bottom of a joint will melt and flow through the joint area, forming a fillet at the top of the joint. The operator can see that the filler metal has flowed through the joint area completely, Figure 4.4.

If the filler metal is applied to the top of the joint, the operator can determine, after heating, that the filler metal has melted and flowed down into the joint, but not whether joint penetration is complete. This approach increases inspectability of the joint. Preforms generally produce joints of

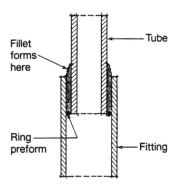

Fig. 4.4 Schematic of braze filler metal preform placement.

good appearance. This feature can reduce or eliminate finish machining, otherwise necessary to improve the appearance of brazed joints.

Wire forms

In terms of the volume of brazing filler metals used in industry today, nonfabricated filler metals are the most widely used. However, while the cost of this material is probably the lowest of the three basic forms, many of its other associated costs are among the highest.

Many companies use nonfabricated filler metals for hand-brazing operations. However, when labor costs are added, hand brazing can be quite expensive compared to automated brazing methods. In addition, research and experience have shown that hand-brazing operations will use anywhere from 10% to 40% more brazing filler metal per joint than operations which use preformed filler metal.

The repeatability of making brazed joints with hand-fed filler metal is generally lower – sometimes significantly lower – than the repeatability of brazed joints made with preforms or paste. Using filler metals in wire form with automatic wire-feeding equipment may sometimes overcome the problems mentioned. Generally, however, problems do exist, and the hidden costs have a negative impact on both productivity and profitability.

When using wire-feeding equipment to braze assemblies, cold and unfluxed wire are applied to components that have been heated. When the brazing filler metal is then introduced, it has a tendency to chill the joint area. The most common way to overcome this potential problem is to overheat the assembly so that it has enough heat present to bring the brazing filler metal in wire form up to the temperature it needs to melt and flow properly.

Because the wire is unfluxed and the assembly is often overheated, many people use liberal amounts of flux to ensure that enough will be left when the brazing actually takes place. This often leads to extensive post-braze cleaning requirements, which is why this type of brazing is most commonly used on simple assemblies which do not require strict cosmetic standards.

Filler pastes and dispensers

Preforming brazing filler metals or paste are mandatory in applications where furnaces are used as the heating method.

For all their attributes, there are situations where preformed brazing filler metals are at a disadvantage. In low-volume runs involving a wide variety of assemblies, for example, the necessary inventory of custom-manufactured preforms can be difficult to manage.

Similarly, the automated dispensing of complex preforms in applications which require total automation can be relatively expensive

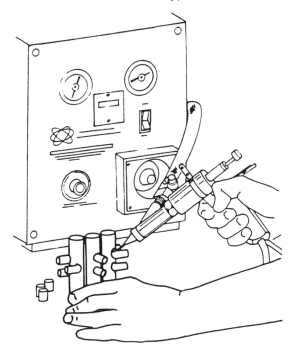

Fig. 4.5 Application of filler metal paste using hand-held applicator gun/dispenser.

compared to other forms, such as paste filler metals. Manufacturers of dispensing systems for preform and paste brazing filler metals may provide the final answer.

Brazing filler metals in paste form also offer users a unique set of advantages. All the elements required to produce a brazed joint – filler metal and flux – are delivered to the joint at one time as a paste deposit. This feature alone can make paste an attractive filler metal form for those metalworking manufacturers who wish to automate the placement of their brazing filler metal. The precise amount of alloy needed in the joint area can be applied with relatively inexpensive equipment, Figure 4.5.

Filler metals in paste form provide versatility. It is possible to use one or two filler metals for a variety of applications, thus reducing filler metal inventory.

Paste filler metals are normally custom-formulated to meet exact requirements. A product can be ideally suited to specific requirements by varying the amount of filler metal, powder size, flux percentages and binder percentages.

Several additional characteristics of paste filler metals should also be considered.

In many applications, paste filler metals are placed on the outside of the joint area. As a result, they are exposed to direct heat throughout the brazing cycle. Careful placement of the filler metal and attention to the heating method and time are necessary to insure that the brazing filler metal in paste form does not melt and flow away from the joint area prior to the components reaching proper temperature.

Many paste filler metal systems exist for heating methods that use torches, induction heat, resistance heating and atmosphere furnaces. There are fewer products that perform reliably in vacuum furnaces. Special care must be taken to evaluate paste filler metals intended for use with vacuum furnaces so as not to damage the equipment.

Transfer tapes

There are a number of factors to be considered when selecting the form of the brazing filler metal for any particular application. In evaluating the total cost of the different filler metal forms, careful consideration of all forms is necessary to make the best choice.

A new approach to controlled application of brazing filler metal is the use of transfer tapes and preforms.

Brazing transfer tapes are produced in either rolls or preforms. Incorporating a prefabricated layer of brazing filler metal, transfer tapes have extreme uniformity in thickness and density as well as ease of application. The tapes consist of four layers (Figure 4.6). A thin plastic foil carrier layer serves to 'carry' the filler-metal layer until actual application takes place. It is backed by a filler-metal layer which can be composed of any powdered filler metal mixed with selected organic binders. The thickness of this layer can be varied from 0.01 to 0.06 mm with a thickness tolerance of ± 5%. The third layer, a pressure-sensitive adhesive film, transfers the filler metal to the metal surface and, during the brazing and firing cycle, decomposes without leaving a residue. A paper fourth layer protects this film and is peeled off before application.

In manual applications, the tape is first cut to the size and shape required. The protective paper is then removed, and the tape is firmly pressed, either by hand or by use of a rubber roller, against the area of

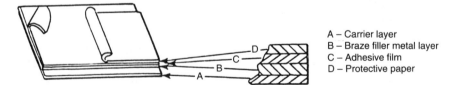

A – Carrier layer
B – Braze filler metal layer
C – Adhesive film
D – Protective paper

Fig. 4.6 The four layers used in transfer tape.

cleaned metal surface to be coated. During this step, the filler metal is transferred to the base metal. The process is completed by simply stripping the carrier layer from the filler metal layer.

Transfer tape has been used to braze turbomachinery seals for gas turbines and has eliminated problems of filler metal distribution and excessive infiltration during brazing.

In another application, use of transfer tape significantly improved a process for assembling sheet-metal turbine blades. Conventional methods of joining a two-piece blade and an internal cooling duct using nickel-base brazing filler metal had resulted in poor control over the location and amount of filler metal applied. Use of transfer tape solved this problem and also permitted simplification of the fixtures required for the brazing operation.

In some cases, gold-nickel transfer tape has replaced filler metal foil. Although prepared and applied similarly to foil, the tape provides more accurate placement and ease of application.

In brazing of honeycomb sections, particularly large, fine-cell honeycomb, the use of transfer tape has solved the problem of migration of brazing filler metal and, in some instances, has reduced the time required to prepare large sections for brazing by as much as 85%.

Plating

When an assembly is made of copper, an excellent method of brazing with the copper-silver eutectic composition in any orientation is to silver plate the joint area to a thickness of 0.0025 to 0.010 mm to produce a tight push fit. Although silver melts at 960°C, heating above the eutectic temperature of 777°C is sufficient to cause the whole joint interface to become molten. The excess plating around the joint provides sufficient additional filler metal to eliminate voids and produce small fillets at the customary joint interface. Larger amounts of filler metal may be obtained by the use of copper–silver filler metal wire rings in conjunction with the silver plating, the latter ensuring excellent and reliable penetration completely through the joint.

Various other plating combinations are possible, e.g., gold plating on nickel, nickel plating on zirconium, and nickel plating on tantalum, all of which produce a lower-melting-point eutectic. 'Electroless' nickel (a nickel–phosphorus alloy) and a whole range of pure metals such as copper, silver, gold, and platinum, may be deposited on joint components of higher-melting-point stainless steels and used as brazing filler metals at their normal melting points without necessarily forming eutectics.

It is worth noting that where electroplating techniques are used, the expertise usually exists to dissolve material from the joint area in a highly controllable manner by either chemical or electrochemical methods. Such techniques have been used to overcome the tight-tolerance problem sometimes encountered with copper components.

Nickel plating may also be used in brazing practice to ensure good wetting of stainless steels by low-melting-point filler metals in protective-atmosphere furnaces, or to prevent oxidation where good control of the dew-point of the furnace atmosphere is not possible. For this purpose, plating at least 0.025 mm thick is desirable to delay diffusion of chromium in the base metal to the surface.

Additionally it should be noted that nickel plating is usually required on the brazed joint areas of stainless, corrosion or heat-resistant steels or superalloys having a percentage by weight equal to, or greater than, the following alloying elements:

(a) Titanium – 0.70%
(b) Aluminum – 0.40%
(c) Titanium plus Aluminum – 0.70%

A 0.01 to 0.02 mm thickness is the nickel plate normally applied.

Foils and sheets

In recent years thin, ductile amorphous foils of braze filler metals have been produced by using **rapid solidification** (RS) technology.[17] Compositions previously could only be utilized in powder form, or as powder-filled pastes.[18, 19] Most of the filler metals used in high temperature brazing are eutectic compositions formed by the transition elements, such as nickel, iron, chromium, etc., and metalloids, such as silicon, boron and phosphorus. The very presence of metalloids at or near the eutectic concentration promotes RS conversion of such alloys into a ductile amorphous foil.

Accordingly, three groups of filler metal compositions have emerged as brittle brazing filler metals which are now being prepared as ductile foils by RS technology. The brazing filler metals fall into three groups; the eutectic-type alloys, the peritectic alloys for vacuum brazing, and the copper-silver eutectics, Table 4.2.[20]

Rapid solidification is characterized by high cooling rates (approaching 10^6 K/s) that enable the stabilization of certain alloys into an amorphous solid state, which has a space distribution of atoms similar to that in liquids. When rapidly solidified microcrystalline materials are formed, the dimensions of their crystalline grains and phases are usually an order of magnitude smaller than those in the material of the very same composition but manufactured using conventional process technology. Because RS micro-crystalline and, in particular, amorphous materials are

Table 4.2 Rapidly solidified (RS) solder brazing filler metals. Courtesy: Metglas ®

No.	Alloy family and type	Metglas prods. designation	AWS designation	Structural state	RS forms	Base metals joined	Major applications, current (potential)
1	Al–Si eutectic	none (in development)	BAlSi–2,3,4,5,6,7	microcrystalline	foil, powder	Aluminum and aluminum alloys, steel to aluminum alloys and aluminum to beryllium	Car radiators, heat exchangers, honeycomb aircraft structures, structural parts
2	Cu–Sn, peritectic	MBF–2004B	none	microcrystalline	foil	Copper and copper alloys, copper to mild steel, copper to stainless steel	Heat exchangers, structural parts, automotive
3	Cu–P eutectic	MBF–2002,2005	none	amorphous	foil, powder	Copper to copper, copper to silver, copper to silver/oxide PM composites	Electrical contacts, bus bars (heat exchangers)
4	Cu–Ag, eutectic	none (in development)	BAg–1,2,2a,3,4 8,13,18,19 20,21	microcrystalline	foil, powder	Most ferrous and nonferrous metals, except aluminum and magnesium	Most widely used utility filler metals
5	TM–Si–B[a], eutectic a) (Ni/Fe + Cr)–Si–B	MBF–10,–15,–17,–20,–30,–35,–50,–60,–65,–75,–80,–90	BNi–1,1a, 2,3,6,7	amorphous	foil, powder	AISI 300 and 400 series stainless steels and nickel- and cobalt-base superalloys. Carbon steels, low-alloy steels & copper	Aircraft turbine components, automotive, heat exchangers, honeycomb structures (electronic components)

Table 4.2 (Continued)

No.	Alloy family and type	Metglas prods. designation	AWS designation	Structural state	RS forms	Base metals joined	Major applications, current (potential)
	b)(Ni,Pd)-Si-B	MBF-1000 series: 1001, -2,-5,-7,-8,-10, -11,-20,-22	none	amorphous	foil, powder, paste	AISI 300 series stainless steels, cemented carbide, superalloys	Honeycomb structures, cemented carbide/ polycrystalline diamond tools, orthodontics, catalytic converters
	c)(Co,Cr)-Si-B	MHF-157,-6M	BCo-1	amorphous	foil, powder	Cobalt-base heat- and high-corrosion-resistant superalloys	Aircraft engines (honeycomb marine structure)
6	Cu-N-Mn-(Si), solid solution	none (in development)		microcrystalline	foil, powder	Nickel-base heat-resistant alloys, steels	(Honeycomb structures, structural turbine parts)
7	Ti-Zr-Cu-Ni, eutectic and peritectic	none (in development)	none	amorphous, microcrystalline	foil, powder	Titanium-base alloys	Titanium tubing (aircraft engineers, honeycomb aircraft structures, aircraft structural parts)

(a) This group includes alloys based on transition metals such as Ni, Fe and Co.

compositionally much more uniform, their melting under transient heating occurs over a narrow temperature range. The resulting 'instant melting' of rapidly solidified materials is an important feature. From the practical point of view, this provides an opportunity to braze at a lower temperature and for a shorter time than in the case of using conventional filler metals. This is particularly important when, for example, brazing fine-gauge honeycomb cores, which have to be protected from erosion by molten filler metals. A shorter brazing time is also beneficial in cases where base metal parts may lose their inherent strength due to annealing during the brazing operation. The joining of cold-formed stainless steels and of dispersion-hardened superalloys are good examples of the cases where a short brazing time can be critically important.

Another convenient method of supplying filler metal is to use brazing sheets which consist of a core of aluminium base metal alloy and a coating of the lower melting filler metal. The coatings are aluminium silicon filler materials and may be applied to one or both sides of the base metal sheet. The coating is roll bonded to the core normally during the mill fabrication. Thus the brazing sheet is a product that can be formed by drawing, bending, or other normal metal working processes without removing this coating. The formed part can be assembled and brazed without placing additional filler metal in the joint. Brazing sheet is frequently used as the member of an assembly with the mating piece made from unclad brazeable alloy. The filler metal on the brazing sheet flows by capillary action and gravity to fill the joint at contact sites. In addition, brazing filler metal clad thin wall tubing has been made by continuous seam welding brazing sheets.

4.1.4 Composition of major filler metals

Brazing filler metals have been divided into eight categories and into various classifications within each category. The eight major categories are:

1. aluminum–silicon;
2. manganese–aluminum–zinc;
3. copper and copper–zinc–tin;
4. copper–phosphorus and copper–phosphorus–silver;
5. gold–copper and gold–nickel–palladium;
6. nickel;
7. cobalt;
8. silver–copper–zinc.

Aluminum–silicon

The aluminum–silicon group of braze filler metals has been used in joining aluminum grades 1060, 1100, 1350, 3003, 3004, 3005, 5005, 5050,

6053, 6061, 6951, and cast alloys A712.0 and C711.0. All types are suited for furnace and dip brazing, while some types are also suited for torch brazing, using lap joints rather than butt joints, see Appendix Table A.7.

The aluminum–silicon brazing filler metals have been used for making brazed assemblies which can be generally subjected to continuous service temperatures up to 150°C. Short-time operating temperatures of 205°C may be permissible, depending on the operating environment. Flux should be used in all cases and removed after brazing, except for vacuum brazing. In joints with overlaps of less than 6.4 mm, joint clearances of 0.15 to 0.25 mm are commonly used, whereas clearances up to 0.64 mm are used for greater overlaps.

Where a choice of aluminum–silicon brazing filler metals is possible, the following general guides may be helpful.

1. For torch brazing, select a brazing filler metal with a liquidus temperature as far below that of the parent metal as possible. Because temperature control is difficult with a torch, a large temperature difference will reduce the chance of accidentally melting the parent metal.
2. When it may be necessary for the operator to push the molten filler metal into a far corner or to torch braze a long joint, a filler metal with a wide spread between its solidus and liquidus will be helpful.
3. When the importance of producing a 'perfect' brazed joint is uppermost in the engineer's mind, brazing filler metals with narrow melting ranges (minimum spread between solidus and liquidus) should be selected. BAl–Si–4, for example, is a filler metal with only 5.6°C between its solidus and liquidus. This brazing filler metal (88% Al–12% Si) is almost eutectic and will change quickly from a solid to a liquid, reducing the time during which the parts must be subjected to brazing heat, reducing filler metal/base metal diffusion, and reducing the possibility of the filler metal solidifying in the joint before it has reached all corners.
4. For furnace or dip brazing, select a filler metal with a narrow melting range. High speed is most desirable.
5. For brazing by torch or induction, filler metals with wide melting ranges are advisable. The filler metal starts to flow at a lower temperature, and better control is possible.

There are nine major aluminum–silicon brazing filler metals.[15] See Appendix Figure A.1. The most common filler metal which is excellent for general-purpose use and work with all brazing processes is BAlSi–4. Several new filler metals, BAlSi –6, –7, –8, –10 and –11 have been developed for use as a bare and/or clad vacuum braze filler metal.[15, 21–23]

Of the nine brazing filler metals designed primarily for joining aluminum alloys or as cladding for aluminum brazing sheet, the aluminum–silicon filler metals contain 86 to 92.5% aluminum. Five of the aluminum braze filler metals mentioned previously are suitable for vacuum brazing of aluminum.[21, 23] Their magnesium content ranges from 1 to 3%. During the vacuum brazing process, magnesium is depleted from these filler metals, causing their solidus temperatures to increase and approach the solidus temperatures of similar nonvacuum brazing filler metals that are low in magnesium

Magnesium–aluminum–zinc

Magnesium filler metal AZ92A (Mg–1) is used to join AZ10A, K1A, and M1A magnesium alloys by torch, dip, or furnace brazing processes. The other magnesium–based filler metal, AZ125A (BMg–2), with a lower melting range, is used for brazing AZ31B and ZE10A compositions. See Appendix Table A.8. Heating must be closely controlled to prevent melting of the base metal. Joint clearances of 0.10 to 0.25 mm are best for most applications. Corrosion resistance is good if the flux is completely removed after brazing. Brazed assemblies are generally suited for continuous service up to 120°C or intermittent service at 150°C, subject to the usual limitations of the actual operating environment.

Aluminum–zirconium

The application of ceramics in technical systems often requires the joining of a **ceramic to a metal** or to itself. In recent years there has been interest in using Al-based braze filler metals for joining oxide and nonoxide ceramics.[24] Aluminum possesses the following favorable properties: it is oxidation-resistant; it wets a variety of ceramics; and it can relax residual stresses that result from the joining process. It has been shown that strong joints can be achieved in the systems Al–Al_2O_3, Al–SiC, Al–Si_3N_4, and Al–PSZ.[25–28] The alloys used in these studies were based on Al with Mg, Si, or Cu additives.

To join zirconium-based systems, one would expect that aluminum filler metals may offer some promise. It has been found that pure Al and Al–Cu alloys can wet calcia-partially-stabilized zirconia surfaces, although temperatures in excess of 1100°C seem to be required.[27, 28] Al–Zr alloys offer some interesting possibilities as a braze material. One would suspect that the zirconium addition to the aluminum may aid in wetting zirconium ceramics, and in addition, these alloys can be dispersion strengthened.[29, 30]

In this preliminary study[30] the use of a two-phase alloy system for joining ceramics may offer some important opportunities. Control of the size, volume fraction, and distribution of the precipitate may allow the optimization of the joint ductility and strength. For the particular case of Y–TZP, it was found that Al–Zr alloys can be a useful braze material.

The Y–TZP was joined with an Al–5.8% wt Zr filler metal at 900°C and above. Large precipitates of the intermetallic phase, Al_3Zr, can aid in strengthening of the joint, especially if they are close to the interface. With decreasing layer thickness, the strengths increased with values as high as 420 MPa.

Copper and copper–zinc–tin

These brazing filler metals are used to join ferrous and nonferrous metals. The corrosion resistance of the copper–zinc alloy filler metals is generally inadequate for joining copper, silicon bronze, copper–nickel alloys, or stainless steel. Typically, lap and butt joints are used with brazing processes.

The practically pure copper brazing filler metals are used to join ferrous metals, nickel-based alloys, and copper–nickel alloys. They are free flowing and often used in furnace brazing with a combusted gas, hydrogen, or dissociated ammonia atmosphere without flux. Copper filler metals are available in wrought and powder forms.[15]

As a result of the work of Wigley, Sandefur, Jr., and Lawring[31] it is possible to understand in general terms the basic mechanisms involved in bond formation. At temperatures in excess of ~1000°C and in the moderate vacuum involved, it appears that the oxide film in the surface of stainless steel is no longer self-repairing and that when copper melts at 1083°C, it is able to wet the metal surface. Heat-treatments involving short times and relatively low temperatures produce essentially a brazed joint between the two surfaces, and if enough copper is present, the bond is void-free. At longer times and higher temperatures, diffusion of copper away from the bond line permits the asperities on the two surfaces to come into contact and allows the formation of diffusion bonds.

The results of these initial experiments show that high strength void-free bonds can be formed by vacuum brazing of stainless steels using copper filler metals. In Nitronic–40, brazed joints have been formed with strengths in excess of the yield strength of the parent metal and even at liquid nitrogen temperatures the excellent mechanical properties of the parent metal are only slightly degraded.[31]

Copper–zinc filler metals are used on steel, copper, copper alloys, nickel, nickel-based alloys and stainless steel where corrosion resistance is not a requirement. They are used with the torch, furnace and induction brazing processes. Fluxing is required and a borax/boric acid flux is commonly used. The copper–zinc brazing filler metals are used extensively in braze welding of low-carbon and low-alloy steels.

The copper–zinc brazing filler metals have melting and brazing temperatures higher than those of silver brazing filler metals. Overheating must be avoided, however, because of their high zinc contents. When these filler metals are overheated, zinc vaporizes (fumes), causing voids in the joint.

Copper–phosphorus

These filler metals are primarily used to join copper and copper alloys. They have some limited use for joining silver, tungsten, and molybdenum. They should not be used on ferrous or nickel-based alloys, nor on copper–nickel alloys with more than 10% nickel. These filler metals are suited for all brazing processes and have self-fluxing properties when used on copper.

Corrosion resistance is satisfactory except where the joint is exposed to sulfurous atmospheres at elevated temperatures. Brazed assemblies can generally be subjected to continuous service temperatures up to 150°C. Short service at 200°C may be permissible, depending on the operating environment. However, flux is recommended with all other metals, including copper alloys. Lap joints are recommended, but butt joints may be used if requirements are less stringent. Joint clearances should be 0.03 to 0.13 mm. The range of clearance depends on the fluidity of the particular filler metal.

Mentioned previously and shown in Table 4.1 is a new family of copper–phosphorus and copper–tin brazing filler metals which have recently been developed by RS (rapid solidification) techniques.[32] Both the 78Cu–10Ni–8P–4Sn and 77Cu–10Sn–7P–6Ni filler metals are silver- and cadium-free, thus enhancing brazing safety and lowering cost. They are potential replacements for the Cu–15% Ag–5%P (BCuP–5) and the 45%Ag–15%Cu–16%Zn–24%Cd (BAg–1) in brazing copper and its alloys. Both filler metals are available as rod, foil, and powder, are self-fluxing, are suitable for both torch and furnace brazing, and have brazing temperatures similar to that of the BAg–1 filler metal and lower than that of the BCuP–5 filler metal. Copper-to-copper brazed joint mechanical properties – tensile, shear and impact strengths – are either similar or superior those of BCuP–5 and BAg–1 brazed joints.

An 80Cu–20Sn filler metal, available as foil only, is capable of brazing ferrous and nonferrous base metals by torch brazing with flux assist and by furnace brazing in nitrogen, argon, or vacuum (0.13Pa).

Copper–manganese–tin

A new copper filler metal system is being developed as an alternative to the expensive silver-based filler metals with cadmium which has been declared as a toxic material and is avoided in many brazing applications. Cu to mild steel (MS) and MS to itself have been successfully joined by a copper–manganese–tin system. Cu–11Mn–16Sn–1Ni produces high-tensile-strength joints of MS to MS at temperatures above 850°C. Cu–12Mn–19Sn and Cu–13Mn–20Sn–1Ce can be used to braze MS to MS at about 800°C with acceptable joint strength. Cu–10Mn–30Sn is capable of joining Cu to MS at about 750°C.

Though the addition of a small amount of nickel improves the MS to MS joint strength considerably, it also raises the effective brazing temperature. The addition of a small amount of cerium improves the flowability of the filler metal but lowers the joint strength, so a balance has to be sought.

Copper–titanium

Titanium added to active braze filler metal has been widely used as the brazing filler for ceramics because of its beneficial effect on the wettability. However, a substantial amount of titanium is required in binary Cu–Ti alloy to improve wettability,[32, 33] and it causes excess interfacial reaction and formation of a brittle intermetallic compound which may decrease the bond strength.

The wettability of ternary Cu–Ti–X (where X = Al, Si, Y) filler metal is enhanced by an increase in titanium activity. In some cases, however, the surface energy of the melt, which depends strongly on the properties of alloying additions and their contents, also influences the wettability. Moreover, interaction of the alloying elements with titanium also greatly affects the titanium activity, and accordingly the wettability and reactivity to Si_3N_4.[34]

Addition of aluminum, which has a high interaction with titanium, decreases the titanium activity, and hence increases the contact angle and decreases the reaction layer thickness. Addition of silicon, which has a strong interaction with titanium, decreases the titanium activity greatly. Hence the contact angle increases and the reaction layer thickness decreases markedly even with addition of small amounts of silicon. Large amounts of silicon depress the melting point of the filler metal and reduce the surface energy of the melt, thus causing the contact angle to decrease significantly.

Addition of yttrium, which has high reactivity with Si_3N_4 and low interaction with titanium, increases the titanium activity (or activity of reactive components) and accordingly decreases the contact angle. While the thickness of the reaction layer increases markedly when the yttrium content is less than 5%wt (total content of yttrium plus titanium is less than 10%wt), the reaction layer thickness decreases sharply with yttrium addition of more than 10%wt (i.e. total content of yttrium plus titanium more than 15%wt).

Addition of tin and silver, which have low interaction with titanium, also causes the titanium activity to increase, and hence the contact angle and reaction layer thickness to decrease.

The bond strength of the Cu–Ti filler metal is affected more by the reaction layer morphology rather than by the wettability.

The effect of alloying elements on the wettability, interfacial reaction and bond strength was very different. Aluminum addition up to 10%wt reduced the reaction layer thickness and increased the strength remarkably, irrespective of the variation of wettability. Therefore, aluminum is regarded to be very effective for the control of interfacial reaction. Since silicon addition reduces the reaction layer thickness and decreases the strength greatly, silicon is considered to be ineffective as a reaction control element. Yttrium addition tends to promote the interfacial reaction by titanium, and decreases the contact angle and the blond strength. But yttrium addition of less than 5%wt to the Cu–Ti filler metal containing less than 5%wt titanium is very effective for the marked improvement of wettability and strength.[34]

Gold–copper and gold–nickel–palladium

Gold filler metals are used to join parts in electron tube assemblies where volatile components are undesirable. They are used to braze iron, nickel, and cobalt-based metals where resistance to oxidation or corrosion is required. They are commonly used on thin sections because of their low rate of interaction with the base metal.[33–36]

Applied under a protective atmosphere, Au–Cu and Au–Ni–Pd brazing filler metals need not be fluxed; however, for certain applications, a borax/boric acid flux may be used. The gold-based brazing filler metals are generally suitable for continuous service at 425°C, and intermittent service at 540°C, depending on the operating environment. Normally, joint clearances of 0.03 to 0.10 mm are used.

Several of the gold filler metals, BAu–1, –2 and –3, see Appendix Table A.9, have different temperatures and are very popular for use in **step brazing**. The significance of step brazing is that when two or more joints are to be brazed in sequence at points not widely separated on the same assembly, brazing filler metals that differ in working temperature are selected, and the higher-melting filler metal is used first. Some overlap of brazing temperature ranges can be tolerated, depending on the work metal, the size of the parts, the proximity of the joints, the presence or absence of heat sinks, and the closeness of temperature control.[37–39]

Two recently-developed gold-based brazing filler metals should also be mentioned. These filler metals (19Au–7Ni–6Pd–25Mn–43Cu and 30Au–10Ni–10Pd–16Mn–34Cu) have lower gold contents and therefore are far less expensive than the other gold-based filler metals listed in the Appendix and previously discussed. They are produced in wire, foil, and powder forms and are used for brazing in the temperature range from 960 to 1010°C. They are much stronger than other high-temperature filler metals of their type, provide better gap-filling capabilities and excellent wetting, exhibit nonaggressive behavior, have good oxidation and salt-spray resistance and are not embrittled in hydrogen environments.

These filler metals were developed for vertical tube-to-tube brazing in the Space Shuttle main engine flight nozzle, but they can be used for many non-space applications.

Nickel brazing filler metals are generally used on 300- and 400-series stainless steels, nickel and cobalt-based alloys, even carbon steel, low alloy steels and copper when specific properties are desired. They exhibit good corrosion and heat resistance properties. They are normally applied as powders, pastes, rod, foil, or in the form of sheet or rope with plastic binders.

Nickel filler metals have the very low vapor pressure needed in vacuum systems and vacuum tube applications at elevated temperatures.

The brazing filler metals that contain nickel as their principal alloying element are most important from the standpoint of high-temperature service. The number of nickel-based brazing filler metals with significant differences in their compositions is confusing to one who must select a filler metal for a specific application, see Appendix Table A.4. However, these filler metals vary extensively in their physical, mechanical and metallurgical properties. These variations are intentional, because it is necessary to provide brazing filler metals that meet varying requirements.[40, 41] Depending on composition, the temperatures at which joints made with the filler metals listed in the Appendix are resistant to oxidation vary from about 540 to 1095°C.

Some nickel-based filler metals are extremely fluid and are used to braze long, close-fitting joints where the filler metal is distributed by capillary attraction. Others are moderately sluggish in their flow properties and must be preplaced on the faying surfaces of the joint to ensure consistent brazing. Still others are extremely sluggish when they are molten; they are used to braze joints whose clearances cannot be controlled readily (e.g., joints such as those encountered in sheet-metal assemblies).

Certain nickel-based filler metals have been developed for nuclear applications that contain elements resulting in a low-capture cross-section to thermal neutrons; others are suitable for exposure in high-temperature water or in liquid metals or their vapors. Nickel-based filler metals developed for fabricating honeycomb sandwich structures have limited reaction rates with thin metals used in such structures.

Nickel-based brazing filler metals have not only the attributes discussed above, but also disadvantages and limitations that must be recognized to ensure proper selection. Some react severely with heat-resistant structural alloys; if the brazing conditions promote prolonged reactions of this nature, erosion and/or penetration of thin metal sections may occur. The diffusion of certain alloying elements, such as boron (and silicon to a lesser degree), into the grain boundaries of the base metal can result in the production of joints with poor mechanical properties. Also, some nickel-based filler metals produce joints that lack ductility. To a large degree, the unfavorable characteristics of nickel-based filler metals (and other filler metals used for high-temperature brazing) can be limited or minimized by proper control of brazing variables.[37, 38] These nickel-based filler metals not only provide oxidation and corrosion resistance but also are suitable for subzero applications down to liquid helium temperatures (–270°C).

The nine major classes of nickel filler metals are listed in Appendix Table A.9.

Improvements in nickel-based brazing filler metals for specialized applications have also been made. For example, several filler metals containing phosphorus instead of boron as a melting point depressant have been developed for applications in the nuclear industry. Among them are Ni–25Cr–10P, Ni–26.3Cr–5.1Si–3P, Ni–14.8Cr–8Si–3P–3Fe, and Ni–20.3Cr–11.5Si–0.5P.[39–41]

The phosphorus-containing filler metals suffer from low ductility because they form nickel phosphides. Other work has concentrated on developing filler metals for brazing mechanically alloyed, (MA), high-temperature nickel-based alloys that are oxide-dispersion-strengthened.[41] Because of their excellent high-temperature strength and corrosion resistance, these MA nickel-based alloy structural materials, such as Incoloy MA–356, which would be serviceable at temperatures up to 1200°C, are potential candidates for various components for future gas-turbine engines. The most promising brazing system has been Ni–Cr–Pd with tungsten, which has improved mechanical properties by solid-solution strengthening.

Another new brazing filler metal, developed especially to wet and flow at 1080°C on superalloys with combined aluminum and titanium contents of 1 to 3% (such as René 41), has a composition of 6% Si, 4.3% Cu, 0.079% Misch metal, remainder Ni (no boron).

The latest development in nickel-based brazing filler metals[20, 42–44] has been the production and commercial availability of homogeneous, ductile foils, with compositions equivalent to those of BNi–1a, –2, –3, –6, and –7, made by liquid-metal quenching and RS technology. The foils have a metallic glass rather than a crystalline structure. They are more easily handled than powders, contain no organic binders and can be bent or punched to exact joint shapes, see Appendix Table A.9 and Table 4.1.

It has been demonstrated that the use of metallic glass foil in brazing results in joints with strengths comparable to those of joints made with existing filler metal products. The joints are characterized by almost complete diffusion of the metalloid elements. Subsequent work has shown that the strengths of joints made with BNi–1a and BNi–3 and several other metallic glass brazing filler metals are comparable to those of joints made with BAu–4.[20, 44]

Developmental work continues on a nickel-based brazing filler metal containing 4.63% Fe, 8.63% Cr, 35.32% Pd, 2.69% B, remainder Ni and having solidus and liquidus temperatures of 945 and 995°C, respectively, as well as on several boron-free filler metals containing 38 to 50% Ni, 41 to 53% Pd, and 6.1 to 8.8% Si and having solidus and liquidus temperatures ranging from 715 to 830°C and 745 to 875°C, respectively.

Finally, some developmental work also continues in the cobalt metallic glass family of brazing filler metals. One of these has a composition of Co–21Cr–4.5W–1.6Si–2.4B–0.07C and solidus and liquidus temperatures of 1100 and 1200°C, respectively.

Si_3N_4 ceramics with Al_2O_3 and Y_2O_3 as additives were joined with an 80%wtNi–20%wtCr filler metal sheet as an insert layer. Joining was performed by hot-pressing between 1000 and 1350°C in argon, and under uniaxial pressures in the range of 50 to 100 MPa. The average strength, evaluated by four-point bending, was large enough (>300 MPa) for some industrial applications.[45] However, the scatter of the joint strengths was large. It was probably due to the formation of pores at the joining interface or near the surface. The influence of nitrogen gas partial pressure in the joining atmosphere and uniaxial pressure on the formation of pores at the joint interfaces was confirmed microscopically. Cr coating the Si_3N_4 ceramic before joining was effective in reducing the scatter of joint strengths. The oxidation resistance of the joint was excellent up to 800°C in air.[45]

Y. C. Wu and associates[46] have been investigating the eutectic bonding of nickel to 3%mol yttria-stabilized zirconia using the eutectic melt of nickel and nickel oxide that exists at 1440°C. This eutectic point is 15°C below the melting point of pure nickel. In fact, it is the liquid eutectic which is used as an intermediate layer to bond the solid nickel to the ceramics without the nickel member losing its original shape. The major advantage of the eutectic bonding technique is the excellent wettability of the nickel–nickel oxide eutectic liquid on the ceramic substrate.[47, 48]

Liquid eutectic bonding can be used to join zirconia and nickel foil. This liquid phase relaxes the stress caused by differential shrinkage between the metal and the ceramics during bonding, and wets the interface well. Upon solidification of the eutectic melt, a strong bond is established. The strength is higher than that without the eutectic layer. The eutectic bonding employed proves to be superior to the pure Ni–YSZ bonding at 1450°C and to NiO–YSZ bonding in air at 1450°C.[46]

Nickel alloy brazes are used for joining high-temperature components, particularly for the aerospace and nuclear industries. They are based on the Ni–P, Ni–Si and Ni–B eutectics so that the melting temperature of the nickel is depressed from 1453°C to the brazing temperature range of 900–1250°C. Although used successfully in demanding applications, systematic characterization of the behavior of these braze filler metals is relatively rare.

However, several recently-published reports have renewed investigations on the wetting and spreading of Ni–P brazes and the wettability of nickel filler metals with boron using RS foils, MBF 35, 1005, 80 and 85, see Table 4.1.[49,50]

Looking toward future reusable metallic **thermal protection systems** (TPS) for entry vehicles[51] a ten-year completed study showed the successful design and fabrication of Inconel 617 honeycomb sandwich panels brazed with Ni–14Cr–4Fe–2B filler metal at 1121°C. The significance is the brazing of 0.13 and 0.06 mm face sheets (foil) without any erosion. The TPS was designed for 100 missions and the vehicles must endure environments where the temperature range is 536 to 1093°C and pressure loads do not exceed 13.8 KPa. The TPS panel design used the requirements for the Space Shuttle Body Point 1300 as representative design criteria.[51]

Cobalt

The cobalt filler metal is used for its high-temperature properties and its compatibility with cobalt-based metals. Brazing in a high-quality atmosphere or diffusion brazing gives optimum results. Special high-temperature fluxes are available for torch brazing. BCo–1 filler metal, see Appendix Table A.15, like the nickel-based filler metals, receives attention when designers need high-temperature resistance, especially for jet engine parts and honeycomb sandwich structures of cobalt-based alloys.

Silver–copper–zinc

The silver-based brazing filler metals are primarily used to join most ferrous and nonferrous metals, except aluminum and magnesium, with all methods of heating. They may be preplaced in the joint or fed into the joint area after heating. Lap joints are generally used with joint clearances of 0.05 to 0.13 mm when mineral-type fluxes are used, and up to 0.05 mm when gas-phase fluxes (atmospheres) are used. However, butt joints may be used if the service requirements are less stringent. Fluxes are generally required, but fluxless brazing with filler metals free of cadmium and zinc can be done on most metals in an inert or reducing atmosphere (such as dry hydrogen, dry argon, vacuum and combusted fuel gas).[15]

Copper forms alloys with iron, cobalt, and nickel much more readily than silver does. Also, copper wets many of these metals and their alloys satisfactorily, whereas silver does not. Consequently, the wettability of silver–copper brazing filler metals decreases as the silver content increases in brazing of steels, stainless steels, nickel-chromium alloys and other metals. Thus, a high silver content filler metal does not wet steel well when brazing is done in air with a flux. Copper wets many of these metals and their alloys satisfactorily, where silver does not. When brazing in certain protective atmospheres without flux, silver–copper filler metals will wet and flow freely on most steels at the proper temperature.

Addition of cadmium to silver–copper–zinc brazing filler metals dramatically lowers their melting and flow temperatures. Cadmium also increases the fluidity and wetting action of the filler metal on a variety of base metals. Cadmium-bearing filler metals should be used with caution. If they are improperly used and subjected to overheating, cadmium oxide fumes can be generated. Cadmium oxide fumes are a health hazard, and excessive inhalation of these fumes must be avoided. Because cadmium-bearing filler metals are not intended for fluxless brazing, an appropriate flux should always be used with these filler metals when brazing in either air or furnace atmospheres.

Zinc is commonly used to lower the melting and flow temperatures of copper–silver brazing filler metals. Zinc is by far the most helpful wetting agent for joining alloys based on iron, cobalt or nickel. Alone, or in combination with cadmium or tin, zinc produces alloys that wet the iron-group metals but do not alloy with them to any appreciable depth.

Tin has a low vapor pressure at normal brazing temperatures. It is used in silver-based filler metals in place of zinc or cadmium when volatile constituents are objectionable, such as when brazing is done without flux in atmosphere or vacuum furnaces, or when the brazed assemblies will be used in high vacuum at elevated temperatures. Tin additions to silver–copper filler metals result in wide melting ranges. Filler metals containing zinc wet ferrous metals more effectively than those containing tin, and where zinc is tolerable, it is preferred over tin.

Generally, as the combined zinc and cadmium content is increased beyond 40%, the ductility of the filler metal decreases. This fact puts a practical limit on how much the flow temperatures of silver-based filler metals can be lowered.

Stellites, cemented carbides, and other molybdenum- and tungsten-rich refractory alloys are difficult to wet with silver–copper–zinc brazing filler metals. Manganese, nickel, and (infrequently) cobalt are often added as wetting agents in filler metals used for joining these materials.[52] An important characteristic of silver-based filler metals containing small additions of nickel is improved resistance to corrosion under certain conditions. They are particularly recommended where joints in stainless steel are to be exposed to salt water corrosion.

When stainless steels and other alloys that form refractory oxides are to be brazed in reducing or inert atmospheres without flux, silver-based filler metals containing lithium as the wetting agent are quite effective. The heat of formation of lithium oxide is very high; consequently lithium is capable of reducing the adherent oxides on the base metal. The resultant lithium oxide is readily displaced by the brazing filler metal. Lithium-bearing filler metals are advantageously used in very pure dry hydrogen or inert atmospheres.

Continuous service temperatures for silver-based brazing filler metals range up to 205°C, with intermittent service up to 315°C, adjusted for the actual operating environment.

Silver–copper plus palladium[53]

A recently-completed study by Hardesty[45] reported that a 65Ag–20Cu–15Pd (Palcusil–15) and two other silver-based filler metals, 92Ag–7.8Cu–0.2Li (BAg–19) and 60Ag–30Cu–10Sn (BAg–18) were tested in edge-wise compression in beryllium panels and carried loads up to 649°C.

Silver–copper plus manganese and gallium

In a study reported in *Welding Production*,[54] Lashko and Sokopov developed a low-silver brazing filler metal for brazing corrosion-resistant steels. To reduce the silver content of the brazing filler metal and retain the high ductility of the brazing joints in the corrosion-resistant steels at cryogenic temperatures, the standard silver brazing filler metals have been replaced by a face-centered cubic metal, i.e., copper, whereas the components forming wide ranges of solid solutions with copper have been represented by manganese, silver, and gallium. The optimum percentage chemical composition of the brazing filler metal was 5Ag, 10Mn, 5Ga, remainder Cu. The brazing filler metal was manufactured in the form of 0.1–2.0 mm thick strips and 0.5–4.0 mm diameter wires.

The brazing filler metal spread satisfactorily on the corrosion-resistant steels heated in vacuum and in an air furnace. The results of the mechanical tests on the braze filler metal showed that in the cast condition the filler metal was sufficiently strong, ductile, and had high impact toughness, both at room temperature and –20°C.

Vacuum-grade filler metals[15]

There are several special noble-metal brazing filler metals which should also be mentioned. Among these are the vacuum-grade filler metals, which are made to high purities and are virtually free of high vapor pressure elements. The vacuum-grade brazing filler metals are used in brazing parts for vacuum tubes and electronic circuits that require durability in demanding applications. Made to be spatter-free, most vacuum-grade filler metals come in two grades. Grade I contains zinc and cadmium to a maximum of 0.001%; grade 2 contains 0.002 max Zn and Cd. These brazing filler metals are also low in Pb, P, C, Hg, Sb, K, Na, Li, Ti, S, Cs, Rb, Se, Te, Sr, and Ca, which are elements that have vapor pressures greater than 0.00013 Pa at 500°C.

The cadmium-free brazing filler metals BAg –22, –24, –25, and –26 are listed in Appendix Table A.14. Due to Federal regulations and the requirements placed on health and safety, the use of cadmium-free brazing filler metals as required is extremely important. Addition of small and carefully-controlled amounts of one or more of the elements tin, nickel, and manganese to alloys of the silver–copper–zinc ternary alloy system provides a series of useful cadmium-free brazing filler metals. These newly-developed and commercially available filler metals have particular potential for production brazing applications in the refrigeration, shipbuilding, cutlery, automobile, and rock-drilling tool industries.

The availability of these materials provides industry with a new range of brazing filler metals which are both technically and commercially attractive and which enable companies who wish to do so to avoid the use of cadmium in their production brazing operations.

Other combinations

There are other groups of special brazing filler metals; those that have been developed for joining refractory metals and their alloys, graphite, and ceramics to themselves and to metals – especially those with brazing temperatures ranging from 1040 to 2300°C. Most work up to the present has involved **metallizing**. In order to obtain wetting of the joint surfaces by the brazing filler metal, the area of the joint on the ceramic side must first be provided with a firmly bonded metal coating – the metallized film. These films can be applied by numerous techniques, including electrolytic precipitation, gas-phase precipitation, thermal and plasma spraying, ionic plating, and electron- and laser-beam coating. At temperatures up to about 1200°C, brazed joints can be used on metallized ceramics. In higher temperature ranges, alloying between the filler metal and the metallizing film can influence the adhesive strength of the joint.[55]

Metallization of the surface of the ceramic is unnecessary when the newly-developed ceramic brazing filler materials are used. These are oxide mixtures which, in the molten state, wet both the surface of the metal and also that of the ceramic base. Additionally, there are 'active' brazing filler metals which contain interfacially-active components that reduce the interfacial energy between the ceramic and the molten filler metal to such a low level that wetting of the ceramic takes place.

The best known interfacial active elements are titanium and zirconium. Even small concentrations of a few percent in copper or silver are sufficient to produce excellent wetting of Al_2O_3 and other oxides. Characteristic of the active metals (Ti and Zr) is their great affinity for oxygen, which permits them to react with the ceramic oxides and form their own oxide phases. It is probably this reaction which makes possible the high degree of adhesion between the components of the joint but which, under certain conditions, can also lead to weakening of the joint. Due to the high degree of affinity for oxygen involved, brazing of such joints can be carried out only under conditions of high vacuum or in a dry, high-purity, inert gas atmosphere.[55]

Three new active silver-based brazing filler metals have been developed which permit brazing of metal parts to high-alumina and other structural ceramics (such as wear-resistant, heat-resistant, and similar parts) without metallizing of the ceramic material. These new ductile filler metals are adaptable for brazing of metals to such materials as silicon nitride, partially-stabilized zirconia, transformation-toughened aluminas, and silicon carbide, as well as many other refractory materials. Foil, wire, and preforms of these filler metals are available.

Silver–copper–hafnium

Due to the increasing applications of nonoxide ceramics in technical structures, new active filler metals are being developed for joining ceramics either to other ceramics or to metals. In commercially-dominant active braze filler metals, titanium as the reactive agent is added to the base filler metal, particularly to the Ag–Cu eutectic. Apart from titanium and other reactive elements, hafnium additives are also known for promoting the wetting of ceramics by conventional braze metals.

Lugscheider and Tillman[56] investigated the Hf-added Ag–Cu active filler metals and the examination of their brazing properties for joining SiC and Si_3N_4 ceramics to themselves and to steel. Active filler metals in a system of Ag–Cu–(In)–Hf, with hafnium contents ranging from 2 to 5%wt and a eutectic Ag–Cu composition has shown that the filler metals have a melting behavior similar to titanium containing Ag–Cu filler metals.

Wettability tests conducted on SiC and Si_3N_4 proved that they exhibit good wetting properties at elevated temperatures (>1000°C). Brazing test joints at different brazing conditions showed that hafnium-containing filler metals are suitable for joining SiC/Si_3N_4 ceramics to themselves, as well as for joining them to steel. The quality of the joint strongly depends on the quality of the vacuum because of the high reactivity of hafnium, even to traces of atmospheric gases.

Regarding the mechanical properties of joints brazed with hafnium-containing filler metals, four-point bending test specimens showed strong joints with bending strengths above 149 MPa in the case of Si_3N_4. SiC ceramics display weaker mechanical properties due to their thicker reaction zone.[56]

Silver–copper–titanium[15, 53, 57]

Nicholas *et al.* conducted a series of experiments[57] to gain information about the wettability of AlN, BN, Si_3N_4 and two SiAlON (Syalon) ceramics by using aluminum, copper-titanium filler metals, and a Ag–28Cu–2Ti filler metal. Wetting by aluminum and the Ag–28Cu–2Ti filler metal was usually good. Both wetting and non-wetting filler metals containing titanium reacted to form TiN and the achievement of wettability was associated with a certain degree of hypostoichiometry. While aluminum should also have reacted, no clear evidence was obtained. In supplementary experiments it was found that bonds formed by brazing with aluminum at 1000°C could have shear strengths as great as 60 MPa. Although the experimental work was preliminary in nature, it suggested that good brazing systems could be developed.

Xian and associates[58] joined Si_3N_4 using a 57Ag–38Cu–5Ti braze filler metal and reported that various temperatures for brazing had an effect on the layer interface structure, Figure 4.7. Hot-pressed Si_3N_4 was joined using 57Ag–38Cu–5Ti brazing filler metal in a vacuum and the maximum bend strength of the joint measured by the four-point bend method was 490 MPa when brazing at 880°C for 5 min. It is important that there is sufficient reaction during brazing between the ceramic and the active element in the brazing filler metal, characterized by forming TiN with an appropriate thickness. On the other hand, insufficient interface reaction will decrease the joining strength.

This work shows that one can expect to obtain a high joining strength using Ag–Cu brazing filler metal with a small quantity of titanium as an active element, without forming a brittle dispersed phase such as silver–titanium or copper–titanium compound after brazing. From the

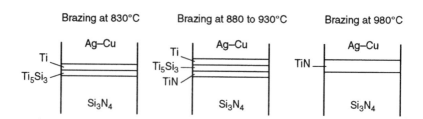

Fig. 4.7 Schematic illustration of layer interface structure in Si_3N_4–Si_3N_4 joints using $Ag_{57}Cu_{38}Ti_5$ brazing filler metal at various temperatures.[58]

above results, in order to obtain high joining strength, the following points are suggested when one designs a brazing filler metal for joining ceramic-to-ceramic or ceramic-to-metal:

1. The selected active element should have a high free energy of reaction with the ceramic, and strong segregation on the ceramic/metal interface.
2. The lower limit of active element content in the brazing filler metal should ensure good wetting of the ceramic by the brazing filler metal, while the upper limit should be such that there is no brittle dispersed phase in the brazing filler metal.
3. The brazing temperature and time should be sufficient to ensure interface reaction of the brazing filler metal with ceramic.

A 27.5Cu–2Ti–62Ag braze filler metal, (Cusil ABA), an 'active braze filler metal', has been successfully used to join hot pressed Si_3N_4 and Incoloy 909, a low-expansion superalloy.[59] Brazing was carried out in-vacuum (about 10^{-3} Pa) at 950°C for 20 minutes.

A 40Ag–5Ti–55Cu braze filler metal was successfully used to join Al_2O_3 to 6Al–4V Ti and 3% yttria-partially stabilized (PSZ) ZrO_2 to 6Al–4V Ti and brazing was performed by heating for 5 minutes in a vacuum furnace (2.6 mPa) at 870°C.[60] The rather good mechanical behavior reported can be related to the very similar thermal expansion coefficients of 6Al–4V Ti and ZrO_2. Another reason for these results can be found in the fact that the PSZ which was used was partially transformed into monolithic ZrO_2, such a transformation accommodating the stresses developed during the brazing procedure.

ZrO_2/6Al–4V Ti brazements exhibit interfacial phases which are very similar to those observed in Al_2O_3/6Al–4V Ti brazements. However, the tensile strength of joints using ZrO_2 ceramics is much higher (150 ± 50 MPa).

The same filler metal was also used to join Al_2O_3 to 6Al–4V Ti by active brazing. The high reactivity of the active braze filler metal formed at the braze/Al_2O_3 interface, a continuous and sinuous layer identified as Cu_2(Ti, Al)$_4$O. The presence of this layer was found to be beneficial to the bonding between the ceramic and the metal.[61]

Al_2O_3 joints of high integrity were produced with a 56Ag–36Cu–6Sn–2Ti (%wt) experimental filler metal by vacuum brazing at 900°C for 20 minutes. It was found by transmission electron microscope (TEM) that the formation of a Ti_3Cu_3O-phase reaction layer in conjunction with a TiO layer can provide a more gradual transition in chemical bonding between the Al_2O_3 and the Ag–Cu metallic filler metal phases than TiO will alone. The Ti_3Cu_3O-phase layer may also provide a more gradual transition in physical properties and help to minimize the effect local strains that develop from thermal expansion coefficient mismatches can have on adhesion.[62–64]

Titanium–zirconium–copper–nickel

Onzawa, Suzumura and Ko[65] investigated the microstructure and mechanical properties of commercially pure titanium (CPTi) and 6Al–4V Ti alloy joints brazed with newly-developed Ti-based amorphous filler metals. Among the developed filler metals were three kinds: Ti–37.5Zr–15Cu–10Ni, Ti–35Zr–15Cu–15Ni and Ti–25Zr–50Cu, whose melting points were approximately 100°C lower than those of conventional Ti-based filler metals, see Table 4.2.

The use of these filler metals makes it possible to braze at below α/β transformation and β transus temperatures of CP Ti and 6Al–4V Ti alloys, respectively. As a result, joints having sufficient tensile properties, as compared to those of the base metals, can only be made by holding for a short time at the brazing temperature. Therefore, in the case of brazing CP Ti and 6Al–4V Ti alloy at below α/β transformation and β transus temperature of each base metal, the original structures of the base metals are completely preserved, and the brazed regions are distinct.

The fatigue properties of 6Al–4V Ti alloy joints brazed at 900°C for 10 minutes and 950°C for 5 minutes approach that of the base metal at maximum stresses below 590 MPa. Brazing at 1000°C above β transus temperature, the joints exhibit less favorable fatigue properties. These brazed joints have excellent corrosion behavior so that no reduction in tensile strength occurs after immersion in a 5% NaCl solution for 1000 hr.

Three other braze filler metals have been developed by Suezawa[66] for wetting 6Al–4V Ti. The three filler metals were 30Pd–60Cu–10Co, 30Pd–40.1Au–39.9Cu, and 65.3Ni–11.1Cr–7.6W–4.3Fe–2.6B–1.5Si. None exhibited any cracks, voids and inclusions in the brazed joint. Brazing took place in vacuum (0.013 Pa), holding time was 5 minutes and 1135°C for the first filler metal listed above, 1050°C for the second and 1448°C for the last filler metal. Filler metal #1 and #2 were the most suitable in brazed 6Al–4V Ti tests mainly because of their lower melting points and minimal interaction and erosion due to brazing.

Cadmium-free systems

Chatterjee and Mingxi[67] initiated a development study of tin-containing filler metals to replace the silver-based metals containing cadmium.

The problem with cadmium-containing filler metals is the toxic fumes they generate during the brazing operation that can be injurious to health if inhaled, or even fatal.[68] Accumulation of sufficient data on the toxic effects of cadmium resulted in the introduction of a **threshold limit value** (TLV) for cadmium in 1970 in most European countries.[69, 70] This has resulted generally in a decision to switch to cadmium-free filler metals rather than installing elaborate fume extraction systems.[70, 71]

In industry, it has been difficult to get a direct replacement filler metal equal in physical properties and intrinsic cost. The Ag–Cu–Zn filler metals are available but have silver contents as high as 55% to maintain a solidus temperature of 630°C, which makes these filler metals very expensive.

Tin has been added to some cadmium-free brazing filler metals to maintain a low working temperature range. A typical filler metal contains 60 Ag–30Cu–10Sn. It can be used in fluxless controlled-atmosphere brazing and vacuum brazing of ferrous and nonferrous alloys. It has a wide working temperature range (600 to 720°C), which aids in filling the joint.

Tin additions improve the wetting characteristics of ferrous alloys over that obtained with binary Ag–Cu filler metals.[69,70] The most effective composition in the range contains 55Ag–21Cu–22Zn–2Sn. This filler metal has a working temperature of 630 to 660°C, and has successfully replaced Ag–Cu–Zn–Cd filler metals in some cases. It is excellent for brazing cutlery, jewelry, hollowwares, etc. [72]

The efficient tin-containing filler metals do not include toxic cadmium, but they have a large amount of silver, which is expensive and influenced by price fluctuations. These two major factors, i.e., the high cost of silver and the toxicity of cadmium fumes, initiated the study for alternative filler metals for brazing copper, mild steel (MS), stainless steel, etc., within the temperature range of 600 to 850°C.

The initial results of the study determined the suitability of Cu/15–40Sn/5–15Mn filler metals as universal brazing materials for joining Cu/Cu, Cu/MS and MS/MS at 750 to 850°C. This series of filler metals can replace expensive and toxic Ag–Cu–Zn–Cd-based filler metals. The working temperatures of these filler metals are slightly higher than silver-based filler metals, but this has to be weighed out against its softening effect on copper joints, economics and the toxicity of Ag–Cu–Zn–Cd filler metals.

Another set of filler metals, Cu–Sn–P and Cu–Sn–P–Ni filler metals can be used successfully to join copper and copper-based filler metals. It has been shown experimentally that it is not advisable to use these filler metals for making Cu/MS or MS/MS joints.

In other experimental work, it was shown that a Cu–30Sn–12Bi filler metal could be used for joining Cu/Cu and Cu/MS at 750 to 800°C. It also appears possible to develop suitable Cu–Sn–Ni filler metals for brazing stainless steel.[67]

Silicate systems

Brazing using a silicate interlayer is similar to metal brazing but differs in that achieving wetting does not generally pose a problem. [73,74] Ceramic brazes often offer better environmental compatibility than metals, but in general still not equal to that of the base material.[73] Ceramic brazes are usually less tolerant of thermal expansion coefficient mismatch.[73,75]

The effect of joint thickness and thermal expansion mismatch on the mechanical properties of joints made with silicate brazes was studied by Kirchner and coworkers.[76,77] This study consisted of finite element analysis (FEA) to estimate the stresses involved[76] and mechanical testing of alumina-to-alumina butt joints with interlayers of various thicknesses and with varying degrees of thermal expansion mismatch.[77] The stress analysis[76] for brazes with thermal expansion coefficients lower than that of the adherend, revealed that axial stresses are tensile and may result in lower fracture stresses. Several glass sealants with joining temperatures in the range of 450 to 1500°C have been developed that are suitable for joining alumina-to-alumina or to sapphire. The high-temperature sealants[78] reported were based on kaolin and on potassium feldspar, while the low-temperature glasses[78] were all borate compositions.

A study by Cawley and Knapp[79] was started to identify and characterize silicate systems as brazes for alumina ceramics. Choosing the proper systems can result in good wetting characteristics, impurity insensitivity and a refractory nature through a postprocessing crystallization anneal. The two silicate systems examined were based on either talc, $Mg_3Si_4O_{10}(OH)_2$, or anorthite, $CaAl_2Si_2O_8$.

Two talc-based brazes were examined in an effort to extend prior work by other researchers. These brazes were pure talc and a 50/50 (by weight) mixture of talc and alumina. Limited mechanical testing using the four-point bend method showed that alumina-to-alumina joints using the 50/50 mixture were relatively strong (~20 to 50 MPa). However, the talc-based brazes were characterized by unacceptable levels of porosity.

Several processing steps were attempted including calcination of powders and raising brazing temperature from 1500 to 1600°C.

These modifications were tried and had some effect on pore morphology but did not significantly reduce porosity. As a result the talc-based systems were abandoned. The anorthite-based system has proven much more promising. A glass with the same chemical composition as anorthite, termed An-glass ($CaAl_2Si_3O_8$), can be readily made by melting together equal molar amounts of whiting, $CaCO_3$, and kaolin, $Al_2Si_2O_5(OH)_4$, at 1600°C, ~50°C above the melting point of anorthite (1553°C) for 0.5 hour, and cooling to room temperature. Additionally, An-glass can be fully crystallized by annealing at 1100°C for 1 hour.

Alumina brazed with An-glass at 1600°C leads to joints with several favorable characteristics. First, a good bond develops between the alumina adherends and the An-glass. In contrast to the talc-based brazes, the porosity problem within the braze layer is minimal. Furthermore, acicular crystals (~100 μm long, 2 μm in diameter) are observed to form within the An-glass matrix. Many of these crystals tend to be aligned roughly perpendicular to the braze layer direction and may be significant contributors to the joint toughness. This system proved successful and average bend strengths of ~145 MPa were achieved with vitreous joints while room temperature strengths were retained to 500°C. Brazing of the same type of alumina used in this study with a filler metal having a composition of 45Cu–26Ag–29Ti (atomic percent), resulted in joints with an average four-point bend strength of 222 MPa at 25°C.[80] Figure 4.8 depicts the sequence of steps in silicate brazing.[79]

Titanium–zirconium–beryllium

Brazing is excellent for fabricating assemblies of refractory metals, in particular those involving thin sections. However, only a few filler metals have been specifically designed for both high temperature and high corrosion applications.

Those filler metals and pure metals used to braze refractory metals are given in the Appendix Table A.11. Low melting filler metals, such as silver–copper–zinc, copper–phosphorus, and copper, have been used to

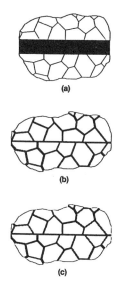

Fig. 4.8 Schematic of the sequence of steps involved in silicate brazing.
(a) Initially, a powdered glass layer of constant thickness is placed between the adherends.
(b) Upon heating, the glass is melted, densifies, and penetrates the grain boundaries drawing the surfaces together.
(c) Crystallization of the interlayer and grain boundary phase is accomplished through a post-joining controlled heat treatment.

join tungsten for electrical contact applications, but these filler metals cannot operate at high temperatures.

Nickel-based and precious-metal-based filler metals have also been used to join tungsten.

Brazing to molybdenum, columbium and tantalum

Various brazing filler metals will join molybdenum. The effect of brazing temperature on base metal recrystallization must be considered. When brazing above the recrystallization temperature, brazing time must be kept short. If high temperature service is not required, copper and silver-based filler metals may be used, see Appendix Table A.11.

Columbium and tantalum are brazed with a number of refractory or reactive-metal-based filler metals. The metal systems Ti–Zr–Be and Zr–Cb–Be are typical, also platinum, palladium, platinum–iridium, platinium–rhodium, titanium, and nickel-based filler metals.

4.2 BASE METAL FAMILY GROUPS

In the ensuing list of base metal groups it is apparent that a prerequisite for ductile joint behavior must be freedom from related hard phase bands in the brazed joint and a matching of elastic/plastic behavior in the brazed joint and base material. It thus follows that not only is the optimal structural development required for a joint capable of deformation, but also the elastic/plastic behavior of the whole system of filler metal and base metal must be considered.[80–82] This results in a ductile behavior in the brazed joint and an elastic and plastic behavior in the base metal. When the elastic and plastic properties of the brazed joint and base metal are well matched, this leads to the demand for brazing filler metals whose mechanical properties in the joint are well matched to those of the base metals.[83]

4.2.1 Aluminum and aluminum alloys

The nonheat-treatable wrought aluminum alloys that are brazed most successfully are the ASTM 1XXX and 3XXX series (e.g. 1350, 1100, 3003, 3004, 3005) and low magnesium alloys of the ASTM 5XXX series (e.g. 5005, 5050, 5052). Available filler metals melt below the solidus temperatures of all commercial wrought, nonheat-treatable alloys.

The heat-treatable wrought alloys most commonly brazed are the ASTM 6XXX series (6151, 6951, 6053, 6061, 6063). The ASTM 2XXX and 7XXX series of aluminum alloys are low melting and, therefore, not normally brazeable, with the exception of 7072 and 7005 alloys, see Appendix Table A.7.

Aluminum alloys that have solidus temperatures above 590°C are easily brazed with commercial binary aluminium–silicon brazing filler metals. Stronger, lower-melting alloys can be brazed with proper attention to filler metal selection and temperature control, and the brazing cycle must be short to minimize penetration by the molten filler metal. High-quality castings are no more difficult to braze than equally massive wrought alloys. Aluminum sand and permanent mold casting alloys 443.0, 356.0, 406, 710, 711, and 850.0 with high solidus temperatures are brazeable and alloys 443.0, 356.0, and 710.0 are the casting alloys most frequently brazed, see Appendix Fig. A.1. Die casting alloys are difficult to braze: the castings are not easily wetted by the molten brazing filler metal and tend to blister when brought to brazing temperature because of their high gas content and entrapped lubricants.

Commercial filler metals for brazing of aluminum and its alloys are aluminum–silicon alloys containing 7 to 12% silicon. Lower melting points are attained, with some sacrifice in resistance to corrosion, by adding copper and zinc. Brazing filler metals for vacuum brazing of aluminum usually contain magnesium, see Appendix Table A.7.

Aluminum can be brazed by most of the standard practices such as the torch, dip, or furnace process. Furnace brazing may be done in air or controlled atmosphere, including vacuum. Other methods, including induction, infrared, and resistance brazing, are used for specific applications. Regardless of the process, the temperature must be closely controlled for successful brazing.

With dip or furnace brazing, automatic proportioning temperature control devices are available that can maintain the flux bath within ±3°C, and the furnace atmosphere within ±6°C, of the desired brazing temperature. In manual torch, induction, or resistance brazing, operator skill and judgement are used to maintain the required temperature range for brazing based on flux color and on melting and flow of the brazing filler metal. On the increase is the use of vacuum furnace brazing because it is done without flux, the joints are free from the corrosion problems commonly associated with residual or entrapped flux. Moreover, brazed assemblies containing inaccessible recesses can be fabricated efficiently. Furnaces operating in the 0.0013 Pa range are used. The success of the operation depends on the use of magnesium vapor as a getter of oxygen on the aluminum surface and magnesium alloyed in the filler-metal (Al-Si) coating.

Furnace brazing

Equipment for furnace brazing of aluminum alloys is much the same as that used for heat treating these materials. Such equipment is designed to operate at temperatures up to 650°C and may be heated electrically or by oil or gas. Circulation of the furnace atmosphere to achieve uniformity of temperature increases the production rate and improves the flow of the flux and the brazing filler metal. Circulation of the furnace atmosphere through radiators and other heat exchangers is necessary to ensure complete brazing at the center of the part. The usual procedure is to adjust the furnace time from 30 s to 2 min longer than is required to bring the assembly to brazing temperature. Too long a time at temperature may result in excessive silicon diffusion, undesirable flow of brazing filler metal, and attack by flux.

Most furnace brazing of aluminum is done in air, but some benefits can be gained by brazing in a controlled atmosphere; by controlling moisture in the furnace at low levels, less flux is needed to achieve brazing (Nocolok process). New types of brazing sheet designed for fluxless brazing require nitrogen or other inert atmospheres.

Both continuous belt-type furnaces and batch furnaces have been used successfully. In some furnace designs, a high rate of heating is obtained by providing a high-temperature zone at the entrance of the furnace, in which the part is brought to within 28 to 55°C of the brazing temperature before being moved to the brazing zone, which is controlled within ±6°C.

Neither the batch furnace nor the continuous furnace is suitable for brazing massive parts. When furnace time approaches 30 minutes or more, braze quality falls off. This is caused by changes in the flux, liquation and diffusion, in decreasing order of importance. These three problems notwithstanding, massive parts are sometimes satisfactorily brazed in furnaces.

Furnaces operating at 0.0013 Pa have been used, and the success has stemmed from the use of magnesium as vapor, which acts as a 'getter' of oxide on the aluminum brazing surface. Work has shown that magnesium in the parent aluminum alloy and in brazing sheet can produce satisfactory brazed joints. Aluminum alloys are sometimes brazed to ferrous materials; in such cases, a nonoxidizing atmosphere is used to protect the ferrous material.

Even though dip brazing of aluminum is still a popular and successful process, the search for improved methods of brazing complex aluminum structures such as heat exchangers which will eliminate or minimize flux removal and corrosive residue problems has continued for many years.[84,85] Aluminum alloys of the 1xxx, 3xxx, 5xxx, 6xxx, and 7xxx series can be vacuum brazed using No. 7, 8, 13, or 14 brazing sheet, which are clad with 4004 brazing filler metal.[86] When additional filler metal is required, 4004 in wire or sheet form can be introduced. The joint designs used for brazing with flux can also be used for fluxless vacuum brazing.

Cold-wall vacuum furnaces with electrical resistance radiant heaters are recommended for aluminum vacuum brazing. Both batch-type and semi-continuous furnaces have been used. In the batch process, the parts are stationary within the brazing chamber during the critical portion of the brazing cycle (temperatures above 400°C) where part temperature uniformity is attained and joint formation takes place with the melting of the brazing filler metal.[87] The entrance vestibule is generally used to preheat and outgas the parts prior to transfer into the brazing chamber. The exit vestibule is used for the initial phase of cooling and filler metal solidification before transfer of the assembly from the vacuum environment of the brazing system to complete the cooling process. Inert gas backfills of dry air or nitrogen are used to bring the vestibule pressures up to one atmosphere before parts are introduced into or removed from the brazing furnace. With this process, the brazing zone temperature and vacuum can be maintained continuously.

In the semi-continuous process (Figure 4.9), the brazing chamber is divided into two or more heating stations so that brazing is brought about by movement of the part from station to station. Entrance and exit vestibules are used in the same way as in the batch process. The feature that distinguishes the semi-continuous process from the batch process is the pressure excursion that occurs in the brazing chamber each time the vestibule doors open to charge and discharge parts during the various heating stages of the brazing process. With this equipment, the braze is completed in three stages, and three carriers of parts, each at a different heating station, are contained in the brazing chamber at all times. In this mode of processing, a given carrier of parts is subjected to environmental changes at three different temperatures during the brazing cycle. Higher

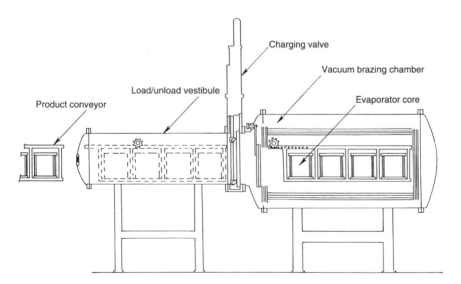

Fig. 4.9 Automatic semi-continuous aluminum brazing furnace.

production rates than for batch processing can be achieved with the semi-continuous system, in which parts are discharged at each transfer. The cycle time for a three-zone chamber is approximately one-third that for the batch process.[88]

The advancements in equipment, brazing filler metals, and controls have reached the point where fluxless aluminum brazing is an acknowledged production process.[89] This process was utilized in fabrication of heat exchangers[90] and cold plate cores for the Apollo Command Module.[91]

Dip brazing

Dip brazing is an excellent method for consistently producing high-quality brazed joints in complicated assemblies such as heat exchangers and wave guides.

Dip brazing of aluminum and its alloys has been used widely in manufacturing for over 40 years.[92, 93] Several pot designs and methods of heating the flux are commercially available. Dip pots can be cylindrical but are more commonly rectangular in shape, with either over-the-wall nickel, stainless steel, or carbon electrodes or through-the-wall carbon electrodes. Alternating current flowing between the electrodes and through the liquid flux provides heating; thermal stirring of the bath achieves temperature uniformity. Resistance heating of the bath is used almost entirely for production applications although externally fired, gas-heated pots have been occasionally used. Figures 4.10 and 4.11 illustrate the principal types of furnaces used for salt bath dip brazing.

In dip brazing the joint surface must be free of grease, oil, paint, oxide, and scale that would prevent the brazing filler metal from wetting the workpiece surfaces. Precleaning of parts is essential for making sound, strong brazed joints in all metals. In working with aluminum nonheat-treatable alloys, vapor or solvent cleaning is usually adequate. For heat-

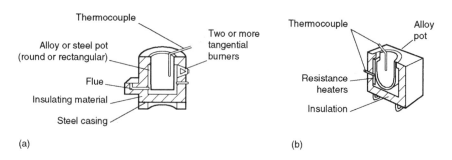

Fig. 4.10 Externally-heated furnaces.[15] (a) gas or oil fired; (b) resistance heated.

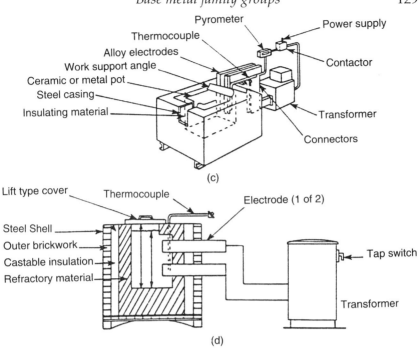

Fig. 4.11 Internally-heated furnaces.[15] (c) immersed electrodes; (d) through-the-wall.

treatable alloys, chemical or mechanical cleaning is necessary to remove the thicker oxide coating.

Parts to be dip brazed are preheated prior to immersion in the molten brazing flux. Preheating temperatures range from 480 to 575°C, depending on the size and complexity of the assembly. Preheating is done to dry the part (eliminating possible rapid steam generation) and prevent solidification of the flux on the assembly, which occurs when a cold part is immersed. Solidification of the flux insulates the part and closes small openings so that the molten flux cannot reach the inside.

If prefluxing is used, preheating dries the flux and vaporizes all moisture from the assembly and the fixture. (Even a slight amount of moisture can cause spattering in contact with molten salt.) Preheating decreases the temperature drop of the salt bath, thus reducing brazing time, and also minimizes the premature melting of externally-placed filler metal. For assemblies consisting of both heavy and light sections, preheating reduces thermal gradients and subsequent distortion, and improves the wetting action on the heavier parts as well. To be effective, the preheating temperature must be at least 55°C lower than the melting temperature of the brazing filler metal. If oven preheating is used, oxidation must be avoided by using temperatures below 480°C. Otherwise, an inert atmosphere is desirable.

During the step of immersion of the assembly/ies to be brazed specific brazing ranges are employed for the individual brazing filler metals. The time in the molten salt bath can differ from one job to another. When only thin section parts are to be brazed, the holding time may be as short as one minute.[94] If assemblies are permitted to remain in the flux longer than three minutes, strength will be substantially reduced due to gradual transformation of the parent material from the 'as wrought' condition to the 'as cast' condition of the joint at the brazed interface.[95]

After the workpieces have been in the bath for the required time, they are carefully lifted from the salt bath. A uniform motion is necessary during removal from the bath; jerky movements can cause the liquid filler metal to be displaced from the joint.

A certain amount of flux adheres to the assembly after brazing, and this must be drained off while the parts are hot.

The parts should be suspended over the salt pot a sufficient length of time to allow the flux to drain back into the pot. This operation not only reduces flux consumption but also reduces the amount of flux that must be removed from the brazed unit. The assembly is allowed to cool and then is immersed in hot running water to remove the frozen salt cocoon. The assembly is normally removed from the fixture and subsequently dipped in a series of acid-to-water rinses to complete the flux removal process.

There are numerous safety precautions for the safe use of the fluorides in the aluminum brazing salt bath, see Chapter 6.

Molten flux must be dehydrated before use to remove chemically-combined moisture. Dehydration is accomplished by immersing coils of aluminum sheet in the flux. As long as moisture is present, hydrogen is evolved and bubbles up to the surface. Often it will ignite on the surface of the bath, showing the characteristic small orange flames. Several hours of dehydration treatment are usually necessary to remove moisture from a freshly-melted bath, depending on size.

Torch brazing is generally used for small parts, short runs, and the attachment of fittings to previously-welded, furnace brazed, or dip brazed aluminum assemblies. Mechanized torch brazing has found application in joining tubes to header joints, return bends on heat exchangers, and other similar joint configurations. All types of air-fuel gas or oxy-fuel gas torches can be used. Oxyacetylene, oxy-hydrogen, and oxy-natural gas are used for brazing of aluminum and its alloys. However, it is also possible to use gasoline blowtorches and gas burners of various sorts.

Brazing flux is mixed with water or alcohol and applied to both the work and the brazing filler metal by brushing, dipping, or sprinkling. The torch should be adjusted so that the flame is slightly reducing. Heat is applied locally to the area to be joined until the flux and the brazing filler metal melt and wet the surfaces of the base metal.

In torch brazing of aluminum, the choice of the torch tip depends on the thickness of the parts and can be most readily determined by trial. Torches with multiple flames or twin tips are extremely useful for heating tubes uniformly. Additionally, brazing with a reducing flame will give a somewhat lower temperature than brazing with a neutral flame.

Oxy-hydrogen torches are often used for brazing of aluminum and other nonferrous alloys. Because the temperature obtained is below that of oxyacetylene, the possibility of overheating the assembly during brazing is reduced.

4.2.2 Beryllium and its alloys

Brazing is the preferred method for metallurgically joining beryllium. From a stiffness/weight standpoint, beryllium is unexcelled by any bulk metal or alloy in existence. However, due to its chemical and metallurgical reactivity, braze joining techniques must be highly specialized, and suitable brazing filler metal systems and brazing temperature ranges include :

- Zinc (427 to 454°C)
- Aluminum–silicon (566 to 605°C)
- Aluminum (645 to 655°C)
- Silver–copper (640 to 904°C)
- Silver (882 to 954°C)

Strictly speaking, zinc, with a melting point of 420°C, does not quite meet the accepted definition for liquidus temperature for a brazing filler metal (425°C). Nevertheless, it is generally accepted as the lowest melting brazing filler metal for brazing of beryllium.

Using rather specialized techniques, zinc brazing of beryllium has found important applications. Its successful use was demonstrated in production joining of over 2000 fuselage longeron components for the beryllium spacer for an ICBM missile. Brazing was selected as the joining process for the longeron assembly because the brazed design had the lowest cost, the least weight, and minimized clearance problems in the assembly.[96] Zinc was selected as the brazing filler metal because it caused no detrimental reaction with the beryllium base material. The dip brazing of beryllium with zinc brazing filler metal offered the following advantages:

- Zinc has no undesirable reaction with beryllium;
- The brazing temperature was low enough that no recrystallization effects took place in beryllium sheet;

- Shear strength at room temperature was consistent and high (128 MPa);
- Shear strength at room temperature decreased linearly to a value of about 17 MPa at 425°C, indicating moderate temperature applications;
- Sophisticated equipment was not required.

Aluminum and aluminum–silicon brazing filler metals that contain 7 to 12% silicon can provide high-strength brazed joints for service at temperatures up to 150°C. These brazing filler metals have been used widely in high-strength wrought beryllium assemblies because joining is performed well below the base metal recrystallization temperature. Fluxless brazing requires stringent processing, but has been used effectively. A significant advantage of aluminum-based versus silver-based filler metals is that metallurgical interaction is minimal. This is of prime concern when thin beryllium sections or foils are to be joined.

When beryllium assemblies are brazed using the aluminum–silicon type filler metal, by virtue of the low furnace brazing temperature (605°C), the wrought sheet beryllium base properties do not deteriorate but maintain the desirable isotropic mechanical properties of beryllium.

Silver and silver-based brazing filler metals, such as Si–7Cu–0.2Li, find use particularly in structures exposed to elevated temperatures. An added advantage with these systems is that atmosphere brazing is used and may be performed in inert gas atmospheres or vacuum. The lithium is added to improve wettability. Silver interacts favorably with beryllium in that hard or brittle intermetallics do not remain upon cooling. Brazing temperatures vary depending on the mass of the parts to be joined, joint precision and pressure applied to the components. Both the silver-based and aluminum–silicon brazing filler metals, however, exhibit poor capillary flow, and preplacement is recommended. The use of the 7%Cu–0.2%Li sterling silver filler metal will markedly improve the ability of the filler metal to make well-defined fillets and bridge gaps of several mils. Copper beryllides, which form generally at the faying surfaces, inhibit rapid penetration of silver into the beryllium components and, hence, a more stable liquid phase will be present at the brazing temperature. Properly executed beryllium brazements may be made to exhibit useful strengths at temperatures up to about 870°C.[96, 97]

4.2.3 Magnesium and its alloys

Brazing techniques similar to those used for aluminum are used for magnesium alloys. Furnace, torch, and dip brazing can be employed, although dip brazing is the most widely used.

Magnesium alloys and those that are considered brazeable are given in the Appendix Table A.8. Furnace and torch brazing experience is limited to M1A alloy. Dip brazing has been successfully used for AZ10A, AZ31B,

AZ61A, K1A, M1A, ZE10A, ZK21A and ZK60A alloys. Other magnesium alloys cannot be brazed with present brazing filler metals and techniques due to their low solidus temperatures.

The two brazing filler metals AZ92A or BMG–1 (9%Al–2%Zn–0.1%Mn–0.00005%Be, reminder Mg) are both suitable for torch, dip or furnace brazing. Because the AZ125A filler metal has a lower melting range, it is usually preferred in most brazing applications. A zinc-based filler metal known as GA432 is an even lower melting composition suitable for dip brazing use only.

The probability of combustion occurring during torch brazing of magnesium is extremely remote because only the joint area, and not the entire structure, is heated, and because the fluxes used prevent burning in the heated area. The dip brazing process presents no combustion problems with magnesium because the entire operation is performed in a flux bath. Furnace brazing offers the greatest potential for magnesium combustion because the entire structure is heated to the brazing temperature. However, because the M1A alloy, with its high solidus temperature (648°C), is the only magnesium alloy suitable for furnace brazing, the risk of combustion is minimal provided that the furnace temperature is controlled within the prescribed limits and brazing time is the minimum necessary to achieve filler metal flow.

Electric or gas heating furnace equipment with automatic temperature controls capable of holding the furnace temperature within ±6°C of the brazing temperature for magnesium base metal should be used to minimize incipient melting of the base metal and to reduce the potential of a magnesium fire. The best results in furnace brazing are obtained if dry powdered flux in sprinkled along the joint. Brazing times depend somewhat on the thickness of the materials used and the amount of fixturing necessary to position the parts.

Torch brazing is accomplished using a neutral oxy-fuel or air-fuel gas flame, because of the close proximity of the initial melting point (solidus temperature) of the magnesium alloy base metal and the flow point (liquidus temperature) of the brazing filler metal. Manual torch brazing is generally preferred using the AZ125A magnesium filler metal. Natural gas is also well suited for torch brazing because its relatively low flame temperature reduces the danger of overheating.

Dip brazing of magnesium alloys is accomplished by immersing the assembly in a molten brazing flux and holding the parts and fixture at the desired brazing temperature. The flux serves the dual function of heating and fluxing. The immersion time in the flux bath can vary from 30 to 45 sec up to 1 to 3 min. Large assemblies with more metal mass and/or fixturing will require the longer times. Because of the large volume of flux and uniform heating, more consistent results are achieved with dip brazing than with other brazing procedures.

The corrosion resistance of brazed joints in magnesium alloys depends primarily on the thoroughness of flux removal and the adequacy of joint design to prevent flux entrapment. Because the brazing filler metal is a magnesium-based alloy, the problem of galvanic corrosion is minimized. Protection of magnesium alloys from corrosion is provided by applying appropriate chromate and other coatings.

Torch brazing has been used to join hydraulic lift floats, whereas dip brazing has been used to produce battery containers and microwave antennas.

4.2.4 Copper and its alloys

The copper alloy base metals include copper–zinc alloys (brass), copper–silicon alloys (silicon bronze), copper–aluminum alloys (aluminum bronze), copper–tin alloys (phosphor bronze), copper–nickel alloys, and several others. They are commonly brazed with copper- and silver-based brazing filler metals.

The brazeability of copper and its alloys is generally rated from good to excellent using one or more of the conventional brazing processes. These processes include furnace, torch, induction, resistance, and dip brazing.

Softening of the base metal frequently occurs during brazing because many copper-based alloys derive their properties as a result of heat treatment at relatively low temperatures, cold working, or both. The degree of softening increases with higher temperatures and longer times of exposure to elevated temperatures. Softening of areas in proximity to the braze can be minimized by cooling the assembly, except for the area to be brazed, by immersion in water, packing with wet rags or wet asbestos, or otherwise providing a heat sink to keep the overall temperature of the part as low as possible. In all cases, brazing with a low-melting filler metal for a minimum time will reduce softening. The importance of uniform and controlled heating cannot be overemphasized, especially when dealing with brasses, cold-worked phosphor bronzes, and cold-worked silicon bronzes. They are especially susceptible to cracking.

Tough pitch coppers are subject to embrittlement when heated in reducing atmospheres containing hydrogen. Consequently, although tough pitch coppers are generally rated as having good to excellent brazeability, they should not be brazed in a furnace that contains hydrogen with a reducing potential, such as dissociated ammonia, or in an exothermic-based or endothermic-based atmosphere. Heating by open flame or by torch also may result in hydrogen diffusion and embrittlement.

Phosphorus-deoxidized and oxygen-free coppers can be brazed without flux in hydrogen-containing atmospheres without risk of embrittlement, provided that self-fluxing filler metals (copper–phosphorus) are used. The use of flux is required, however, when silver-based filler metals that contain additives such as zinc, cadmium, or lithium are used to braze these coppers to each other or to copper alloys or other metals.

The coppers, including those that contain small additions of silver, lead, tellurium, selenium, or sulfur (generally no more than 1%), are readily brazed with the self-fluxing, copper–phosphorus brazing filler metals, but wetting action is improved when a flux is used and when a sliding motion between components is provided while the filler metal is molten.

Another copper group includes those with enhanced mechanical properties due to the addition of small amounts of alloying elements. These **precipitation–hardenable (PH) copper alloys** contain beryllium, chromium, or zirconium and form oxide films that impede the flow of brazing filler metal. To ensure proper wetting action of the joint surface by the filler metal, beryllium–copper alloys should be freshly machined or mechanically abraded before being brazed. Removal of beryllium oxide from joint surfaces requires the use of a high fluoride content flux.

Brazing PH coppers in the aged condition reduces their mechanical properties. The high-strength beryllium–copper alloys (2% beryllium) can be furnace brazed and simultaneously solution treated at 790°C. The temperature is lowered to 760°C to solidify the filler metal, and then the assembly is rapidly quenched in water and aged at 315 to 345°C, which develops adequate hardness in the base material. The 72Ag–28Cu eutectic filler metal is generally used with flux. When the sections to be brazed are thin and can be cooled very rapidly, solution-heat-treated beryllium copper may be brazed in the temperature range of 620 to 650°C. Other beryllium–copper alloys (0.5% beryllium) can be silver brazed rapidly with filler metals containing 45 to 50% silver, 15% copper, 16% zinc, and 18 to 24% cadmium.

The chromium coppers are usually brazed with silver-based brazing filler metal and fluoride flux and simultaneously solution-treated at 900 to 1010°C. Cold work and ageing can take place as a subsequent operation to develop improved mechanical properties. However, zirconium coppers do not precipitation-harden without the benefit of prior cold working, a sequence that is incompatible with brazing. In the absence of cold working, the strength of zirconium coppers is not improved by ageing treatments.

Brasses

The copper–zinc alloys (brasses) are produced with varying ratios of the two elements to provide desired properties and casting characteristics.

Other elements occasionally are added to enhance particular mechanical or corrosion properties; these special brasses are identified by the added element – for instance, lead, which gives leaded brass. Additions of manganese, tin, iron, silicon, nickel, lead, and aluminum, either singly or collectively, rarely exceed 4%.

Red brasses (low zinc content) contain up to 20% Zn and are readily brazed with a variety of filler metals. Flux is normally required for best results, especially when the zinc content is above 15%. All the brasses can be brazed with the silver and copper–phosphorus filler metals, and the higher melting-point (low-zinc) brasses can also be brazed with the copper–zinc filler metal.

It is recommended that protective atmospheres be used in furnace brazing of brasses; however, even in protective atmosphere, flux should be used to promote good wetting by the brazing filler metal and to reduce zinc fuming. When heated above 400°C, brass tends to lose zinc by vaporization. This loss can be reduced by fluxing the parts during furnace brazing or by using an oxidizing flame during torch brazing. Brasses are also subject to cracking and should therefore be heated carefully and uniformly.

Yellow Brasses (high zinc content) contain 25 to 40% Zn and are readily brazed, but low-melting filler metals should be used to avoid dezincification of the base metal.

Leaded brasses are formed when lead is added to red or yellow brass in amounts up to 5% while alloys containing more than 5% lead are usually not brazed. The lead forms a dross on heating that can seriously impede wetting and the flow of the brazing filler metal. Consequently, in brazing of leaded brasses, the use of a flux is mandatory to prevent dross formation in the joint area.

Naval brass, leaded naval brass, and admiralty brass contain up to 1% tin and may contain other alloying elements such as lead, manganese, arsenic, nickel and aluminum; except for the aluminum-containing alloys, these brasses are readily brazed; they have greater resistance to thermal shock and are less susceptible to hot cracking than the high-lead brasses. For proper wetting, brasses that contain aluminum require a special flux.

Bronzes

Alloys of copper and tin are properly termed tin bronzes. However, these copper-tin alloys contain small amounts of phosphorus (up to 0.25%) added as a deoxidizer. The tin bronzes have come to be known commercially as **phosphor bronzes**. Although susceptible to hot cracking in the cold worked condition, phosphor bronzes have good brazeability and are adaptable to brazing with any of the common brazing filler metals that have melting temperatures lower than that of the base metal.

To avoid cracking, phosphor bronzes should be stress relieved at 290 to 345°C before brazing. After this stress relief or anneal, the parts should be supported in a stress-free condition during brazing, and slow heating cycles should be used to avoid thermal shock. When the tin content is high or when there are appreciable lead additives, adequate flux protection during brazing is necessary. All the phosphor bronzes can be brazed with the silver–copper–zinc–cadmium and copper–phosphorus filler metals while the copper–zinc filler metals are appropriate for brazing the low tin varieties.

Aluminum bronzes are copper–aluminum alloys with high copper contents and 3 to 13% aluminum, with or without varying amounts of iron, nickel, manganese, and silicon. Aluminum bronzes are generally considered difficult to braze because of their aluminum content, which results in the formation of refractory aluminum oxide at brazing temperature in alloys containing more than 8% aluminum. However, alloys containing 8% aluminum or less are brazeable provided that appropriate fluxes are used to dissolve the aluminum oxide. The oxide, which inhibits the flow of brazing filler metal, cannot be reduced in dry hydrogen.

Electroplating of copper on the surface to be brazed is one technique used to enhance wetting of the low-melting silver filler metal, while use of flux and a protective atmosphere in a furnace is a technique employed if unplated surfaces are used.

Silicon bronzes which contain up to about 3.25% silicon and are in a highly stressed condition are susceptible to hot shortness and stress cracking by molten brazing filler metal and should be well fixtured during heating and brazing to prevent excessive stressing. To avoid cracking, these alloys should be stress relieved at about 290 to 345°C before brazing. For best brazing results, joint surfaces of copper–silicon alloys should be freshly machined, cleaned, and flux coated or copper plated before brazing to prevent the formation of refractory silicon oxide. Mechanical cleaning is recommended, or for light oxide the material can be pickled. Silver-bearing brazing filler metals and flux generally are used.

Other copper alloys

Copper nickels which are available commercially may contain from 5 to 40% nickel, and are susceptible to both hot cracking and stress cracking by molten brazing filler metal. To prevent cracking, copper–nickel alloys should be stress relieved before brazing. Stresses should not be introduced during brazing. The silver–copper–zinc–cadmium filler metals and flux are preferred for brazing the copper nickels.

Nickel silvers, which are brasses (copper–zinc–nickel) that contain up to about 20% nickel but do not contain silver, are highly susceptible to hot cracking and should be stress relieved at about 290°C before being brazed. The nickel silvers can be brazed readily with the same procedures used for brazing brass. When copper–zinc brazing filler metals are used, however, care is required because of the relatively high brazing temperatures.

Copper brazing processes

Typical brazing processes used for copper and its alloys are compared in Figure 4.12. The advantages of furnace brazing that are applicable to the joining of copper and copper alloys relate to the furnace as a source of heat and to the cooling chamber that is provided on conveyor-belt furnaces as a means for cooling assemblies from the brazing temperature to 150°C or below. To a lesser degree, the protective atmospheres used most conveniently in furnace brazing – notably the exothermic-based and endothermic-based atmospheres – constitute another advantage of brazing deoxidized copper alloys in a furnace.

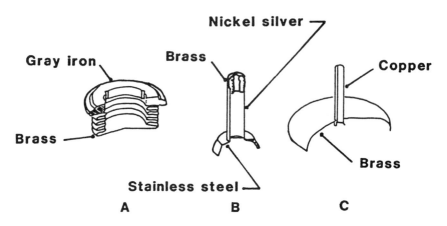

Assembly	Brazing process	Production rate, assemblies/hour
A	Induction	720
	Torch (manual	120 (a)
B	Furnace	140
	Torch (manual)	60 (b)
C	Induction	60 (b)
	Furnace	12 (c)

(a) Assemblies brazed one at a time; heating time, 30 s. (b) Assemblies brazed one at a time; heating time, 60 s. (c) Assemblies brazed one at a time; heating time, 5 min.

Fig. 4.12 Comparison of production rates for copper and brass brazing processes.

The fuel-gas mixtures generally used in **torch brazing** of copper and copper alloys are oxyacetylene, oxy-natural gas, oxypropane; oxyhydrogen, and air-natural gas. Oxy-fuel gases are highest in cost, flame temperature, and heating rate, with oxyacetylene being the highest in the group in each respect. Oxy-fuel gases are widely used in manual torch brazing where temperature can be controlled by torch manipulation. The cost advantage of air-natural gas is exploited particularly in mechanized and automatic high-production torch brazing, where the lower flame temperature offers protection from damage caused by overheating.

Although high heat input is necessary to overcome high thermal conductivity, many copper alloys cannot withstand the rapid heating rates that can be used on steel. Under too-rapid heating, high thermal expansion can cause local stresses, resulting in distortion and cracking in some alloys. Accidental overheating can cause damage to copper and copper alloy assemblies more readily than to those made of steel and in particular phosphor bronzes, leaded brasses, nickel silvers, and silicon bronzes.

Although the use of oxyacetylene flame requires more care to avoid overheating, it is often preferred over other types of flames because of its high rate of heating.

Precautions should be taken in torch brazing certain coppers and copper alloys. Where it is necessary to braze oxide-containing coppers, a reducing atmosphere in the flame must be avoided because it can promote hydrogen embrittlement. For these coppers, a neutral or slightly oxidizing flame and a short brazing cycle are necessary. Brasses are subject to volatilization of zinc when over-heated or when held too long at brazing temperature and therefore, application of flux suppresses zinc volatilization.

Alloys containing elements that readily form refractory oxides (aluminum, beryllium, chromium, and silicon) must be protected by flux and should not be exposed to an oxidizing flame.

The efficiency of **heating by induction** varies directly with the electrical resistivity of the alloy. Brass, because it has higher electrical resistivity, can be heated more efficiently than copper; steel, which has even higher resistivity, can be heated more efficiently than brass. In terms of the high-frequency power input and the time required to heat 0.45 kg of metal in a joint assembly at a brazing temperature of 705°C and a power input (450 kHz) of 15 kW, steel required 16 s, whereas brass required 30 s and copper about 55 s. The main advantages of induction brazing of copper and its alloys over torch brazing are minimized warpage, reduced post-braze cleaning, and less required operator skill. Induction heating is suited for mass producing brazed assemblies, primarily because inductors can be designed that heat a line of assemblies as they are carried through the induction field by conveyor belt or turntable. One of the general limitations of induction brazing is the cost of induction heating equipment, which far exceeds the cost of torch brazing equipment and usually exceeds the cost of equipment for resistance brazing or dip brazing in molten salt. Because the efficiency of heating copper and copper alloys by induction is generally low, costs to achieve given production rates are high. The low long-term operating and maintenance costs of induction units, however, may out-weigh the higher initial costs.

Other general limitations of induction brazing may relate to the size and shape of assemblies that can be brazed, the design of inductors (cooling coils used to convey heat to the assembly), and requirements for matching impedances which apply to brazing of copper and copper alloys.

In a recently completed study[98] on the usage of advanced particulate-reinforced composites and dispersion-hardened materials as candidate materials to replace pure annealed copper for the electromagnetic windings of the compact ignition Tokomak nuclear fusion reactor, it was found that direct brazing of such materials may lead to brittle joints and that indirect brazing may need to be employed for joining such materials.

As a result 0.25% vol alumina dispersion-hardened copper (ADHC) was brazed satisfactorily using induction brazing in an argon atmosphere using 72Ag–28Cu eutectic and pure Ag electroplate as interlayer materials. Both interlayers completely wet ADHC. For both interlayers, the application of additional stress (200 MPa) on the joint improved the joint quality. By plating a $17.5\mu m$ layer of copper on the ADHC prior to brazing, other researchers have obtained shear strengths as high as 223 MPa.

Resistance brazing is often used in joining copper conductors, terminals, and other parts in lap joints for electrical connections where heating must be localized and closely controlled during brazing and where the brazed joint must have low electrical resistance. Heat in the brazing filler metal and nearly complete filling of the joint with a thin layer of the filler metal help meet all these objectives. Normally the brazing filler metal most frequently used is copper–phosphorus, which is used without a flux – especially in brazing of copper.

The success of **resistance brazing** for joining small copper electrical conductors to massive copper assemblies has led to the use of the process for attaching armature leads to commutator bars on large electric motors and generators.

The use of conventional resistance welding equipment and high-resistivity electrodes with specially contoured tips makes it possible to concentrate the heating at the joint, to keep the heating time to a minimum, and to obtain efficient handling and comparatively high production rates.

Another use of portable resistance welding machines for resistance brazing is in attaching copper busbar terminals or similar strip connectors to large electrical equipment that cannot be brought to, or positioned for brazing in, a conventional fixed-position resistance welding machine. Electrical connections to such equipment can often be made more economically by resistance brazing than by mechanical means and are made more readily by resistance brazing than by arc welding.

Resistance brazing done with portable resistance welding machines is a convenient and economical way of interconnecting large copper electrical busbars or of attaching either large or small copper busbars to motor-generators, transformers, and other electrical equipment.

The low melting temperature of the brazing filler metal helps to avoid overheating and excessive annealing of the work, and the usual selection of self-fluxing copper–phosphorus filler metal avoids corrosion problems and the need for flux removal.

In some applications, filler metal for resistance brazing can be provided in the form of a coating already present on one or both members to be joined, eliminating not only the use of flux but also the operation of placing filler metal at the joint. This is done when copper wire is joined by high-speed resistance brazing to copper terminals clad with copper–phosphorus filler metal.

Salt bath furnaces have been used for brazing of copper and copper alloys with silver filler metals. The same neutral salts, operating temperatures, and brazing procedures used for steel are used for dip brazing of most copper alloys. Applications of dip brazing of copper alloys in molten salt include waveguides and waveguide hardware, flowmeter hardware, and capillary tube and bellows assemblies.

4.2.5 Nickel, cobalt and heat-resistant alloys

Nickel and the high-nickel alloys may be divided into the following classes:

1. commercially pure nickel;
2. nickel–copper alloys;
3. nickel–chromium–iron alloys;
4. nickel–chromium–molybdenum alloys:
5. thorium-dispersed nickel alloys.

Alloys in groups (1) and (2) are used primarily for applications where corrosion resistance is important, or where product purity must be maintained. Alloys in groups (3), (4) and (5) also have good corrosion resistance in many media and, in addition, have high strength and oxidation resistance at elevated temperatures. To utilize these properties, it is important to determine the effects of service conditions on the brazed joint.

Nickel and the high nickel alloys are embrittled by sulfur and low-melting metals present in brazing filler metals, such as zinc, lead, bismuth, and antimony. Base metal surfaces must be thoroughly cleaned prior to Brazing to remove any substances that may contain these elements and sulfur and sulfur compounds must also be excluded from the brazing atmosphere.

Nickel and its alloys are subject to stress cracking in the presence of molten brazing filler metals. Parts should be annealed prior to brazing to remove residual stresses, or carefully stress relieved during the braze cycle.

In selection of a brazing process and a brazing filler metal for a nickel-base alloy, the characteristics of the alloy must be carefully considered. Nickel alloys differ significantly not only in physical metallurgy (precipitation-strengthened versus solid-solution-strengthened) but also in process history (cast versus wrought). These characteristics can have a profound effect on brazeability.

The brazing filler metals normally used for ferrous metals are suitable for joining nickel and high-nickel alloys. It also is important to consider any heat treatments that may be required for the base metal, because the brazed joint must withstand the temperatures involved. In corrosive environments, high-silver brazing filler metals are preferred while cadmium-free brazing filler metals are chosen to avoid stress corrosion cracking.

Most **high-nickel alloys** are capable of being brazed with pure copper filler metal, which is similar to the brazing of carbon and low-alloy steel except that the copper filler metal will characteristically alloy to a greater extent with nickel than with iron. Alloying during brazing makes capillary flow difficult. The copper will not flow far before it has picked up enough nickel to raise its liquidus and reduce its fluidity. To eliminate this problem, the brazing filler metal should be placed as close to the joint as possible, and there should be a sufficient reservoir to fill the joint. Secondly, the assembly should be heated as rapidly as practicable to the brazing temperature.

Phosphorus combines with many metals to form brittle compounds known as phosphides. For this reason the copper–phosphorus brazing filler metals usually are not used with any iron or nickel-based alloy; however, the BNi–6 and BNi–7 filler metals are used for brazing nickel-based alloys. Nickel based brazing filler metals offer the greatest corrosion and oxidation resistance and elevated temperature strength.

Nickel-based brazing filler metals containing palladium or platinum, and gold-based, palladium-based, and platinum-based filler metals, have also been used successfully to braze high-nickel alloys. These brazing filler metals generally have good wetting and flow characteristics. They have a low interaction rate with most nickel-based metals and are used to advantage in many special applications where joints are required to have good ductility, high strength, and good oxidation resistance. Brazing filler metals consisting of base material plus additions of silicon and boron have been used successfully for brazing the precipitation-strengthened alloys of nickel.

When appreciable amounts of aluminum and titanium appear (greater than 1%) in the precipitation-strengthened nickel-based alloys, the oxides of aluminum and titanium are almost impossible to reduce in a controlled atmosphere (vacuum or hydrogen). Therefore, nickel plating or the use of a flux is necessary to obtain a surface that allows wetting by the filler metal.

Most commercial brazing processes may be used on nickel and high-nickel alloys. The most common of these are torch, furnace, induction, and resistance brazing; salt bath dip brazing has limited application. Whereas the silver-based brazing filler metals may be used in torch brazing, the copper and nickel filler metals are usually used in controlled-atmosphere brazing.

Attempting to braze over the **refractory oxides** of titanium and aluminum that may be present on precipitation-hardenable nickel-based alloys must be avoided. Procedures to prevent or inhibit the formation of these oxides before and during brazing include special treatments of the surface to be joined and brazing in a highly-controlled atmosphere. Surface treatments include electrolytic nickel plating and reduction of the oxides to metallic form. As stated earlier, a typical practice is to nickel plate the joint surfaces of any alloy that contains aluminum and/or titanium. For vacuum brazing, when aluminum and titanium are present in trace amounts, plating 0.01 mm thick is considered optional. Alloys with up to 4% aluminum and/or titanium require plating 0.015 mm thick, whereas alloys with aluminum and/or titanium contents greater than 4% require plating 0.02 to 0.03 mm thick. When brazing is done in a pure dry hydrogen atmosphere, thicker plating (0.03 to 0.04 mm) is desirable for alloys with greater than 4% aluminum and/or titanium contents.

Brazing of base materials containing more than a few percent of aluminum, titanium, zirconium or other elements that form very stable oxides requires vacuums of 0.13 Pa or lower. Vacuum furnaces for this type of brazing usually employ a diffusion pump usually backed by a mechanical pump.

Consideration must be given to the effect of the **brazing thermal cycle** on the base metal. Brazing filler metals that are suitable for brazing nickel-based alloys may require relatively high thermal cycles. This is particularly true for the filler metal alloy systems most frequently used in brazing nickel-based alloys – the nickel–chromium–silicon and nickel–chromium–boron systems.

It is significant to point out the effects of thermal cycles on solid-solution-strengthened and precipitation-strengthened nickel-based alloys. An example of the former is Inconel 600, a high-temperature nickel-based alloy containing 15.5% Cr, 76% Ni-Co and 8% Fe which resists oxidation up to 1175°C and which is used in furnace parts and fixtures, heat exchangers, chemical-handling equipment and turbine and reactor parts. This alloy may not be adversely affected by nickel-based filler metal brazing temperatures of 1010 to 1230°C. An example of the latter is Inconel 718, a nickel-based superalloy containing 50 to 55% Ni, 17 to 21% Cr, 4.75 to 5.5% Nb and Ta, 1% Co, approximately 0.20% Fe and greater than 2% Al and Ti. This alloy is oxidation-resistant up to 980°C and is used in aircraft turbine parts, pumps, and rocket motors. This alloy may, however, display adverse property effects when exposed to brazing cycles higher than its normal solution heat treatment temperature. Inconel 718, for example, is solution-heat-treated at 955°C, for optimum stress-rupture life and ductility. Brazing temperatures of 1010°C or above result in grain growth and an attendant decrease in stress-rupture properties, which cannot be recovered by subsequent heat treatment.

Consideration of base metal property requirements for service enables selection of an appropriate brazing filler metal. For lower melting temperatures (below 1040°C), filler metals are available within the nickel-based filler metal family and within other filler metal systems, see Appendix Tables A.4 and A.5.[99–101]

Brazing is a preferred method for joining dispersion-strengthened nickel alloys that must function at elevated temperatures. High strength brazements have been made with special nickel-based brazing filler metals and then tested up to 1300°C.

Procedures have been developed for brazing this new family of materials, TD–Ni and TD–NiCr foils, for high-temperature service using the candidate filler metals shown in the Appendix Table A.10.

Heat-resistant alloys are suitable for use under moderate to high loading in the temperature range from 540 to 1100°C. These metals are complex austenitic alloys based on nickel or cobalt or both. They have often been termed 'superalloys' and their greatest use is in the construction of gas turbine engines and hot airframe components.

Heat-resistant alloys are generally brazed in hydrogen atmosphere or high-vacuum furnaces with nickel-based or special brazing filler metals. Because the brazing temperatures are high, the effect of the brazing thermal cycle on the base metals should be taken into account. The non-heat-treatable alloys will suffer moderate strength losses due to grain growth during brazing. Cold-worked alloys should not be brazed unless the severe loss in strength from annealing during brazing is considered in the design.

Brazing of **cobalt-based alloys** is readily accomplished by the same techniques used for nickel-based alloys. Because most of the popular cobalt-based alloys do not contain appreciable amounts of aluminum or titanium, brazing atmosphere requirements are less stringent. These materials can be brazed in either a hydrogen atmosphere or a vacuum. Brazing filler metals are usually nickel- or cobalt-based alloys or gold–palladium compositions. Silver or copper brazing filler metals may not have sufficient strength and oxidation resistance in many high-temperature applications. Although cobalt-based alloys do not contain appreciable amounts of aluminum or titanium, an electroplate or flash of nickel is often used to promote better wetting of the brazing filler metal. Nickel-based brazing fillers such as BNi–3 (Appendix Table A.5) have been used successfully on cobalt-based alloys[102,103] for honeycomb structures. After brazing, a diffusion cycle was used to raise the brazed joint remelt temperature to 1260 to 1315°C. The BCo–1 brazing filler metal (Appendix Table A.5) appears to offer a good combination of strength, oxidation resistance, and remelt temperature for use on cobalt-based alloy foil such as Haynes 25.

Cobalt alloys, much like nickel alloys, can be subject to liquid metal embrittlement or stress corrosion cracking when brazed under residual or dynamic stresses. This frequently is observed when silver or silver–copper brazing filler metals are used. Liquid metal embrittlement of cobalt-based alloys by copper brazing filler metals occurs with or without the application of stress; therefore, copper filler metals should be avoided in the brazing of cobalt-based alloys.

Superalloys

Superalloys can be subdivided into two categories: conventional cast and wrought alloys and powder metallurgy (P/M) products. P/M products may be produced in conventional alloy compositions and oxide-dispersion-strengthened (ODS) alloys.

Oxide-dispersion-strengthened alloys are P/M alloys that contain stable oxide evenly distributed throughout the matrix. There are two commercial alloy classes of ODS alloys; the dispersion-strengthened nickel chromium and dispersion-strengthened nickel discussed previously; and mechanically alloyed (MA) alloys. An example is Inconel MA 754, a turbine vane material made from powder. The material has extremely good creep and oxidation resistance at temperatures up to 1050°C and is used in gas turbines (vanes, nozzle, burners, combustors), chemical plant (liners, heat exchangers, combustion chambers), etc. They owe their oxidation resistance to a stable and tenacious oxide film, and their creep resistance to the combination of a large and directional grain structure plus a dispersion of submicron-sized particles consisting of chiefly yttrium oxide.

There is no joining method[104,105] which can preserve the coarse and elongated grain structure of the recrystallized parent metal across a joint line. However, some sacrifice of the maximum properties may be allowable if such a joint were to be sited outside the area of maximum stress; alternatively a joint parallel to the major stresses might be acceptable in certain cases.

Although brazing introduces a change in composition across the joint line, and although it does melt some of the parent metal and thus causes agglomeration of the dispersoid, nevertheless these effects can be minimized by a suitable choice of composition and by close control of brazing and postbrazing conditions. Furthermore, brazing does not introduce mechanical distortion and it can be carried out at temperatures well below the recrystallization range.

Inconel MA 754 is one of the easiest alloys to braze in the family of ODS alloys. Vacuum, hydrogen, or inert atmospheres can be used for brazing. Pre-braze cleaning consists of grinding or machining the faying surfaces and washing with a solvent that evaporates without leaving a residue. Generally, brazing temperatures should not exceed 1315°C unless demanded by a specific application that has been well examined and tested. The brazing filler metals for use with these ODS alloys usually are not classified (see Appendix Table A.9). In most cases, the brazing filler metals used with these alloys have brazing temperatures in excess of 1230°C. These include proprietary filler metals containing nickel, cobalt, gold, or palladium.

Bucklow[105] recently completed a series of braze filler metal evaluation tests where he found that brazing fine-grained MA 754 at 1200°C by induction heating for less than 30 s with a thin β-free coating equivalent to BN1–5 gave joints of excellent appearance and without recrystallization.

A second nickel-based ODS alloy is Inconel MA 6000. Like Inconel MA 754, it is γ-strengthened. Inconel MA 6000 has a solidus temperature of 1300°C; therefore, the brazing temperature should be no higher than 1250°C. Ni and Co brazing filler metals have been used to join this alloy. Other MA ODS alloys include Fe-based creep-resistant MA 956 which was also recently examined by Bucklow.[106] Although it was possible to produce a reasonable brazed joint in MA 956, by means of sputter-coated brazes based on the nickel filler metal BNi–1a (72.5Ni–16.5Cr–4Fe–4Si–3B), an agglomeration of the dispersed oxide at the braze interface was unavoidable. Furthermore, as in the brazing of Ni-based ODS alloys, the presence of boron in the braze causes local recrystallization of the parent material. This has led to future work whereby diffusion bonding was to be examined as the only joining process which did not require either a foreign material or a melted or disrupted region in the joint. Diffusion bonding therefore offered the opportunity to achieve a bond in an Fe–ODS alloy in which the massive grain structure continued unbroken across the bond line.

4.2.6 Molybdenum, niobium, tantalum, tungsten and their alloys

The refractory group of metal elements have a history of use in applications where structural integrity at high temperatures is required. These applications include filaments and lead wires for incandescent lamps and vacuum tubes and grids for cathode structures. Because of its limited effect on base metal properties, brazing has been accepted widely as a method of joining assemblies fabricated from refractory metals. The characteristics that affect the brazeability of refractory metals include ductile-to-brittle transition behavior, recrystallization temperature, and reactivity with oxygen, nitrogen, hydrogen, and carbon.

The refractory metals include those metals that have melting points in excess of 2205°C, i.e., niobium, iridium, molybdenum, tantalum, ruthenium, tungsten, osmium, and rhenium. Sometimes vanadium, hafnium, rhodium, and chromium are included in this group of metals. The most important refractory metals, from a structural standpoint, are molybdenum (Mo), tantalum (Ta), niobium (Nb), and tungsten (W).

Niobium, molybdenum, tantalum, and tungsten have much in common; however, there are differences that have a bearing on the manner in which these metals are used and joined. All of the refractory metals have a body-centered cubic crystal structure, with very high melting temperatures, high to very high densities, low specific heats, and low coefficients of thermal expansion. Mechanical strength and structural integrity at high temperatures are excellent.

The mechanical properties of refractory metals are affected markedly by ductile-to-brittle transition behavior, recrystallization temperature, and reactions with carbon and various gases. These characteristics must be considered when brazing procedures are established.

Refractory metals, because of their body-centered cubic structure, have a well-defined transition from ductile to brittle behavior. The transition temperature ranges for the pure refractory metals are as follows:

- Niobium, –200 to –75°C
- Molybdenum, 150 to 260°C
- Tantalum, <–195°C
- Tungsten, 260 to 370°C

Therefore, molybdenum and tungsten are brittle at room temperature and must be handled carefully to avoid damage. Also, these metals must be brazed in a stress-free condition. The strength and ductility of refractory metals are adversely affected by microstructural changes that occur when the recrystallization temperatures of these metals are exceeded. Recrystallization temperature ranges for unalloyed refractory metals are :

- Niobium, 985 to 1150°C
- Molybdenum, 1150 to 1200°C
- Tantalum, 1100 to 1400°C
- Tungsten, 1200 to 1650°C

Some applications permit brazing with filler metals that melt below the recrystallization temperature range. Other applications require the use of filler metals that melt above this temperature range. The joint also must be designed to accommodate the loss in mechanical properties associated with recrystallization.

Research has improved the high-temperature mechanical properties of the refractory metals and has led to an increase in recrystallization temperatures through alloying. For example, titanium, zirconium, and hafnium can be used to strengthen molybdenum and increase its recrystallization temperature. The recrystallization temperature of unalloyed molybdenum is about 1150 to 1200°C. In contrast, the recrystallization temperature of Mo–0.5Ti–0.7Zr molybdenum alloy (TZM) is about 1480°C and the stress rupture strength of this alloy at 980 to 1095°C is several times that of unalloyed molybdenum.

The third consideration is reactions with gases and carbon. The environment in which refractory metals are brazed is determined by the reactivity of these metals with oxygen, hydrogen, and nitrogen and the effects of these elements on mechanical properties.

Tantalum and niobium are embrittled by the presence of hydrogen at relatively low temperatures. In contrast, molybdenum and tungsten can be brazed in a hydrogen atmosphere. Molybdenum, tantalum, and niobium are embrittled by nitrogen at high temperatures. All of the refractory metals form carbides in the presence of minute quantities of carbon and its compounds.

Based on the various characteristics of refractory metals, the following principles are applicable to brazing of refractory metals:

1. If maximum joint strength is required, refractory metals must be brazed at temperatures below those at which recrystallization occurs.
2. However, brazing at much higher temperatures may be necessitated by service requirements, and some decrease in joint properties must be anticipated.
3. All of the refractory metals can be brazed in a vacuum or in an argon or helium atmosphere with a very low dewpoint.
4. Graphite fixturing should not be used to position refractory metal parts during brazing, because these metals readily form carbides.

However, graphite tooling may be acceptable if coated with a refractory material.

5. Ceramics can also be used for fixturing, but care in their selection must be exercised. Some ceramics cannot be used in a vacuum because of outgassing, whereas others react with refractory metals at high temperatures.
6. Refractory metals with higher melting temperatures than the one being brazed also can be used for fixturing.

The filler metals used for brazing of refractory metals must be selected on the basis of :

(a) the service conditions;
(b) the specific application;
(c) their compatibility with the base metal and coating (if a coating is used), see Appendix Table A. 6.

Molybdenum and its alloys

Copper- and silver-based brazing filler metals can be used to braze molybdenum for low-temperature service while for high-temperature applications, molybdenum can be brazed with gold, palladium, and platinum filler metals, nickel-based filler metals, reactive metals and refractory metals that melt at lower temperatures than molybdenum. The nickel-based filler metals have limited applicability for high-temperature service, since nickel and molybdenum form a low-melting eutectic at about 1315°C.

There are three basic limits on brazing of molybdenum and its alloys for use above 980°C:

1. Recrystallization of molybdenum;
2. Formation of intermetallics between refractory metals and brazing filler metals;
3. Relative weakness of brazing filler metals at elevated temperatures.

The formation of intermetallics between molybdenum and brazing filler metals is detrimental to joint soundness because the intermetallics become brittle and may fracture at relatively low loads when the joint is stressed.

The relative weakness of brazing filler metals at elevated temperatures limits the use of brazed molybdenum assemblies. Most of the nickel-based elevated-temperature brazing filler metals melt between 980 and 1150°C, where the superior elevated-temperature strength of molybdenum begins to manifest itself.

Two binary brazing filler metals, V–35Nb and Ti–30V, have been evaluated for use with the Mo–0.5Ti molybdenum alloy. Tee-joints with the Ti–30V filler metal were brazed in a vacuum for 5 min at 1650°C and 1870°C for the V-35Nb filler metal. The brazing filler metal had excellent metallurgical compatibility with the molybdenum alloy, and minimum erosion of the base metal occurred during brazing.[107]

Brazements made on Mo–0.5Ti alloy with the diffusion sink brazing filler metal, Ti–8.5Si, which melts at approximately 1330°C, exhibited excellent filleting and wetting, joint ductility, and freedom from cracks. Specimens with molybdenum powder added to the filler metal powder were also evaluated and brazed at 1400°C.[103]

Because of its excellent high-temperature properties and compatibility with certain environments, molybdenum is a prime candidate for use in isotopic power systems for components of nuclear reactors and chemical processing systems and the best filler metal base on overall performance has been Fe–15Mo–5Ge–4C–1B.[108,109]

The molybdenum alloy TZM (0.5Ti–0.08Zr–Mo) has also been brazed successfully at 1400°C with molybdenum powder added to the Ti–8.5Si brazing filler metal[102,103] and Gilliland and Slaughter[110] have shown that other brazing filler metals can be used. A Ti–25Cr–13Ni filler metal with a brazing temperature of 1260°C has produced the highest remelt temperature on TZM.

Remelt investigations conducted to develop vacuum brazements of molybdenum and tungsten which can be used in seal-joint applications up to 1600°C were completed with the following brazing filler metals:

1. Ti–65 wire
2. V wire
3. MoB–50MoC powder mixture
4. V–50Mo powder mixture
5. Mo–15MoB$_2$ powder mixture
6. Mo–49V–15MoB$_2$ powder mixture.

Brazing temperatures ranged from 1625 to 2255°C. Molybdenum joints made with Ti–65V, pure vanadium wire, V–50Mo, and MoB–50MoC were as strong, or stronger, than the base metal at elevated temperature. The most resistant joints tested were brazed with Ti–65V, V–50Mo, and MoB–50MoC. Without further testing, it probably would be best to limit the maximum service temperature to 1395°C.

In a more recently completed wettability study conducted by McDonald, Keller, Heiple and Hoffman[111] they found that:

1. Brazing filler metals generally wet molybdenum significantly better than they wet TZM. This is seen in Tables 4.3 and 4.4. The wettability index (WI) shown for each filler metal indicates that indices > 0.05 are

indicative of good performance during brazing, and WI > 0.10 are indicative of excellent performance during brazing.

2. The decrease in wettability of the brazing filler metals on TZM versus molybdenum is believed to be due to the presence of titanium in the oxide on the surface of TZM.

3. Some nickel-containing brazing filler metals have been shown to embrittle both molybdenum and TZM. Liquid–metal embrittlement is believed to be the mechanism by which this embrittlement occurs.

Torches, controlled-atmosphere furnaces, vacuum furnaces and induction and resistance heating equipment can be used to braze molybdenum with the variety of brazing filler metals described above.

Niobium and its alloys

Niobium is used mostly for nuclear and aerospace applications. One aerospace application required the development of manufacturing procedures for fabricating flat and curved niobium alloy sandwich heatshield panels.[103] Titanium and the titanium-based alloy Ti–11Cr–13V–3Al (B120VCA) produced joints with excellent ductility and with base metal reaction.[112,113]

The flow characteristics of the B120VCA brazing filler metal were more sluggish than those of pure titanium; as a result, the filler metal could be used to bridge wide joint clearances. The previously-mentioned honeycomb sandwich panels were successfully brazed with pure titanium at 1815°C. Heat shield panels have been brazed with B120VCA brazing filler metal at 1650°C in the same vacuum brazing environment. Simulated re-entry environment tests have shown that the heat shield panel could be used at temperatures up to 1315°C. Room temperature and elevated-temperature tests of the structural honeycomb panel indicated that it possessed useful properties up to 1260°C.[103]

Other developments have evaluated the high remelt-temperature brazing techniques based on diffusion-sink and reactive brazing concepts. The diffusion-sink technique involves either permitting the brazing filler metal to react with the base metal under controlled conditions or adding base-metal powder to the brazing filler metal powder. One diffusion-sink filler metal is Ti–33Cr, requiring brazing temperatures of 1455 to 1480°C.[107] The other is a Ti–30V–4Be reactive filler metal that requires brazing temperature of 1290 to 1315°C.

Test results of Ti–33Cr brazed joints with D–36 niobium alloy indicate an increase in lap shear strength from approximately 17 MPa to more than 31 MPa at 1370°C and 7 MPa at 1650°C.

Table 4.3 Wettability index for commercial brazing alloys on molybdenum WI

Common name	Temperature (°C)											
	750	800	825	850	875	900	950	975	1000	1025	1050	1075
Palcusil 5				0.012		0.034	0.045					
Palcusil 10				0.011		0.030	0.045					
Palcusil 15						0.044	0.045	0.152				
Palcusil 20				0.015		0.037						
Palcusil 25							0.102	0.119	0.418			
Nicusil 3		0		0.024		0.053						
Nicusil 8				0.020		0.052	0.096					
T50				0.023		0.052	0.069					
T51		0		0.023		0.049						
T52				0.024		0.057	0.075					
Cusiltin 5	0.030			0.008	0.012	0.057						
Cusiltin 10			0.021	0.037	0.031							
Braze 580		0.037		0.045								
Braze 630			0	0.066		0.080						
Braze 655			0.061	0.075	0.069							
Nioro								0.158				
Nicoro 80							0.024			0.071		
Palniro 7										0.023	0.119	0.102
Gapasil 9						0.017	0.061					

Table 4.4 Wettability index for commercial brazing alloys on TZM WI

Common name	Temperature (°C)																				
	750	800	825	850	875	900	925	950	975	1000	1025	1050	1075	1100	1125	1150	1175	1200	1225	1275	1300
580		0.007		0.009																	
Nicusil 3		0.008		0.009	0.018	0.015		0.022													
T51		0.007		0.009		0.010	0.020	0.018													
630		0.015		0.018		0.016															
Cusiltin			0	0	0	0															
655			0.009	0.009	0.025																
Palcusil 5				0		0.011		0.027													
Palcusil 10				0.007		0.016		0.028													
Nicusil 8				0.008		0.018		0.024													
T50				0.014		0.020		0.025													
T52				0.016		0.017		0.026													
Cusiltin 5				0	0	0															
Palcusil 20					0.009	0.017	0.034														
Ticusil					0.024	0.101															
Cusil						0	0.009	0.009													
Palcusil 15						0.023		0.039		0.50											
Gapasil 9						0		0.025		0.041											
Silcoro 60						0	0.020	0.023													
Palcusil 25							0.020	0.096		0.204											
Nioro								0	0.105	0.225											
Nicoro 80								0.020	0.070	0.029											
Ticuni								0.093				0.049									
Silver												0	0.013	0.013							
Incuro 60									0.008	0.012	0.013										
Silcoro 75									0	0	0										
Palnicusil									0.089		0.318										
Altizirbe									0.034												
Palniro 7												0.107	0.133								
Gold											0	0.088	0.107								
Palniro 1																0.037	0.053	0.070			
Palniro 4																	0.037	0.044	0.047		
Palco																				0.041	0.048

Other niobium alloys – D43 (10W–1Zr–0.1C–Nb), Nb–752 (10W–2.5Zr–Nb), and C–129Y (10W–11Hf–0.07Y–Nb) have been brazed successfully with two brazing filler metals, B120VCA and Ti–8.5Si whose brazing temperature was 1455°C.

Tantalum and its alloys

The brazing filler metals for moderate temperature applications of tantalum and its alloys are chosen on the basis of the intended application. Nickel-based filler metals (such as the nickel–chromium–silicon filler metals) have been used in successful brazing of tantalum. These filler metals are satisfactory for service temperatures below 980°C. Copper–gold alloys having less than 40% gold can also be used as brazing filler metals, but gold in amounts between 40 and 90% tends to form brittle, age-hardening compounds. Copper–tin, gold–nickel and copper–titanium filler metals also have been used in brazing tantalum and its alloys.

Because most uses of tantalum and its alloys are in elevated-temperature applications (at or above 1650°C, there is a lack of high-temperature brazing filler metals for use with tantalum. Investigations have examined conventional brazing, reactive brazing and diffusion-sink brazing concepts for new tantalum filler metals. The reactive brazing concept uses a filler metal containing a strong melt-temperature depressant. The depressant is selected to react with the base material or powder additions to form a high-melting intermetallic compound during a post-braze diffusion treatment. As the depressant diffuses into the base material, the joint remelt temperature is increased. Successful application of this concept appears highly dependent on controlling the intermetallic compound reaction to form discrete particles.

Diffusion-sink brazing with titanium and Ti–30V filler metals, whose brazing temperatures range from 1675 to 1760°C, has produced remelt temperatures exceeding 2095°C in tee and lap joints. Diffusion-sink brazing with 33Zr–34Ti–33V filler metal with a brazing temperature of 1425°C has produced remelt temperatures exceeding 1760°C. Therefore, these remelt temperatures indicate that service temperatures could be 1925°C for the titanium and Ti–30V filler metals and 1650°C for the 33Zr–34Ti–33V filler metal.

Most of the brazing filler metals currently available for tantalum are in powder form, which is difficult to work with at elevated temperatures. New powder filler metals, such as Hf–7Mo, Hf–40Ta and Hf–19Ta–12.5Mo have been used successfully but require further development and refinement.

Special techniques are necessary to braze tantalum satisfactorily. All gases that have any reactivity must be removed. Oxygen and carbon monoxide must be eliminated. Tantalum forms oxides, carbides, nitrides, and hydrides very readily with these gases. A loss of ductility ensues. At high temperatures, tantalum should be protected from oxidation. One method is to electroplate the surfaces with copper or nickel, in which case it is necessary for the filler metal to be compatible with the plating. Controlled-atmosphere and both hot and cold wall vacuum furnaces have been used for brazing tantalum and its alloys. Induction, resistance, and torch brazing are not recommended.

Tungsten and its alloys

Tungsten has been successfully brazed in vacuum. The variety of filler metals which have been used to braze tungsten as well as the other refractory metals are listed in the Appendix Table A.6. These brazing filler metals and pure metals having liquidus temperatures ranging from 650 to 1925°C are potentially useful for brazing.

Tungsten can be brazed in much the same manner as can molybdenum and its alloys, using many of the same brazing filler metals. Brazing can be accomplished in a vacuum or in a dry argon, helium, or hydrogen atmosphere. To some extent, the selection of the brazing atmosphere depends on the filler metal used. For example, filler metals that contain elements with high vapor pressures at the brazing temperature cannot be used effectively in a high vacuum.

Care must be exercised in handling and fixturing tungsten parts because of their inherent brittleness; these parts should be assembled in a stress-free condition. Contact between graphite fixtures and tungsten must be avoided to prevent the formation of brittle tungsten carbides. Although nickel-based brazing filler metals have been used successfully to braze tungsten, a reaction between nickel and tungsten that results in base metal recrystallization can occur; this reaction can be minimized by short brazing cycles, minimum brazing temperatures, and the use of small quantities of filler metal.

Data on brazing of tungsten are more limited than data on brazing of other refractory metals. Unalloyed tungsten has been vacuum brazed with two experimental filler metals, Nb–2.2B and Nb–20Ti. The lap shear strength of joints brazed with the Nb–2.2B filler metal was about 35 MPa at 1650°C and 55.2 MPa at 1370°C. The strength of joints brazed with Nb–20Ti was somewhat lower, about 21 MPa at both test temperatures. The flow of Nb–20Ti filler metal was more sluggish than that obtained with Nb–2.2B.

Brazing filler metals based on the platinum–boron and iridium–boron systems were developed to braze tungsten for service at 1925°C. They contained up to 4.5% boron and could be used for brazing below the recrystallization temperature of tungsten. Tungsten lap joints were brazed and diffusion treated in a vacuum at 1095°C for 3 h. This cycle resulted in the production of joints with remelt temperatures of about 2040°C. A slight increase in joint remelt temperature was noted when tungsten powder was added to the brazing filler metal. The highest remelt temperature of 2170°C was obtained when joints were brazed with Pt–3.6B plus 11%wt tungsten powder.[114]

Studies have been conducted to develop and evaluate brazing filler metals that could be used to braze tungsten for nuclear reactor service at 2500°C in hydrogen atmosphere. Butt joints have been brazed using a gas tungsten arc as the heat source and the filler metals W–25Os, W–50Mo–3Re, and Mo–5Os.

Tungsten can be brazed by furnace (inert gas or reducing atmosphere and vacuum), torch resistance or induction brazing. Furnace brazing operations usually are used when the parts are larger than those considered practical for the induction and resistance processes.

Tungsten–copper electrode tips are materials made by powder metallurgy processes. A normal analysis of one of the materials is 80W–20Cu. These tips may be used for resistance welding electrodes that are usually brazed to the copper–chromium alloys, which constitute the main current-carrying portion of the electrode. Any of the silver–copper and/or zinc series of brazing filler metals are suitable for this application. Tungsten–copper tips usually are applied by induction or torch heating in a neutral or reducing atmosphere. The most suitable brazing filler metal is 72Ag–28Cu. To hold parts in alignment, pins are incorporated in the design, and the brazing filler metal is preplaced in the form of shim stock located between the mating parts.

In brazing of resistance welding electrodes with pure tungsten tips, precoating of the tungsten surface aids uniform flow of the brazing filler metal between the faying surface of the joint. Usually one of the silver-based filler metals is used as a precoating. Without pretreatment, poor wetting of the tungsten by the brazing filler metal often results, causing premature joint failure under the high pressures and temperatures of resistance welding.

4.2.7 Cast iron

Brazing of gray, ductile, and malleable cast irons differs from brazing of steel in two principal respects; special precleaning methods are necessary to remove graphite from the surface of the iron, and the brazing temperature is kept as low as feasible to avoid reduction in the hardness and strength of the iron.

The processes used for brazing of cast irons are the same as those used for brazing of steel, which include furnace, torch, induction, and dip brazing. As with other metals, selection of the brazing process depends largely on the size and shape of the assembly, the quantity of assemblies to be brazed, and the equipment available.

In recent years, brazing of ordinary gray cast irons using silver-based filler metal has become practical commercially; the development of a satisfactory surface pretreatment has opened up many design possibilities. Intricate forms can be built from simple castings or by joining castings to standard wrought forms, such as tubing and rolled shapes. Foundry work may be simplified by making intricate castings in several parts to reduce coring. Gray, ductile, and malleable irons commonly are brazed to steels or to other cast irons.

In brazing of ductile or malleable irons, certain precautions are imperative. If ductile or malleable irons are heated above 760°C, the metallurgical structure may be damaged; brazing thus should be done below this temperature.

The effects of graphite are significant which is present in all gray, ductile, and malleable cast irons. Although gray, ductile, and malleable irons all have lower brazeability than carbon or low-alloy steels, the three types of iron are not equal in brazeability.

Malleable iron is generally considered the most brazeable of the three types of cast iron, largely because the total carbon content is somewhat lower (seldom over 2.70%) and, therefore, graphitization is lower. Brazeability is also enhanced because the graphite occurs in the form of approximately round nodules and thus is easier to remove or cover up (as by abrasive blasting). Also, malleable iron is lower in silicon than the other types of cast iron and thus is less graphitized, which makes it better suited for brazing.

Ductile iron can have a composition nearly the same as gray iron, but the graphite particles are spheroidal rather than flake-shaped. The spheroidal shape is more favorable for brazing. Shot or grit blasting is effective in rolling metal over graphite particles that are exposed at the surface.

Gray iron, which is characterized by large flakes of graphite, is the most difficult type of cast iron to braze. Until the development of electrolytic salt bath cleaning, brazing of gray iron was considered impractical.

Abrasive blasting with steel shot or grit has proved reasonably successful for preparing surfaces of ductile and malleable iron castings, but is seldom suitable for preparing surfaces of gray iron castings. Electrolytic treatment in a molten salt bath, alternately reducing and oxidizing, has been the most successful method for surface preparation and is applicable to all graphitic cast irons. Ordinary chemical cleaning methods such as degreasing, detergent washing, and acid pickling have the distinct disadvantage of not removing surface carbon, which interferes with bonding.

Before any procedure for cleaning is adopted, tests should be made by cleaning samples of the iron intended for use in the castings to be brazed, fluxing the samples, and applying brazing filler metal (preferably on a smooth, flat surface). The samples are then heated to the pre-established brazing temperature, cooled, and examined visually. If the samples show an indication that the brazing filler metal has not uniformly wetted the test piece, the surface is not sufficiently clean.

Alkaline-based salts (sodium hydroxide) operating at 400 to 480°C are extremely effective in removing surface oxides and sand from iron castings. At these temperatures, salt baths exhibit the required high chemical activity. This action is further enhanced in the cleaning operations by the introduction of electrical energy. The effective use of these salts and the removal of their by-products to produce an excellent metallurgical surface has resulted in quality bonds and has permitted direct babbitting of cast iron bearings and silver brazing fittings to cast iron surfaces. With proper cleaning and preparation of the cast iron, it has been possible to produce a brazed bond between metal surfaces that exceeds the strength of the parent metal.

With proper surface preparation, the usual variety of brazing filler metals suitable for steel brazing is satisfactory for cast irons. In selection of the brazing filler metal, the metallurgical effects of the brazing cycle on the base metal matrix (hardening, softening, etc.) should be considered. Various brazing filler metals in the silver–copper series are commonly used. Copper filler metal has been used successfully in furnace brazing. Some cast irons have low melting (solidus) temperatures, suggesting that care must be exercised to keep the brazing temperature below 1120°C. Brass filler metals (copper–zinc series) frequently are used to braze cast irons while the silver-based brazing filler metals have been used to join copper to cast iron.

Most brazing of cast iron is performed to join assemblies at lower cost than is possible by another process or to fabricate parts that are difficult to produce – one-piece castings, for example. In some applications, two or more cast iron components are brazed together; in other applications, one or more components of a brazed assembly are made of another metal – most often steel, but also copper alloys. Localized heating methods, such as torch and induction, have the advantages of restricted heat effects and limited metallurgical phase changes in base metal components. Furnace brazing may be desirable for production lots due to cost savings and/or consistency of results. Brazing with copper and nickel filler metals require atmosphere controls during heating, usually accomplished by furnace brazing. The use of dry hydrogen as a protective atmosphere promotes wetting by reducing oxides such as silica. Electric-arc braze welding processes have successfully been used with bronze–brass brazing filler metals on cast iron and require special preheating techniques to prevent cracking.

Cast irons that contain pearlite or free carbide graphitize and decrease in strength at elevated temperatures. Because graphitization is a function of both time and temperature, some experimentation is usually necessary to develop a brazing cycle that produces acceptable joints without excessive graphitization and decreases in strength. Two identical assemblies were heated in different plants for 1 h. In the first plant, the assemblies were heated to about 705°C with no significant decrease in strength. In the second plant, the assemblies were heated to about 790°C again without significant decrease in strength. In the latter application, however, heating was done by induction instead of in a furnace, and the heating time was much shorter than would have been required in a furnace.

In some plants using furnace brazing, the furnace is operated at a considerably higher temperature than is desired for brazing (sometimes as high as 870°C), but the time cycles used are so short that the assemblies never reach furnace temperature.

For applications in which little or no decrease in strength of the cast iron can be tolerated, it is mandatory to use a brazing filler metal with as low a flow temperature range as possible, thus permitting a low brazing temperature, and to keep the time at brazing temperature to the minimum.

The temperature required for wetting the base metal and for flow of brazing filler metal having a melting range of 620 to 645°C may vary from about 690 to 845°C, depending on the complexity of the joint design – especially the distance the brazing filler metal must flow. Simple joints with short flow distances can be brazed at lower temperatures than more complex joints.

4.2.8 Low-carbon, low-alloy, and tool steels

Brazing of low-carbon and low-alloy steels is a highly-developed, low-cost production process and these steels are brazed without difficulty. Brazing filler metals can be either manually or automatically applied or preplaced in the joint. The heat treatment considerations of low-alloy steels are factors in determining the specific brazing filler metals and brazing temperatures to be used.

The steels covered in this section include low-carbon (less than 0.30% C) and low-alloy steels. The low-alloy steels include the SAE/AISI 23xx nickel steels, 31xx nickel–chromium steels, 41xx chromium–molybdenum steels, 43xx nickel–chromium–molybdenum steels, and several other types containing less than 5% total alloy content.

Leaded steels can be torch brazed using brazing filler metal and flux combinations normally recommended for steel. Tests conducted on low-carbon free-machining steel containing 0.25 to 0.35% lead show detrimental effects after torch brazing with silver-based brazing filler metals; satisfactory joints have been made in these leaded steels using copper and nickel-based brazing filler metals.

All the silver-based brazing filler metals can be used for brazing low-carbon and low-alloy steels (see Appendix Table A.12). The silver-based filler metals containing nickel usually provide better wettability and are preferred for brazing low-alloy steels wherein joint strength is most important.

Copper-based filler metals are used mainly for preplacement in controlled-atmosphere furnaces. Copper is also generally preferred because of its low cost and the high strength of the joints produced. Copper–zinc brazing filler metals often are face-fed into the joint but can also be preplaced for furnace and induction heating. The high solidus temperatures (1095 to 1150°C) necessary when copper-based filler metals are used often allow simultaneous brazing and heat treating of certain low-alloy steels. Nickel-based filler metals have been used for joining low-carbon and low-alloy steels when special joint requirements so dictate. Brazing usually is restricted to controlled-atmosphere furnaces.

Low-carbon and low-alloy steels can be brazed using virtually all known processes. Acetylene, natural gas, propane, and proprietary gas mixtures are the types of fuel gas most often used in torch brazing of carbon, low-alloy, and stainless steels. Torch, furnace, and induction heating techniques are the most common. Brazing filler metals in the form of continuous wire or strip can be automatically applied using electromechanical wire feeders; powder filler metals blended with flux and paste-forming ingredients are applied automatically with pressurized dispensing equipment. For torch brazing, the equipment would include standard oxyacetylene or similar torches. Furnaces can be of the batch or conveyor type, with or without atmosphere control; they can be electric, gas, or oil-fired and should provide accurate temperature control.

The principal advantage of furnace brazing over other brazing processes is that it permits the use of a variety of prepared protective atmospheres – notably, the rich exothermic-based, endothermic-based, and some prepared and commercial nitrogen-based atmospheres. These atmospheres are among the least expensive; they can be generated in the plant in large volume, or, in the case of commercial nitrogen-based atmospheres, they can be stored in liquid form outside the plant. They provide excellent protection against oxidation, and they can be prepared with any carbon potential in the range of about 0.2% to more than 1.0% C, depending on the atmosphere. This range of carbon potential is sufficient to accommodate all carbon and low-alloy steels, including those carburized before brazing. By selecting an atmosphere with a carbon potential that matches the carbon content of the work metal, brazing can be accomplished without carburizing or decarburizing the work metal.

Because the protective atmospheres used for furnace brazing are sufficiently reducing to iron oxide, they usually eliminate the need for fluxes in brazing carbon steel with copper filler metal. An oxide-free surface normally promotes wetting of the workpiece by the molten brazing filler metal. However, some low-alloy steels that contain a total of more than 2 or 3% chromium, manganese, aluminum, and silicon form more stable surface oxides and they require highly reducing atmospheres (such as dry hydrogen or dissociated ammonia), a flux or nickel plating to obtain adequate wetting.

Most of the limitations of furnace brazing are directly related to the high temperatures required for brazing of steels with copper filler metal. These temperatures exceed the average brazing temperature required for brazing with silver-based filler metals by 280°C or more. They are high enough to cause grain coarsening in medium carbon, high carbon, and low alloy steels; however, grain refinement can be obtained by subsequent heat treatment.

Steels brazed with copper filler metal will develop lower tensile and yield strengths and increased ductility as the brazing time or temperature, or both, are increased.

These changes in properties are a result of decarburization in some types of atmospheres and alteration of grain size. Original grain size can be restored by subsequent heat treatment below the remelt temperature of the copper brazing filler metal. Loss of carbon through decarburization is generally unimportant in low-carbon steels. However, surface hardness of some low-alloy steels may be substantially lowered. Such loss of surface hardness can be very deleterious when thin-gage material is involved.

For alloy steels, the filler metal should have a solidus well above any heat-treating temperature to avoid damage to joints that will be heat-treated after brazing. In some cases, air-hardening steels can be brazed and then hardened by quenching from the brazing temperature.

A filler metal with brazing temperature lower than the critical temperature of the steel can be used when no change in the metallurgical properties of the base metal is wanted.

In torch brazing, a neutral or slightly reducing flame usually is preferred, because the brazing filler metal is face fed into the prefluxed joint. Flux-coated filler metal is often beneficial. As in all brazing processes, it is important to avoid overheating in brazing to prevent undesirable effects on the base metal, the brazing filler metal, or the flux.

Time at temperature is an important consideration, especially when the brazing filler metal contains volatile elements such as zinc and cadmium. Excessive heat also might affect the integrity of the braze and reduce the mechanical properties of the joint. Automated torch and burner-type production equipment is available for high-production applications. These units usually have brazing fluxes and silver-based filler metals, Figures 2.1 and 3.2.

In production applications, the brazing filler metal (usually copper) is preplaced in, or adjacent to, the joint before the preassembled parts are charged into controlled-atmosphere batch- or conveyor-type furnaces for brazing. Induction heating or brazing is also advantageous in that maximum temperature rise is restricted to the immediate joint area by selective coil design and the use of an appropriate frequency for the induction heating circuit.

In joining certain hardenable low-alloy steels, it usually is desirable to use the lower-melting silver-based brazing filler metals, which can be applied below the lower transformation temperature of the steel. Some local annealing, however, may occur. When postheat treatments are required, the higher solidus filler metals are necessary to preclude the possibility of joint impairment. Low-alloy steels are sometimes brazed and heat treated simultaneously using brazing filler metals of the copper–zinc and silver classifications. The solidus temperature of the filler metal chosen must be above the austenitizing temperature recommended for the base metal prior to quenching. In an application of this type, the joint is made at normal brazing temperature, removed from the heat source to permit a drop to the hardening temperature, and quenched. The procedure is satisfactory only for base metals compatible with rapid cooling.

Dip brazing has also been used to join carbon and low-alloy steels with silver-based, copper-zinc, and other copper-based brazing filler metals.

The types of salts used in dip brazing of carbon and low-alloy steels are neutral chloride salts, plus a fluxing agent such as borax or cryolite, and carburizing and cyaniding salts, which are also fluxing-type salts. Types and compositions of brazing salts and temperatures used for brazing of carbon and low-alloy steels with various brazing filler metals are given in the Appendix Table A.13.

Neutral salts, so called because normally they do not add or subtract anything from the surface of the steel being treated, protect the surface from attack by oxygen in the air. Oxide on the workpiece, however, cannot be reduced by the salt, and a flux must generally be provided.

The neutral salts are mildly oxidizing to steel when they are used at recommended austenitizing temperatures. The oxides produced by heating steel in molten salt are largely soluble; hence, the steel is scale-free after heating. The accumulation of oxide in the molten salt, however, progressively makes the salt more decarburizing, and for this reason salt baths may require periodic rectification.[115]

Carburizing and cyaniding salts provide their own fluxing action. In addition, they supply carbon or carbon and nitrogen to the surface of the steel assembly as it is being brazed. Although silver-based brazing filler metals have been used successfully, copper–zinc filler metal is generally preferred.

Various applications involve the use of brazing filler metals in brazing low-carbon and low-alloy steels in everyday production. These include many components of such vehicles as automobiles, trucks, bicycles, motor cycles, snowmobiles, all-terrain vehicles and the like. Other common brazements include window and door frames, ducts, tanks, containers of all types, perforated and expanded steel panels, steel partitions and shelving. A great variety of tubular steel furniture is production brazed, and brazing is also used in production of cutting tools and industrial knives, hydraulic oil tanks, reservoirs, electronic chassis and supports, hand tools, honing appliances, certain instruments and steel assemblies of all types.

Tool steels

High-carbon steels contain more than 0.45% carbon. High-carbon tool steels usually contain 0.60 to 1.40% carbon. In discussing the brazing of tool steels, it is convenient to group them in two broad classifications; carbon steels and high-speed tool steels.

Brazing of high-carbon steels is best accomplished prior to or during the hardening operation. Hardening temperatures for carbon steels range from 760 to 820°C. Filler metals having brazing temperatures above 820°C should be used. When brazing and hardening are done in one operation, the filler metal should have a solidus at or below the austenitizing temperature.

Tempering and brazing can be combined for high-speed tool steels and high-carbon, high-chromium alloy tool steels which have tempering temperatures in the range of 540 to 650°C. Filler metals with brazing temperatures in that range are used. The part is removed from the tempering furnace, brazed by localized heating methods, and then returned to the furnace for completion of the tempering cycle.

Localized heating for brazing may decrease the hardness of heat-treated steels when the brazing temperature is above the tempering temperature of the steel. Except for thin sections, these steels must be quenched drastically during heat treatment to achieve optimum properties. Alloying elements may be added to carbon steels to impart special properties, such as reduced distortion on heat treatment, greater wear resistance, and toughness or better high-temperature properties.

Such steels are referred to as alloy tool steels; they are known by various trade names and grades, and their properties are adequately covered in manufacturers' published information and in various handbooks.

The alloy steel in question should be studied carefully to determine its proper heat treating cycle, the kind of quench necessary (water, oil, or air), the best brazing filler metal, and the proper technique for combining the heat treating and brazing operations to achieve maximum properties in service life.

High-speed steels, although logically falling in the alloy steel group, are classified separately because their properties depend on relatively high percentages of such alloying elements as tungsten, molybdenum, chromium, and vanadium. Their carbon contents normally are much lower than those of carbon tool steels. High-speed steels are widely used in industry as metal cutting tools. A common type known as 18–4–1 contains 18% tungsten, 4% chromium, and 1% vanadium.

The choice of the brazing filler metal to be used depends on the properties of the tool steel being brazed and the heat treatment required to develop its optimum properties. Practically all brazing filler metals of the BAg, BCu, and RBCuZn classifications are used at various times. The best filler metal to use should be determined for the specific application.

Torch, furnace and induction heating are the three most commonly-used processes in brazing tool steels. Available equipment is frequently the main factor in process selection.

4.2.9 Stainless steels

Stainless steels include a wide variety of iron-based alloys containing chromium which are used primarily for applications demanding heat or corrosion resistance. Tighter process controls are required than for brazing of carbon steels. These more rigorous requirements are imposed by the inherent chemical characteristics of stainless steels and the generally more arduous service environments in which they are used. Success in the fabrication of stainless steel components by brazing depends on knowledge of the characteristics of the various types of stainless steels, and rigid adherence to certain items of process control required by these characteristics.

All stainless steels are difficult to braze because of their high chromium content. Brazing of these alloys is best accomplished in a purified (dry) hydrogen or in a vacuum. Dew points below – 50°C must be maintained because wetting becomes difficult following the formation of chromium oxide. Torch brazing requires fluxing to reduce any chromium oxides present.

Most of the silver, copper and copper–zinc filler metals are used for brazing stainless steels. Silver filler metals containing nickel are generally best for corrosion resistance. Filler metals containing phosphorus should not be used on highly stressed parts because brittle nickel and iron phosphides may be formed at the joint interface.

Boron-containing nickel filler metals are generally best for stainless steels containing titanium or aluminum or both, because boron has a mild fluxing action which aids in wetting these base metals.

Stainless steels may be grouped into four categories:

1. austenitic nonhardenable
2. ferritic nonhardenable
3. martensitic hardenable
4. precipitation-hardening.

All these alloys are iron-based and contain chromium, the basic element that imparts corrosion resistance. The corrosion resistance of stainless steels varies widely from one alloy to another, and for any given alloy varies widely from one corrosive medium to another. If doubt exists regarding the proper stainless steel to use in a given environment, standard reference works[9, 10] or manufacturers' representatives should be consulted.

Austenitic nonhardenable stainless steels

Austenitic nonhardenable stainless steels contain sufficient nickel or nickel plus manganese additions to stabilize austenite down to room temperature, causing these alloys to be nonmagnetic and nonhardenable by heat treatment. Stainless steels of this class possess the highest heat and corrosion resistance. They are designated as AISI 300 and 200 series alloys. One commonly used alloy is type 302, which contains about 18% chromium and 8% nickel. In the 200 series stainless steels, some of the nickel is replaced by manganese. For example, type 202, the parallel to type 302, contains 18% chromium, 5% nickel, and 9% manganese.

The 300 series stainless steels are used widely for both torch and furnace brazed assemblies. These alloys have relatively high thermal expansion and low thermal conductivity. This combination of properties makes thermal distortion a major concern in furnace brazing of large or complex assemblies or assemblies in which dissimilar materials are brazed to stainless steel. In design of fixtures, heat shields, and thermal cycles, the requirement to provide uniform heating and cooling must be considered.

In brazed assemblies for which corrosion resistance is important, precautions must be taken to avoid sensitization to intergranular corrosion. The problem occurs when an unstabilized grade of austenitic stainless steel, such as type 302 or type 304, is held at temperatures in the range from 425 to 815°C or slowly cooled through this range. The excess carbon combines with chromium and precipitates as chromium carbide along grain boundaries of the austenite. The region around the precipitate is depleted of chromium and thus becomes susceptible to corrosion.

The 300 series stainless steels have been used widely for brazed components; the 200 series alloys, which are only 25 years old, have not seen as much use. Due to their high manganese content, the 200 series stainless steels are more difficult to furnace braze in hydrogen atmospheres than the 300 series. Manganese forms an oxide that is not reduced easily by dry hydrogen at the furnace brazing temperatures normally used for stainless steels. It is important to start with thoroughly cleaned surfaces and maintain a low dew point in the hydrogen atmosphere.

All the chromium-nickel steels are subject to stress-corrosion cracking in the presence of molten brazing filler metals. This phenomenon occurs when the base metal is under stress – either residual or resulting from applied loads while the braze is being made. One form of applied load results from thermal gradients during brazing. The brazing filler metal penetrates the base metal along the grain boundaries at the points of stress, producing a greatly weakened base metal. Best results, therefore, are obtained with stress-relieved material. This stress relief may be done prior to or during the brazing cycle. If it occurs during the cycle, it must be done below the solidus temperature of the filler metal.[116,117]

Ferritic nonhardenable stainless steels

Ferritic nonhardenable stainless steels are basically low-carbon alloys of iron and chromium in which sufficient chromium has been added to the iron to stabilize ferrite, the low-temperature phase in steels, over a wide temperature range. The more common AISI stainless grades in this category are types 405, 406, 430, and 446. Type 430 is a widely-used grade that is particularly subject to a form of interface corrosion when brazed with some silver-based brazing filler metals. This corrosion apparently is caused by electrochemical action whereby the bond between the base metal and the brazing filler metal is destroyed. In many cases this action has been found to occur in the presence of tap water. It has been found that addition of small percentages of nickel to the silver-based filler metal prevents interface corrosion of brazed joints in most stainless steels. The nickel-containing silver-based filler metal is not completely effective, however, with type 430, even though its use greatly reduces the rate of attack. For type 430, a special silver-based filler metal has been developed, (63Ag–28.5Cu–6Sn–2.5Ni).

The ferritic stainless steels (405, 406, and 430) cannot be hardened and their grain structure cannot be refined by heat treatment. These alloys degrade in properties when brazed at temperatures above 980°C, because of excessive grain growth. They lose ductility after long heating times between 340 and 600°C. However, some of the ductility can be recovered by heating the brazement to approximately 790°C for a suitable time.[116, 117]

Martensitic hardenable stainless steels

Martensitic hardenable stainless steels are iron–carbon–chromium alloys of two basic types: the low-chromium, low-carbon grades (types 403, 410, and 416) and the high-chromium, high-carbon grades (types 440A, B, and C). These steels are closely related to the ferritic nonhardenable grades, but their alloy compositions are so balanced that they air-harden upon cooling from brazing, which occurs above their austenitizing temperature range. Therefore, they must be annealed after brazing or during the brazing operation. These steels are also subject to stress cracking with certain brazing filler metals.

The primary precaution in brazing components made of these alloys is that the brazing thermal cycle must be compatible with the required heat treatment.

If high-temperature nickel-based brazing filler metals are used, it is possible to reaustenitize the assembly after brazing. Although this procedure would increase costs, it may be desirable to develop optimum properties in critical components.

The thermal expansion properties of martensitic stainless steels are relatively low – in the same range as those of ferritic alloys.[118]

Precipitation-hardening stainless steels

These steels are basically stainless steels with additions of one or more of the elements copper, molybdenum, aluminum and titanium. Such alloying additions make it possible to strengthen the alloys by precipitation hardening heat treatments.

Some of the designations of the precipitation-hardening stainless steels are 17–7PH, PH14–8Mo, PH13–8Mo, 15–5PH, PH15–7Mo, AM350, AM355, 17–4PH, and A–286. As in the case of the martensitic hardenable stainless steels, brazing thermal cycles used in joining these alloys must be compatible with their heat treatments. Because the heat treatments vary rather widely, specific brazing procedures are required for each alloy.

Precipitation-hardening alloys that contain aluminum or titanium are difficult to wet in the usual furnace brazing atmospheres. Nickel plating generally is used as a surface treatment before furnace brazing.

In brazing of stainless steels, base metal inclusions and surface contaminants are even more deleterious than in brazing of carbon steels. Base metal inclusions, such as oxides, sulfides, and nitrides, interfere with the flow of brazing filler metal. Flow is also impeded by surface contaminants, which may include lubricants such as oil, graphite, molybdenum disulfide and lead that are applied during machining, forming and grinding or by aluminum oxide particles produced by grit blasting or by grinding with aluminum oxide wheels or belts.

Some brazing filler metals in powder form contain or are mixed with an organic binder to form a paste that is subsequently applied to stainless steel for brazing; acrylics and other plastics are often used for this purpose. Although some binders form a soot residue, this residue does not usually interfere with filler metal flow.

The brazing characteristics of stainless steels can also be severely impaired by unsuitable fixturing materials, such as graphite, or by protective atmospheres with nitriding potentials. Carbon in graphite fixtures unites with hydrogen to form methane (CH_4), which carburizes stainless steel and impairs its corrosion resistance. Dissociated ammonia unless sufficiently dry and completely (100%) dissociated, nitrides stainless steel.

A wide variety of brazing filler metals, including silver-based, nickel-based, gold-based and copper, are commercially available for brazing stainless steel parts, see Appendix Table A.14. The factors to consider in selecting a filler metal for a particular application include the following:

1. Service conditions, including operating temperature, stresses, and environments;
2. Heat treatment requirements if martensitic or precipitation-hardening steels are involved;
3. Brazing process to be used;
4. Cost;
5. Special precautions, such as sensitization of unstabilized austenitic stainless steels at certain temperatures.

Commercial brazing filler metals are available that have copper, silver, nickel, cobalt, platinum, palladium, manganese, and gold as the base or as addition elements; these are grouped conveniently according to service temperature.

The most widely-used filler metals for brazing stainless steels are the silver-based family (BAg group). BAg-3, which contains 3% nickel, is probably the most frequently selected silver-based brazing filler metal, although several other silver-based filler metals can also be used successfully. Where improved corrosion resistance is needed, BAg-3 and BAg-4 are recommended.

Silver brazed joints cannot be used for high-temperature service; the recommended maximum service temperature is 370°C, which is the maximum temperature for BAg–13 and BAg–19 filler metals. Of the silver-based brazing filler metals all except BAg–19 and possibly BAg–13 are used at brazing temperatures that fall within the effective sensitizing range for austenitic stainless steels, 540 to 870°C. Chromium carbide precipitation occurs in the sensitizing temperature range, resulting in impairment of the corrosion resistance of the base metal. Carbide precipitation, however, depends on time as well as temperature, and exposure within the sensitizing temperature range for only a few minutes is unlikely to produce a significant amount of precipitate. Nevertheless, the lower melting temperatures of the silver-based brazing filler metals prohibit re-solution treatment of the base metal after brazing, and if corrosion resistance in service is sufficiently critical, an extra-low-carbon, titanium-stabilized (321) or niobium–tantalum-stabilized (347) austenitic stainless steel should be selected instead of a nonstabilized type.

Ferritic and martensitic stainless steels that contain little or no nickel are susceptible to interface corrosion in plain water or moist atmospheres when brazed with nickel-free silver-based filler metals using a flux. Brazing filler metals containing nickel (BAg–3) help to prevent interface corrosion. However, for complete protection, special filler metals containing nickel and tin should be used, and brazing should be done in a protective atmosphere without flux.

The nickel-based brazing filler metals usually rank second in frequency of use as filler metals for brazing of stainless steels. Nickel-based filler metals provide joints that have excellent corrosion resistance and high-temperature strength. These filler metals alloy with stainless steel, however, and form phases with two undesirable characteristics: the phases are considerably less ductile than either the base metal or the filler metal even at elevated temperatures, and thus are a potential source of rupture; and the alloys formed with stainless steel are higher-melting alloys that are likely to freeze and block further flow into the joint during brazing.

Because of the relatively high brazing temperatures required for nickel-based brazing filler metals, their use is generally restricted to furnace brazing in a controlled atmosphere (including vacuum), although there are occasional exceptions.

The nickel-based brazing filler metals normally are supplied as powders; however, they can be obtained as sintered or cast rods, preforms, plastic bonded sheet, plastic bonded wire and tape. Many of these filler metals now are available as metallic foils produced by ultrarapid cooling from the molten state. The BNi brazing filler metals commonly are used on stainless steels for oxidation resistance at temperatures up to 980 to 1095°C. Filler metals BNi–1, BNi–2, BNi–3, and BNi–4 (see Appendix Table A.15) tend to erode thin sheet metal because of their interaction with many base metals. Time at brazing temperature and quantity of filler metals should, therefore, be controlled carefully when these filler metals are used. Boron-free brazing filler metals BNi–5,

BNi–6, and BNi–7 are suitable for use in nuclear reactor components where boron cannot be tolerated because of its absorption of neutrons.

The gold-based brazing filler metals (see Appendix Table A.15) are sometimes used for brazing stainless steels, although their high cost restricts their use to specialized applications such as heat exchangers for manned space-flight vehicles. When a gold-based filler metal is used, alloying with the stainless steel base metal is minimized, and, as a result, joints exhibit good ductility.[119] In addition to the gold family of brazing filler metals there are palladium and platinum filler metals such as gold–nickel–palladium, copper–platinum, silver–palladium–manganese and palladium–nickel–chromium, which are useful for brazing heat- and corrosion-resistant components for jet and rocket propulsion systems and the nuclear energy fields.

The cobalt-based brazing filler metal BCo–1 (see Appendix Table A.15) is very rarely used for brazing stainless steels, but it has been included in the table for its availability.

Stainless steels can be brazed with any of the various brazing processes. A large volume of controlled atmosphere brazing is being performed on stainless steel, and the success of this type of brazing can be attributed to the development of reliable atmospheres and vacuum furnaces.

The protective atmospheres most often used in furnace brazing of stainless steels are dry hydrogen and dissociated ammonia. These atmospheres are effective in reducing oxides, protecting the base metal, and promoting the flow of brazing filler metal. The low-cost exothermic atmospheres that are widely used in furnace brazing of low-carbon steels are not suitable for stainless steels. An inert gas such as argon, or vacuum, may be used to satisfy special requirements and provide protection in applications for which hydrogen or hydrogen-bearing gases are unsatisfactory.

A dry hydrogen atmosphere is preferred for many stainless steel brazing applications. Hydrogen, the most strongly reducing of protective atmospheres, reduces chromium oxide and provides for excellent wetting by some brazing filler metals without the need for flux. The principal disadvantages of hydrogen are high cost, difficulty in drying sufficiently, need for special furnace equipment, and danger involved in storing and handling hydrogen.

Dissociated ammonia, when it is free of moisture and 100% dissociated, is a suitable atmosphere for brazing of stainless steel with some brazing filler metals without the need for a flux. Dissociated ammonia is strongly reducing, but less so than pure dry hydrogen. Consequently, although it will promote wetting by reducing chromium oxide on the surface of the stainless steel, dissociated ammonia may not be sufficiently reducing to promote the flow of some brazing filler metals, such as copper oxide powders. Because of its high (75%) hydrogen content, dissociated ammonia forms explosive mixtures with air and must be handled with the same precautions as those required for handling hydrogen.

Unless the atmosphere used in brazing of stainless steel is completely decomposed (100% dissociated), even minute amounts of raw ammonia (NH_3) in the atmosphere will nitride stainless steels, especially those containing little or no nickel. Finally, to avoid oxidation of base metal and brazing filler metal, the dissociated ammonia atmosphere must be kept dry and pure while it is inside the furnace.

Argon is occasionally used as a furnace atmosphere in brazing stainless steels to other stainless steels or to reactive metals such as titanium. Argon has the advantage of being chemically inert in relation to all metals; thus, it is a useful protective atmosphere for metals that can combine with or absorb reactive atmospheres, such as hydrogen. An argon atmosphere has the disadvantage of being unable to reduce oxides; consequently, the surfaces of stainless steel components must be exceptionally clean and free of oxides when brazed in argon.

The principal advantages of furnace brazing are high production rates and the means for using controlled protective atmospheres at controlled dew points, which often make it unnecessary to use a flux to obtain satisfactory wetting. In most furnace brazing applications, both of these advantages are exploited. Occasionally, however, furnace brazing is selected solely on the basis of production rate, and brazing is performed without a protective atmosphere but with a suitable flux. The lower-melting filler metals are generally selected for brazing under these conditions.

Vacuum brazing of many structural configurations made of austenitic stainless steels offers excellent heat and corrosion resistance for high-temperature service applications.

For stainless steels, the fundamentals of torch brazing, as well as its advantages and limitations, are basically the same as carbon steels. Because of the metallurgical characteristics of stainless steel and its requirements for corrosion resistance, however, best results are obtained when special consideration is given to the type of flame at the torch and to the brazing filler metal composition.

To aid in reducing the oxide already present, as well as to prevent further oxidation of the work-metal surfaces, a strongly reducing flame should be used for torch brazing stainless steel to itself. A reducing flame is also satisfactory for brazing stainless steel to nickel alloys or carbon steels. In brazing stainless steel to copper alloys, however, some compromise is necessary. Although a slightly oxidizing flame is normally best for brazing copper, for brazing stainless steel to copper a slightly reducing flame usually provides a satisfactory compromise in flame adjustment.

The silver-based filler metals that flow at relatively low temperatures are used almost exclusively for torch brazing of stainless steels. BAg–3 is most often used, because it flows well in the temperature range from 705 to 760°C and provides joints that have greater resistance to corrosion than those brazed with filler metals such as BAg–1 or BAg–1a, although these filler metals have been used.

Depending on the metallurgical and physical properties of particular stainless steels, their behavior in heating by electrical induction may differ considerably from that of carbon and low-alloy steels, and from that of the more widely-used nonferrous metals. In addition, depending on whether a stainless steel is magnetic or nonmagnetic at room temperature, the response of the steel to induction heating varies considerably. Differences in specific heat and electrical conductivity markedly affect response to heating by induction.

Ferritic and martensitic (400 series) stainless steels are ferromagnetic at all temperatures up to the **Curie temperature**. The Curie temperature is the temperature of magnetic transformation below which a metal or alloy is magnetic and above which it is paramagnetic.[117] Thus, given the same power input, these steels generally heat faster than austenitic stainless steels, which are nonmagnetic in the annealed condition.

Stainless steels may be induction brazed in an air atmosphere, using a suitable flux, although for critical applications induction brazing is sometimes done in a protective atmosphere or in vacuum, whereas in other applications an inert gas such as argon may be used as a backing gas to minimize oxidation.

Brazing of stainless steel by immersion of all or a portion of the assembly in molten salt offers essentially the same advantages that would apply to brazing similar assemblies of carbon steel. Similarly, the same limitations are applicable.

4.2.10 Titanium and zirconium (reactive metals)

Titanium and zirconium combine readily with oxygen, and react to form brittle intermetallic compounds with many metals and with hydrogen and nitrogen. Parts must be cleaned before brazing and brazed immediately after cleaning.

Silver and silver-based filler metals were used in early brazing of titanium, but brittle intermetallics were formed and crevice corrosion resulted. Type 3003 aluminum foil as a braze filler metal will join thin, lightweight structures, such as complex honeycomb sandwich panels. Electroplating various elements on the base metal faying surfaces will let them react *in-situ* with the titanium/titanium alloy base material during brazing to form a titanium alloy eutectic. That transient liquid phase flows well and forms fillets, then solidifies due to interdiffusion.[11, 15]

Other brazing filler metals with high service capability and corrosion resistance include Ti–Zr–Ni–Be, Ti–Zr–Ni–Cu and Ti–Ni–Cu filler metals. The best braze processing is obtained in high vacuum furnaces using closely controlled temperatures in the range of 900 to 955°C. Selection of filler metals for use in brazing of reactive metals is critical to avoid formation of undesirable intermetallic compounds.

Selection of brazing filler metals and brazing cycles that are compatible with the heat treatment required for (α–β) and β-titanium-based metals may present some difficulty. Ideally, brazing should be conducted at temperatures from 55 to 83°C below the β transus, because the ductility of (α–β) based metals may be impaired if this temperature is exceeded. The β transus can be exceeded when β-titanium-based metals are brazed; however, if the brazing temperature is too high, base metal ductility after heat treatment may be impaired by considerable interaction between the properties of the heat-treatable titanium alloys and may be affected adversely by brazing, unless the assembly can be heat-treated afterwards. For example, the (α–β) titanium alloys must be solution-treated, quenched, and aged to develop optimum properties. It is not easy to select a brazing filler metal that permits brazing and solution treating in a single operation. Similarly, it is not always possible to quench a brazed assembly at the desired cooling rate, and certain configurations (e.g., honeycomb sandwich structures) do not lend themselves to rapid quenching without distortion. Brazing at the aging temperature is impractical, because few brazing filler metals melt and flow at these temperatures.

The possibility of galvanic corrosion must be considered when filler metals are selected for brazing titanium base metals. Although titanium is an active metal, its activity tends to decrease in an oxidizing environment because the surface undergoes anodic polarization in a manner similar to that of aluminum. Thus, brazing filler metals must be chosen carefully to avoid preferential corrosion of the brazed joint.

When titanium is brazed, precautions must be taken to ensure that the brazing retort or chamber is free of contaminants from previous brazing operations. As a precaution against any existing contaminants in the brazing furnace, a loose cover of pure titanium foil (0.03 to 0.08 mm thick) should be put over the workpieces. This will act as a 'getter' of any remaining contaminants. Care should be exercised to ensure that the foil does not come into direct contact with the workpieces.

The choice of materials to be used in fixtures must be carefully considered. Nickel or materials containing high amounts of nickel generally should be avoided; nickel and titanium form a low-melting eutectic (28.4% nickel) at about 940°C. Should the titanium workpieces contact fixtures or a retort made from nickel-base alloy, the parts may fuse together if the brazing temperature is in excess of 940°C. If a fixture material containing a high nickel content (such as stainless steel) is used, it should be oxide coated. In most applications, coated graphite or carbon steel fixture materials are used.

Brazing filler metals initially used for brazing of titanium and its alloys were silver-based alloys containing lithium, copper, aluminum, or tin.[120–122] Most of these filler metals were used in low-temperature applications (540 to 595°C). Later developments produced a number of successful commercial brazing filler metals including silver–palladium, titanium–nickel, titanium–nickel–copper, titanium–zirconium–beryllium[123] and titanium–zirconium–copper–nickel.[124, 125] Additionally, these filler metals could be used at temperatures from 870 to 925°C. For joining applications requiring a high degree of corrosion resistance, the 48Ti–48Zr–4Be and 43Ti–43Zr–12Ni–2Be brazing filler metals are outstanding.[123] A silver–palladium–gallium filler metal (Ag–9Pd–9Ga), which flows at 900 to 915°C, is another excellent brazing filler metal which will fill large gaps.[126]

The following four filler metals are new developments and have been tested in laboratory work.[124]

1. Ti–Cu–Ni–2 (Ti–20Cu–20Ni)
2. Ti–Zr–Cu–Ni–1 (Ti–38Zr–12Cu–12Ni)
3. Ti–Zr–Cu–Ni–2 (Ti–38Zr–15Cu–15Ni)
4. Ti–Zr–Cu–Ni–Pd (Ti–37Zr–12Cu–12Ni–2Pd)

Initial tensile and corrosion testing of joints with the latter filler metals shows that the joints exhibit nearly the same mechanical and chemical characteristics as Ti–6Al–4V.

Brazing of titanium honeycomb sandwich structures using aluminum brazing filler metal is an achievement stemming from supersonic transport materials and process technology.[120] Such aircraft structures up to 7 m in length have been successfully fabricated using Al 3003 brazing foils as filler metal which provides satisfactory strength up to about 315°C. High-strength, corrosion-resistant Ti–Zr–Be and Ti–Zr–Ni–Be filler metals are recommended for applications from 540 to 595°C.[123]

Zirconium alloys

Zirconium-based metals of commercial importance are the pure metal and several alloys. The most commonly used zirconium alloy is Zircalloy, which contains small percentages of tin, iron, chromium, and nickel. These structural alloys are used for corrosion resistance in nuclear applications, especially in pressurized water nuclear power reactors.

Like titanium and beryllium, zirconium reacts readily with oxygen, hydrogen, and nitrogen and is embrittled. It also reacts with many metals and alloys to form intermetallic compounds. Therefore, as a result of this reactivity, zirconium must be brazed in a vacuum or in a dry atmosphere of argon or helium. Zirconium joint members must be cleaned carefully before brazing to remove oxides and other surface contaminants, and brazing should be done immediately after cleaning.

Compared with the other reactive metals, very little research has been done to develop filler metals and brazing methods for joining zirconium and zirconium alloys.[127] To a degree, this lack of research on brazing can be attributed to the availability of other joining methods. However, most commercial brazing filler metals do not wet or flow well on zirconium-based metals, nor are they metallurgically compatible with zirconium. In addition, many of these filler metals do not possess the corrosion resistance required in reactor environments.

Research was conducted to develop filler metals for producing brazed joints in Zircalloy 2 that possessed good resistance to corrosion in pressurized water at 360°C. The data from these studies and from metallographic examinations of the brazed joints indicated that the following brazing filler metals most nearly met the service requirements: Zr–5Be, Cu–20Pd–3In, Ni–20Pd–10Si, Ni–30Ge–13Cr, and Ni–6P.[127]

Additional studies to develop improved filler metals for brazing Zircalloy base metals for use in water-cooled reactors were completed recently.[128] Candidate zirconium-based and non-zirconium-based filler metals were formulated, used in vacuum brazing under pressures of 0.0013 to 0.00013 Pa, and screened by wetting tests and corrosion tests in pressurized, high-temperature water (315°C). The following filler metals had acceptable corrosion resistance and mechanical strength:

- Zr–50Ag (brazing temperature, 1520°C)
- Zr–29Mn (brazing temperature, 1380°C)
- Zr–25Sn (brazing temperature, 1730°C)

The Zr–5Be brazing filler metal previously mentioned has been used extensively to braze zirconium base metals to themselves and to other metals (e.g., stainless steel). For example, zirconium sheet stock has been brazed with this filler metal using the following cycle: 10 min at 1005°C, followed by 4 to 6 h at 800°C. Because of its ability to wet ceramic surfaces, Zr–5Be has also been used to braze zirconium to uranium oxide and beryllium oxide.

As stated earlier, titanium and zirconium are both highly reactive metals. The brazing process, therefore, must not allow the joint surfaces to come in contact with air during heating. Induction brazing and furnace brazing in inert gas or vacuum atmospheres can be used successfully. Torch brazing of these base metals is difficult, requiring special precautions and techniques. Induction brazing of small, symmetrical parts is very effective because the speed minimizes reactions between filler metal and base metal. For large, precise assemblies, furnace brazing is favored because it allows uniformity of temperature throughout the heating and cooling cycle to be controlled readily. Titanium and zirconium assemblies frequently are brazed in high-vacuum, cold-wall furnaces.

4.2.11 Carbides and cermets

Carbides of the refractory metals tungsten, titanium, and tantalum, bonded with cobalt, have been used for years in cutting tools and dies. It is necessary to join these carbides to metal parts, particularly for cutting tools. Closely related materials called cermets have been developed and are ceramic particles bonded with various metals. Their high-temperature strengths are intermediate between those of the ceramic materials and the binder metals employed. Their greatest disadvantage is their brittleness.

Brazing carbides and cermets is generally more difficult than brazing metals. Torch, induction, or furnace brazing is used, often with a sandwich brazing technique: a layer of weak, ductile metal (pure nickel or pure copper) is interposed between the carbide or cermet and a hard metal support. The cooling stresses cause the soft metal to deform instead of cracking the ceramic.[129]

Silver-based braze filler metals, copper–zinc filler metals, and copper are often used on carbide tools. Although it is possible to use any of the BAg–1 through BAg–7 brazing filler metals, those which contain nickel (BAg–3 and BAg–4) are generally recommended because nickel improves wettability. The RBCuZn–D and BCu brazing filler metals also have been used, particularly where a post-braze heat treatment is required. The BCu filler metal retains practically all of its strength up to a temperature of 540°C, however, it requires a hydrogen atmosphere furnace for best brazing results. The 85Ag–15Mn and 85Cu–15Mn filler metals are used where the brazed joint will be subjected to elevated temperatures in service and also for wetting the titanium-based or chromium-based carbides. Additionally, the BNi filler metals with boron and a 60Pd–40Ni filler metal have been used successfully for brazing nickel- and cobalt-bonded cermets of tungsten carbide, titanium carbide, and niobium carbide.

Tungsten-based carbides generally are readily wetted by the BAg and BCu brazing filler metals. However, titanium-based carbides are somewhat more difficult to wet. Where it is necessary to mount titanium carbides by brazing, either the joint must be made in an inert or vacuum atmosphere, or the surfaces must be specially treated by the carbide manufacturer, or by nickel plating. In each case, the filler metals just mentioned will then wet the titanium carbide surface.

In specifying a brazing filler metal for a given job, first consideration should be given to the temperature range of anticipated application. This and the other considerations of corrosion and mechanical properties will dictate the alloy composition selected, the brazing temperature required, the equipment to be used, and the joining atmosphere. Generally speaking, the brazing filler metals mentioned previously are considered for the majority of applications which require simple equipment, fluxes, and brazing temperatures below 980°C.

The carbides, in general, are not wetted as readily by brazing filler metals as are most base metals; thus it is preferable, when possible, to preplace shims in the joint rather than to face-feed the filler metal in the form of wire. For larger surfaces, shims having a core of copper or nickel with a coating of the brazing filler metal on both surfaces are frequently used. The core of the shim generally accounts for about 50% of the total thickness, with the brazing filler metal providing another 25% on each side.

4.2.12 Ceramics

The increasing use of ceramic-to-ceramic and ceramic-to-metal joints in industrial and developmental applications is due to the unique combination of properties of ceramic materials.[9, 130–133] The large usage of ceramic-to-metal joints in vacuum tubes in the electronics industry stems from the following properties:

(a) ceramic tubes can be outgassed at higher temperatures than glass tubes;
(b) ceramic tubes can withstand higher temperatures than glass tubes of similar dimensions;
(c) ceramic tubes are mechanically stronger and less sensitive to thermal shock than glass tubes;
(d) ceramic components can be ground to the precise tolerances required for vacuum-tube construction;
(e) ceramic materials have very low electrical losses at high frequencies.

Because of their inertness in many corrosive environments, ceramics are used as seals in fuel cells and other devices that convert chemical, nuclear, or thermionic energy to electricity. Ceramics are also used as friction materials for brakes, clutches, and other energy-absorbing devices; coatings for nuclear fuel particles; constituents in high-temperature adhesives; radomes used to enclose antennae; and ablative materials.

Glass-to-metal seals have been made for many years in the vacuum-tube industry, and the experience thus obtained gives a general insight into the problem of fabricating ceramic-to-metal joints.[134] More recently, the fabrication of refractory-tipped tools, vacuum tubes and various experimental devices has added to the available knowledge and techniques.

Ceramic materials are inherently difficult to wet with conventional brazing filler metals. Most of these filler metals merely ball up at the joint, and little or no wetting occurs. When bonding does occur, it can be either mechanical or chemical. The strength of a mechanical bond can be attributed to interlocking particles or penetration into surface pores and voids, whereas a chemical bond derives strength from material transfer between the filler metal and the base material. Discussions of bonding mechanisms can be found in the literature.[8, 135–140]

Another basic problem in brazing of ceramics results from the differences in thermal expansion between the base material and the brazing filler metal and, in the case of ceramic-to-metal joints, between the two base materials. In addition, ceramics are poor conductors of heat, which means that it takes them longer to reach equilibrium temperature than it does metals. Both of these factors may lead to cracking in the joint. Because ceramics generally have lower tensile and shear strengths, crack propagation occurs at lower stresses in ceramics than in metals. In addition, the low ductilities permit very little distribution of the stresses set up by stress raisers. Alumina, zirconia, beryllia, thoria, forsterite (Mg_2SiO_4), silicon carbide and nitride, are the leading ceramic materials which can be joined by brazing.

If the ceramic is premetallized to facilitate wetting, copper, silver–copper, and gold–nickel filler metals can be used. Titanium or zirconium hydride can also be decomposed at the ceramic–metal interface to form an intimate bond.

Nonmetallized ceramics have been brazed with silver–copper-clad or nickel-clad titanium wires and other useful titanium and zirconium filler metals Ti–Zr–Be, Ti–Zr–V, Zr–V–Nb, Ti–V–Be, and Ti–V–Cr.[141–144]

Silicon nitride (Si_3N_4)

One process for brazing Si_3N_4 with metallic alloys involves vapor coating the ceramic with a 1.0 μm thick layer of Ti before the brazing operation. The coating improves wetting of the Si_3N_4 surfaces to the extent that strong bonding between the solidified braze filler metal and the ceramic occurs. Braze joints of Si_3N_4 are made with Ag–Cu, Au–Ni, and Au–Ni–Pd filler metals at temperatures of 790°C, 970°C and 1130°C.[145]

The study by Santella also showed that Si_3N_4 surfaces are not easily wet by common precious metal-based brazing filler metals, but that vapor coating the surfaces with Ti prior to brazing improved their wetting characteristics and permitted braze filler metals to adhere strongly to the ceramic. The ability of vapor coatings, particularly of Ti, to improve the wetting characteristics of oxide ceramics has been known for some time.[146, 147] Using a metallic vapor coating rather than incorporating an active element like Ti directly into the brazing filler metal permitted readily available commercial braze filler metals to be used for joining.

Si_3N_4 ceramics have recently been joined by an oxynitride brazing method. It was useful for large-sized parts or ceramics with complex geometries in which the high-temperature mechanical properties of the original ceramic components must be maintained.

Joined parts can withstand temperatures that are 300 to 400°C higher than conventionally joined parts can. Joint strength is comparable to the 400 MPa of the Si_3N_4 ceramic material. If joint strengths do not need to exceed 200 MPa, then these joints can be used at temperatures as high as 1250°C.

These large or geometrically complex parts are used in gas turbine and adiabatic diesel engines, magnetohydrodynamic generators, pumps, and numerous other applications that require high strength, corrosion resistance, and use temperatures exceeding 750°C.[148]

SIALON

Oxynitrides were initially referred to by the acronym SIALON, which stands for the Si–Al–O–N system. Both terms, oxynitride and SIALON, are used extensively throughout the literature to refer to specifically Si–Al–O–N materials as well as compounds derived from silicon nitrides or oxynitrides by simultaneous replacement of Si and N_2 by Al and O_2. The term has become a generic one applied to materials where the structure involves (Si, Al), $(O_2,N_2)_4$ or (Si, M) $(0_2,N_2)_4$ tetrahedra.[140, 149, 150]

Silicon carbide

Joining of SiC to SiC was investigated for advanced heat engine applications.[151] The SiC–SiC joints were produced by cosintering β-SiC green forms with and without the use of pressure (HIP and vacuum sintering). The joints attained tensile strengths equal or > 138 Pa at 1530°C and no glassy phases were used for the joint.

A program was undertaken to evaluate active brazing applied to SiC because there is no sufficient metallizing process available for this ceramic.[152]

The employment of metallic interlayers is one way of reducing thermally generated stress. Ductile metal, like Cu, can be employed as an interlayer as well to reduce thermal stress by plastic deformation of the interlayer.

There were several commercially-available active braze filler metals that were evaluated in a program (Table 4.5). Resultant strengths using Cu are reflected in Figure 4.13.

Table 4.5 Commercially available active braze filler metals[152]

Braze filler metal	Chemical composition, weight %				Solidus, (°C)	Liquidus, (°C)	Brazing temperature, (°C)
	Cu	In	Ti	Ag			
1. Ag–Cu–Ti	27.5	—	2	70.5	780	795	840
2. Ag–Cu–In–Ti	23.5	14.5	1.25	60.75	605	715	760
3. Ag–Cu–Ti	26.5	—	3	70.5	803	857	950
4. Ag–Ti	—	—	4	96	960	960	1050
5. Ag–Cu–Ti	6	—	3	91	875	917	970
6. Ag–Cu–In–Ti	19.5	5	3	72.5	732	811	950

a°C = 5/9(°F – 32)

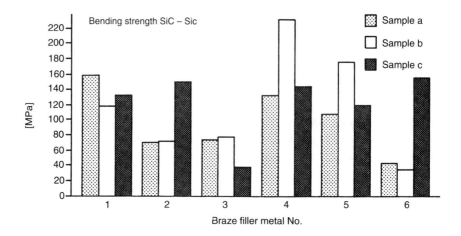

Fig. 4.13 Four-point bending strength of active brazed SiC–SiC joints.[152]

Alumina

A major problem with brazing an oxide ceramic is the resistance to wetting caused by the oxides on the surface of the ceramic. A means of rectifying the problem is to apply pressure to the braze filler metal with sufficient force to counteract the repelling force of the oxides. A study was undertaken to assess the strength of the joints of metals brazed to Al_2O_3 with a Cu braze filler metal.

Results of tests indicated that the strength of the joint steadily increased as the pressure increased up to 5 MPa and then leveled off with any additional pressure. Strength increased as the brazing temperature rose to 1100°C and then dropped with further increases in temperature. Length of holding time under pressure also affected strength. The amount of Al_2O_3 in the ceramic also played a role in joint strength, with those of 100% Al_2O_3 being only 50 to 60% as strong as those with 94% Al_2O_3.

Titanium-containing braze filler metals (Table 4.6) were used to join both 94 and 99+% Al_2O_3 compositions. Resulting tensile strengths of 76.8 to 109.8 MPa compared favorably with conventional Mo–Mn metallizing of Al_2O_3 surfaces and subsequent brazing.[153] The two different Ti-containing braze filler metals which were used were essentially Cu–Ag eutectic compositions which contain small (1 to 3) weight percentages of Ti. The basic difference between the two filler metals was that one contained approximately 10%wt In whereas the second filler metal contained no In. The test results lead to two general conclusions concerning the strength data[153]:

1. For a given ceramic, higher strengths are observed (a) when brazing in a vacuum and (b) when brazing with the In-containing filler metal.
2. All sets of variables examined (filler metal, atmosphere, and ceramic composition) yielded comparable tensile strength to conventional metallizing/brazing techniques.

In order to verify the above conclusions concerning the braze filler metal as an active braze alloy (ABA), a component was selected (Figure 4.14) to see how well the component (94% Al_2O_3) performed when brazed with the In-containing filler metal. The component chosen was a 94%wt Al_2O_3 ceramic header with two Mo/Al_2O_3 cermet electrical feedthroughs. Attached to the connector end of this header were two Cu contacts, which were subsequently brazed to the Mo/Al_2O_3 cermet surface.

The results of the shear tests are contained in Table 4.7. The braze filler metal preforms with 10% In yielded the highest shear strength. As a result it was found that ABAs provide a simplified method of joining Al_2O_3 ceramics. Second, comparable tensile strengths can be obtained from ABAs as from conventional Mo–Mn metallizing techniques. ABA sealing results in the migration of Ti from the bulk braze to the ceramic (Al_2O_3) surface.

Table 4.6 Summary of braze filler metal compositions (%wt) for brazing alumina

Braze filler metal	Filler metal				Braze temperature range
	Copper	Silver	Titanium	Indium	
Cusil ABA	27.5	70.5	2.0	—	820–860°C (1508–1580°F)
Incusil 10 ABA	27.0	62.25	1.25	9.5	770– 800°C (1418–1472°F)

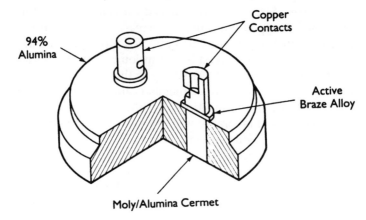

Fig. 4.14 Ceramic header used to test Ti-containing braze filler metals.[9,153]

Table 4.7 Braze test results in evaluation of titanium-containing filler metals[153]

Braze material	Atmosphere	Temperature, (°C)[a]	Time, (min)	Shear strength, (lb)[b]
Incusil–10 ABA(0.002 in.)[c]	vacuum	810	5	118
Cusil ABA (0.002 in.)	vacuum	840	5	110
Cusil ABA (0.001 in.)	vacuum	840	5	32
Cusil ABA (0.001 in.)	H₂	840	5	42
No metallize/nickel/cusil	H₂	820	6	120[d]

[a]°F = 9/5°C + 32. [b]0.45 × lb = kg. [c]25.4 mm = 1 in. [d]Specification requirement = 65 lb.

Additionally, a braze filler metal, 25Cr–21V–54Ti, has been successfully used to join Al_2O_3 to itself without any metallizing coatings (Mo–Mn or Ti hydride pretreatment).[154] Table 4.8 lists some of the systems which have held out promise; however, only numbers 1 and 2 have successfully wet Al_2O_3. Compositions with at least 25% Cr readily wet and flow on Al_2O_3; however, less Cr results in limited or no flow. The ductility of the braze filler metals also depends on composition. Filler metals with <25% Cr are ductile and can easily be rolled into sheet. The success of the above type of filler metals in wetting ceramics is through the employment of active metals such as Ti and Zr as components of the braze filler metal.

Table 4.8 Ternary systems of braze filler metals [154]

Filler metal system	Approximate brazing temperature, (°C)[a]	Materials		
		Refractory Metals	Graphite	Al_2O_3
Ti–V–Cr	1550–1650	x	x	x
Ti–Zr–Ta	1650–2100	x	x	x
Ti–Zr–Ge	1300–1600	x	x	
Ti–Zr–Nb	1600–1700	x	x	
Ti–Zr–Cr	1250–1450	x		
Ti–Zr–B	1400–1600	x		
Ti–V–Nb	1650	x		
Ti–V–Mo	1650	x		

[a]°F = 9/5°C + 32.

Table 4.9 Braze filler metals in the Ag–Cu system [157]

Composition, (wt = %)	Brazing range, [a,b] (°C)
Cu–27 Ag–26 Ti	900–950
Cu–26 Ag–29 Ti	920–1000
Ag–38 Cu–1 Ni–4 Ti	950–1050
Ag–37 Cu–0.75 Ni–7.25 Ti	850–950
Ag–35 Cu–0.7 Ni–10.3 Ti	950–1050

[a]Brazing temperature depends to some degree on substrate material and on the extent of flow desired.
[b]°F = 9/5°C + 32

While the joining of ceramics (Al_2O_3) has been successful without coatings prior to brazing, other investigators have examined other metallic coatings in lieu of Mo–Mn processing and the Ti hydride treatment.[155–157]

Finally, in a comprehensive study of braze filler metals and several engineered ceramic materials for uncooled diesel engines Moorhead *et al*.[157, 158] evaluated two alloy systems Cu–Ag–Ti (Table 4.9) and Cu–Au–Ti (Table 4.10), three types of Al_2O_3, and two types of ZrO_2 (MgO–stabilized PSZ and Y_2O_3–stabilized TZP).

From the braze filler metals based on the Ag–Cu eutectic with additions of Ni and Ti, the best results were achieved with the Cu–26Ag–29Ti (%wt) composition. This filler metal produced wetting angles less than 30° on all the ceramics, and flexural strengths at 400°C of >165 MPa for Al_2O_3 and >130 MPa for a PSZ brazement.

Table 4.10 Experimental braze filler metals with copper and gold as major elements[157]

Composition, (wt = %)	Brazing range,[ab] (°C)
Cu–14 Au–4 Ni–6.5 Ti	1090–1190
Cu–13 Au–3.5 Ni–14 Ti	1050–1150
Cu–22 Au–10 Ti	1350–1450
Cu–20 Au–18 Ti	1050–1150
Cu–18 Au–26 Ti	1050–1150
Au–36 Ni–11 Ti	1050–1200
Au–30 Ni–21 Ti	1150–1250

[a]Brazing temperature depends to some degree on substrate material and on the extent of flow desired.
[b]°F = 9/5°C + 32

Of the Au-bearing filler metals, the Cu–20Au–18Ti (%wt) had superior properties. Flexural strengths at 400°C of brazements made with this filler metal ranged from 106 MPa for one Al_2O_3 to 218 MPa for another Al_2O_3. The room-temperature flexural strength of PSZ brazed with this material was 258 MPa. Finally the fracture toughness of composite specimens of PSZ brazed with a Cu–27Ag–26Ti (%wt) filler metal averaged about the same as that of the bulk ceramic, 6 MPa m$^{1/2}$.

Zirconia

Braze filler metal systems have been developed not only for joining Al_2O_3 but also for joining partially-stabilized ZrO_2.

The contact angles of molten Al–Cu filler metals and their wettability on CaO-stabilized ZrO_2 have been measured and it was found that a ZrO_2 joint brazed with Al–1.7Cu%wt filler metal has produced the maximum fracture strength of 105 MPa compared to 52 MPa for that brazed with pure Al at room temperature. This improved strength of ZrO_2 is maintained at elevated temperatures up to 513°C. The joining strength of a ZrO_2 joint brazed with an Al–Cu filler metal is dominated by the mechanical properties of the Al–Cu in addition to the wettability of the Al–Cu filler metal against ZrO_2.[159]

Carbides

Not considered in the same structural group of ceramic carbides (SiC, B_4C, and the like) tungsten carbide (WC) and others are a very important group of hard carbides. Brazing was one of the first successful methods of mounting carbides to steel or other base alloys.

Although it is possible to braze to any of the BAg–1 through BAg–7 filler metals, those which contain Ni are generally recommended especially BAg–3 which contains 3% Ni and possesses the wettability needed to braze carbide tool bits to shafts.

BCu–1 is unalloyed copper and is used in joint clearances from press fits to 0.05 mm and RBCuZn–D as well as BCu–1 used to braze WC.

The 85Ag–15Mn and 85Cu–15Mn filler metals are used where the braze is subject to elevated temperatures in service and also for wetting the Ti-based or Cr-based carbides. W-based carbides are generally readily wetted by the BAg and BCu filler metals.[160]

Developments in the past few years have yielded new brazing filler metals containing active metals[161–168] as well as improved processes.

Brazing of ceramics is now usually carried out in high-purity inert-atmosphere or in vacuum furnaces. High-frequency induction heating can also be used for brazing ceramics to metals. In the induction brazing of ceramics to themselves, it is necessary to interpose a graphite or metal tube (susceptor) between the inductor and the work. This tube then becomes the heat source.

Ceramic joining methods

Several techniques for joining ceramics have been reported in the literature.[130, 141] The sintered metal powder technique is a widely used brazing method. This process requires several steps to produce a joint:

1. the firing of metal powder held in suspension on the ceramic;
2. plating or deposition of a thin copper or nickel film;
3. brazing by conventional methods to make the ceramic-to-metal joint.

The affinity of titanium and zirconium for ceramics is the principle behind the active-alloy process. Highly active titanium or zirconium can be made available at the ceramic/metal interface by hydride decomposition of a powder slurry on the ceramic surface. The reaction of the titanium or zirconium with the ceramic, or with additional metals placed at the interface, forms an intimate bond. In some cases, the titanium or zirconium is merely painted on the ceramic surface and placed in contact with a suitable brazing filler metal and the base metal to which the ceramic is to be joined. This process has an advantage over some others in that only one firing operation is required.

The affinity of reactive and refractory metals for ceramics is also the basis for the direct brazing approach. In this case, active metals are incorporated as one or more of the constituents in the brazing filler metal, which is placed at the joint as in conventional brazing. An interesting means of promoting flow of the filler metal involves vapor-deposited coatings of titanium, zirconium, or other metals on the ceramic substrate. Electron-beam heating has provided a unique means for producing the metal vapor.[9, 169–171]

Active metal brazing (ABA) has been practiced since the 1940s, when Ti hydroxide in an organic solvent was coated onto ceramic seals, which were then sandwiched around a Ag (or Au or Cu) filler metal and brazed together in vacuum. The trouble was that under high vacuum the excess Ti in the filler metal formed a brittle joint, while under low vacuum most of the Ti formed brittle oxides and nitrides and little was left for the actual active brazing.

One way around the problem was to take Ti wire or foil and clad it with a braze filler metal. While that protects the Ti from the furnace atmosphere until the filler metal starts to melt, so more Ti is available for brazing, the filler metal itself is not ductile enough to yield to stresses caused by the different rates of thermal expansion for the ceramic and metal.

Braze filler metal manufacturers attacked the problem by developing a series of braze filler metals. The series consisted of Cu–Ag–Ti and Cu–Ag–In–Ti. Like cladding, alloying protects the active elements (Ti) until the filler metal starts to melt. Since the Ti is protected, less is needed (1.25 to 2.0%wt), and the resulting braze is less hard and ductile enough to form strong, reliable ceramic-to-ceramic and ceramic-to-metal hermetic bonds without prior metallization. Table 4.7 and Figure 4.14.[9, 10, 157, 158, 172–177]

The system works best when the ceramic surface is free of surface fractures. Researchers found it took more than twice as much energy to peel apart a Fe–Ni–Co (Kovar) strip brazed to a lapped ceramic surface as from a ground ceramic surface. Joint strength for Al_2O_3 ceramic has been shown to be equal whether prepared and joined by the Mo–Mn braze system or the ABA system,[10, 172–174] Figure 4.15.

A unique ceramic-to-metal seal system was developed[170] based on a composite Al_2O_3 insulator and a 49%Ti–49%Cu–2%Be braze filler metal. Pulsed laser beam, furnace, and induction joining techniques were all successful in brazing the sensor electrodes, ceramic insulator, and end seals in a number of brazing sequences and operations.

Pure metal coatings have also been evaluated. A Nb coating film has been electrolytically precipitated at a temperature of about 740°C out of a fluoride melt onto sintered Al_2O_3. The electrical conductivity of the surface that is essential for electrolysis was provided by a thin film of Nb produced by gas phase precipitation. Metallizing coatings can also be created completely by means of gas phase precipitation, for example by the thermal dissociation of halides or organic metal compounds (chemical vapor deposition, CVD). Tungsten coatings, for example, can be produced from WCl_6 in a He atmosphere at a temperature of about 1000 to 1300°C.

Brazed joints on this type of metal coating have created high-vacuum seals; at ambient temperature they have a transverse tensile strength of between 40 and 100 MPa. In cases where high-temperature-resistant metals such as W, Mo, or Nb are involved, their thermal resistance

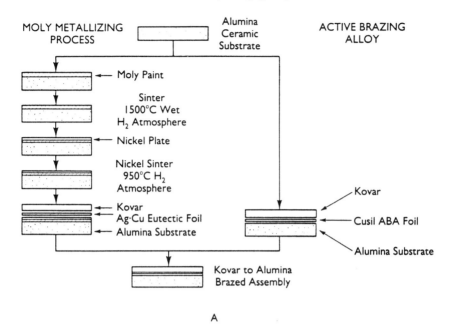

Fig. 4.15 Comparison of two methods of joining Kovar to Al_2O_3.[172]

depends upon the behavior of the brazing filler metal used. The alloying effect between the brazing filler metal and the metal coating can give rise to joint failure. Operational temperatures of between 800 and 1000°C have, however, been obtained. A disadvantage of this metallizing process is the cost of the expensive manufacturing processes involved. A mixture of metal powder (usually Mo or W) and a vitreous or crystalline solidifying oxide bonding phase is first spread over the surface of the ceramic and then fused at a temperature above that of the melting point of the bonding phase, usually under slightly oxidizing or reducing conditions. The oxide melt wets both the ceramic and the metal powder and the latter becomes firmly bonded to the ceramic. In order to increase the wetting power of the brazing filler metal, such coatings are usually Ni coated prior to use.

The coating is a Mo–Mn–silicate coating.[156] This particular coating enables joints having a transverse tensile strength of between 70 and 120 MPa to be effective. In contrast to the conventional Mo–Mn processes[178] which require SiO_2 from the ceramic in order to create a bonding phase, Meyer's method permits strongly bonded metallization of very pure Al_2O_3 ceramics (99.8% Al_2O_3) and other ceramic oxide materials such as ZrO_2.[156]

Considerable success is being achieved at present in the application of reactive metal brazing.[179] The next few years, however, could see a considerable expansion in the market for reactive metal brazes if their performance and associated fabrication procedures are optimized. Current braze filler metals for ceramics employ titanium as their reactive component, but it should be possible and perhaps be technically preferable to use other reagents for some applications. Thus the capability of titanium to form intermetallic compounds with many workpiece metals can be detrimental and replacement by other multivalent metals, such as chromium, which can form ceramic reaction products with wide ranges of stoichiometry could be advantageous – although substantial development work will be required to optimize and produce the new brazes in commercially usable forms. Improving the useability of existing braze alloys is also of importance, and developments could come from the application of rapid solidification technique to brittle alloys to produce the thin sheets needed for stamping out preforms. Of equal importance with the development of materials is the production of joint design codes and the acquisition of engineering performance data. These combined with the tailoring of brazes for use with specific materials should enhance the acceptability of brazing as a joining process for ceramics and make it possible to exploit it more efficiently in advanced engineering projects.[180]

Finally aluminum nitride (AlN) is currently under investigation as a potential candidate for replacing Al_2O_3 as a substrate material for electronic circuit packaging. The requirements for such a material are that it can be metallized and joined to produce hermetic enclosures for semiconductor devices as well as its coupling with its expansion matching that of silicon. A technique for brazing AlN using a nonactive metal braze has been developed.

The study found that active metal brazes, namely titanium, wet the AlN whereas nonactive ones are nonwetting. This method has been used to successfully join AlN to a low expansion lead frame alloy.

The interfacial reactions lead to the formation of titanium and zirconium nitrides and these interfacial reactions enable the ceramic surface to be wet by the molten braze in the development of AlN packages for microcircuit applications.[180]

4.2.13 Graphite

The joining problems associated with brazing of graphite are very similar to those encountered in brazing the ceramic materials previously discussed.[181, 182] Wetting of graphite is more difficult than wetting of metals, and the differences in coefficients of thermal expansion between graphite and conventional structural materials are pronounced. It should be noted that carbon and graphite can also be brazed both to themselves and to metals. These materials vary widely in degree of crystallinity, in degree of orientation of the crystals, and in size, quantity, and distribution of porosity in the microstructure. These factors are strongly dependent on the starting materials and on processing and, in turn, govern the physical and mechanical properties of this product.

Carbons and graphites can be manufactured by several processes that yield materials with a wide range of crystalline perfection and properties. In the most widely-used process, polycrystalline graphites are made from cokes produced as by-products of the residium from the manufacture of petroleum or from natural pitch sources.[183]

The wetting characteristics of all the carbons and graphites are strongly influenced by impurities, such as oxygen or water, that are either absorbed on the surface or absorbed in the bulk material. Moisture absorption always occurs to some extent, with levels as high as 0.25%wt. Brazeability also depends on the size and distribution of pores, which can vary significantly from one grade to another.

A major consideration in brazing of carbon and graphite is the effect of the coefficient of thermal expansion of these materials, which can range from $2 \times 10^{-6}\,°C^{-1}$ up to $8 \times 10^{-6}\,°C^{-1}$ between 25 and 1000°C. Before carbon or graphite is brazed, the type and grade of the material must be established so as to ascertain its expansion characteristics. This information is also important for brazing of carbon or graphite to itself. Joint failure, particularly during thermal cycling, may occur if too great a difference exists between the coefficients of thermal expansion of the graphite and the brazing filler metal.

If the joint gap increases significantly on heating because of a large mismatch in coefficient, the brazing filler metal may not be drawn into the joint by capillary flow. On the other hand, if the materials and joint design cause the gap to become too small, the filler metal may not be able to penetrate the joint. In conventional brazing of dissimilar materials, the material having the greater coefficient of expansion is made the outer member of the joint. Joint tolerances are used that do not allow the gap between the surfaces to become too great for capillary flow.

Additional problems occur in brazing of dissimilar materials when one part of the joint is a carbon or graphite. Carbons and graphites have little or no ductility and are relatively weak under tensile loading. These adverse conditions are usually compensated for in graphite-to-metal joints by brazing the graphite to a transition piece of a metal such as molybdenum, tantalum, or zirconium, with a coefficient of expansion near that of the graphite. This transition piece can be subsequently brazed to a structural metal if required. This minimizes shear cracking in the graphite by transposing the stresses resulting from the large difference in thermal expansion to the metallic components. Thin sections of metals, such as copper or nickel, that deform easily when stressed have also been successfully used for brazing dissimilar metals.

Metals that have strong tendencies to form carbides (titanium, zirconium, silicon, chromium) have been found to wet graphite when they are molten. A commercial brazing filler metal frequently used for brazing of graphite is silver-copper-clad titanium wire. Graphite is also readily wetted by molybdenum disilicide, titanium, and zirconium. In recent years, the requirements of the aerospace and nuclear industries have resulted in the development of several additional brazing filler metals. In general, these filler metals incorporate substantial quantities of carbide-forming elements. They include 48Ti–48Zr–4Be, 35Au–35Ni–30Mo, 70Au–20Ni–10Ta, nickel-clad titanium, 54Ti–21V–25Cr, 43Ti–42Zr–15Ge, and 47Ti–48Zr–5Nb.[182] Additionally, the brazing filler metal 49Ti–49Cu–2Be has been recommended for brazing of graphite as well as oxide ceramics.[184] These filler metals mentioned previously wet graphite and most metals well in either a vacuum or inert atmosphere (pure argon or helium) and span a fairly wide range in brazing temperatures from 1000°C for 49Ti–49Cu–2Be to 1350°C for 35Au–35Ni–30Mo.

At least two commercially-available brazing filler metals wet carbon and graphite as well as a number of metals. The first has a composition of 68.8Ag–26.7Cu–4.5Ti, a solidus of 830°C, and a liquidus of 850°C. This filler metal is suitable for low-to-medium-temperature applications but appears to have only moderate oxidation resistance. The second commercially-available filler metal for graphite brazing has a composition of 70Ti–15Cu–15Ni. With a somewhat higher melting range (solidus, 910°C; liquidus, 960°C), and by virtue of its greater titanium content, it has better oxidation resistance than the silver-base filler metal. Some workers in Europe and the USSR have reported successful joining of graphite to steel using a brazing filler metal with a composition of 80Cu–10Ti–10Sn at 1150°C.[185] In other work, using a technique called 'diffusion brazing', a metallic interlayer was placed between the graphite components: the components were pressed together at a specific pressure and heated to the temperature of formation of a carbon-bearing melt or eutectic. On heating to higher temperatures, the melt dissociated with the precipitation of finely divided crystalline deposits of carbon that interacted with the graphite base material to form a strong joint.[186, 187]

Researchers[188] have developed a procedure for brazing a special grade of graphite to a ferritic stainless steel for a seal in a rotary heat exchanger. It seems apparent that the selection of type 430 stainless steel was based at least partly on its lower coefficient of thermal expansion compared with that of a typical austenitic stainless steel. In addition, a joint geometry was developed that minimized the area of the brazed joint, thereby reducing thermally-induced stresses to acceptable levels.

Specimens of graphite brazed in a vacuum furnace to thin type 430 stainless steel sheet with either Ni–20Cr–10Si or Ni–18Cr–8Si–9Ti filler metal at 1125 to 1175°C performed well in tests at 650°C.

4.2.14 Composites and aluminides

SiC fiber-reinforced SiC matrix composites (SiC–SiC) produced by chemical vapor infiltration (CVI) are being developed for use in structural applications at temperatures approaching 1000°C. These composites contain about 40%vol SiC fibers (Nicalon®) and are infiltrated to about 85% of the theoretical density with SiC. In order to fully realize the advantages of these materials, practical joining techniques are being developed. Successful joining methods will permit the design and fabrication of components with complex shapes and the integration of component parts into larger structures. Joints must possess acceptable mechanical properties and exhibit thermal and environmental stability comparable with the composite which is being joined.

Studies have focused on joints produced using TiC+Ni and SiC+Si interlayers. The microstructures of joints have been characterized and the results appear promising. In one approach, hot pressing elemental titanium, carbon and nickel powders formed a TiC+Ni joint interlayer and in a second approach, silicon infiltration of SiC and carbon powders, or carbon cloth, produced SiC+Si interlayers.[189]

Morimoto, Tanaka and associates[190] investigated the effects of brazing temperature on joint properties of continuous SiC fiber in reinforced aluminum alloy matrix composite and the behavior of the MMC during the heating process. The following results were obtained.

1. There is a brazing temperature at which maximum joint strength is obtained in the brazing process. A brazing temperature lower than this causes the brazed joint to fail in shear at the brazed interface. A brazing temperature higher than this makes the brazed joint fail in the MMC. The matrix was Al–10Si–1Mg alloy foil. The maximum joint strength in this study was 77.7 MPa. The brazing condition was 580°C for 1800 seconds.
2. MMC showed degradation in tensile strength at room temperature after heat treatment higher than 550°C.
3. MMC began to melt at a temperature of about 580°C and completely melted at about 600°C. Voids were produced owing to the solidification shrinkage of the matrix metal.
4. When the melting point of the matrix metal was close to the brazing temperature, the reaction between fibers and matrix metal occurred easily. As a result, accurate temperature control was needed in the brazing of the MMC using the aluminum alloy matrix.[191]

Trial tests have demonstrated that joining the various titanium aluminides which are ordered intermetallic alloys, and are materials which are arousing considerable interest in the aerospace industry for use in gas turbine engines and in high-performance military airframes, is feasible.

Like most materials of this type, (ordered intermetallic alloys), they are very brittle at ambient temperature, although they exhibit good hot ductility. Intermetallic alloys generally have high strength and modulus which are retained to very high temperatures, and often have good resistance to oxidation at elevated temperatures too.

There has recently been much effort to improve the low-temperature ductility of these materials by suitable alloying, as well as to develop improved high-temperature mechanical properties and oxidation resistance. This work has reached the point where two types of titanium–aluminide-based alloys can be seriously considered for use in aerospace applications.

The two alloy systems of immediate interest are based on the intermetallic compounds Ti_3Al (α_2 aluminides) and TiAl (γ aluminides).

The α_2 alloys are made more ductile by alloying with β stabilizers – such as Nb, Mo or V – which improve the number of slip systems, refine the microstructure, and permit a small amount of β phase to be retained at low temperatures. Unfortunately, this is at the expense of the density of the alloy. The key to joining lies in controlling the phase transformation from the high temperature β to the α_2 phase.

The TiAl-based γ alloys are less well developed than the α_2 alloys, but their potential is greater because of their lower density and improved high-temperature performance. Again, β stabilizing elements are added to improve the ductility, and the alloys are usually designed to give a predominantly γ microstructure, containing laths of α_2.

The Welding Institute (TWI) has been very active in developing joining techniques for both α_2 and γ alloys. Figure 4.16 shows a diffusion bond in an α_2 alloy, from which it can be seen that a good quality bond could be obtained at a temperature below the β transus thus avoiding the complications of the β to α_2 transformation. The comparatively high alloy content does not apparently compromise the ability of titanium to be joined by this process.

4.2.15 Dissimilar material combinations

Many dissimilar metal combinations may be brazed, even those with metallurgical incompatibility that precludes welding.

Important criteria to be considered start with differences in thermal expansion. If a metal with high thermal expansion surrounds a low expansion metal, clearances at room temperature which are satisfactory

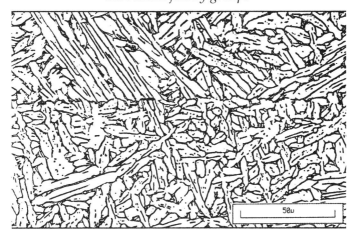

Fig. 4.16 A diffusion bond in Ti–24A1– 11Nb alloy.

for capillary flow will be too great at brazing temperature. Conversely, if a low expansion metal surrounds a high expansion metal, no clearance may exist at brazing temperature. For example, when brazing a molybdenum plug in a copper block, the parts must press fit at room temperature; if a copper plug is to be brazed in a molybdenum block, a properly centered loose fit at room temperature is required.

In brazing tube-and-socket type joints between dissimilar base metals, the tube should be the low expansion metal and the socket the high expansion metal. At brazing temperature, the clearance will be maximum and the capillary will fill with brazing filler metal. When the joint cools to room temperature, the brazed joint and the tube will be in compression.

For a tongue-in-groove joint one should place the groove in the low expansion material. The fit at room temperature should be designed to give capillary joint clearances on both sides of the tongue at brazing temperature.

As noted above the joining of dissimilar metals has become increasingly important during the past two decades because of the service requirements for structures used in missiles and rocket, supersonic aircraft, nuclear equipment, marine systems, and chemical processing equipment. Although certain dissimilar metals have been routinely joined for many years, the advent of space and nuclear requirements has produced a need for sophisticated methods of joining the new structural materials that have been developed for these demanding applications. These new alloys possess exceptional mechanical properties and resistance to corrosive media under extreme operating conditions. However, such alloys are frequently used only for sections of a structure where their specific properties are required; conventional alloys are used for the remainder of the structure for reasons of economy, weight, ease of fabrication, etc. Thus, there is a need for procedures for producing reliable joints between dissimilar metals. The ability to design and fabricate such joints is essential to many segments of our industrial economy.

The selection and use of dissimilar metals in structural applications is governed by the service requirements of the structure and by material and fabrication costs. For example, a relatively inexpensive grade of steel may be used in fabricating the shell of a vessel for the chemical industry for reasons of economy, whereas the corrosion requirements are satisfied by lining the vessel with thin-gage tantalum or titanium.

Corrosion of a dissimilar-metal joint is inevitable if there is an electrolyte present. The degree of corrosion which takes place is dependent on the type of electrolyte and the difference in electromotive potential between the dissimilar metals and/or materials.

Young and Smith[192] discuss the problem of galvanic corrosion between dissimilar metals. The magnitudes of the solution potentials of metals and intermediate compounds depend on the nature of the electrolyte and the difference in potential between the alloys being joined.

Applications requiring the joining of dissimilar metals are discussed briefly below:

1. The lunar module contains 26 pressure vessels in its descent and ascent stages.[193] Depending on their functions, these vessels are fabricated from various alloys plus titanium. Coextruded titanium–stainless steel transition joints are used to connect the titanium pressure vessels to the stainless steel feed system.
2. Joints between beryllium and such metals as aluminum, stainless steel, and titanium are encountered in space-vehicle design where beryllium is an attractive structural metal because of its low density, its stiffness under load, its resistance to damage by impact with meteors, and its high heat capacity, Figure 4.17.
3. Procedures are required for joining dissimilar metals in nuclear-reactor construction. These applications range from the cladding of fuel elements with zirconium alloys to fabrication of dissimilar metal piping joints.
4. Dissimilar metal joints are encountered in aircraft hydraulic and ducting systems, as well as in engine and airframe construction. Because significant quantities of titanium alloys are used in new aircraft, there are occasions to join titanium to other structural alloys to meet specific design requirements,[194] Figure 4.18.

In general, dissimilar metals are used in structures to provide:

● high-temperature or low-temperature strength;
● resistance to oxidation, corrosion, or wear;

Fig. 4.17 Stainless steel-to-beryllium joint brazed with the 49Ti–49Cu–2Be (wt%) filler metal.[194]

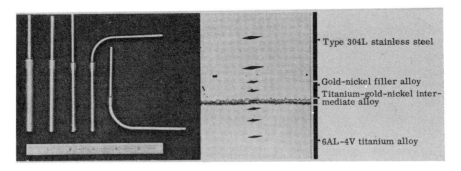

Fig. 4.18 Transition tubes of Ti-6Al-4V joined to type 304L stainless steel.[194]

- resistance to radiation damage;
- other required properties.

Also, the use of dissimilar metals is often attractive from the standpoint of cost.

Joints between ferrous and nonferrous metals are of interest to industry because they combine the strength and toughness of steels with the special properties – such as oxidation resistance, corrosion resistance, etc. – provided by nonferrous metals. Joining of ferrous to nonferrous metals is far more complicated than joining of dissimilar ferrous metals, because of the wider variation in the physical, mechanical, and metallurgical properties of the metals being joined. The extent of these property differences is an excellent indication of the difficulty to be anticipated in joining such metals.[195–198]

In brazing steel to copper, the steel heats much more rapidly than the copper, unless provision is made to equalize the heating rates. In the practice of induction brazing the inductor is coupled more closely to the copper than to the steel or has additional turns to that portion of the inductor heating the copper. Similar provision must be made in brazing carbon steel to brass or austenitic (nonmagnetic) stainless steel. Carbon steel heats faster than either of these materials, although the differential in heating rates is less than that between carbon steel and copper.

The problems inherent in joining dissimilar nonferrous metals are similar to those encountered when ferrous and nonferrous metals are joined, because of the differences in the physical and metallurgical properties of the base metals. Some dissimilar nonferrous metals have been joined routinely for many years; others, such as aluminum to titanium, titanium to nickel, titanium to alumina, Figure 4.19, and aluminum to uranium, are new combinations that owe their existence to their applications in aerospace and nuclear hardware.[199]

Fig. 4.19 Aluminum oxide brazed to titanium bearing assembly with 49Ti–49Cu–2Be (%wt) filler metal by vacuum brazing at 1000°C for 10 min.[142]

Brazing and metallizing of ceramics to form joints with metals finds many uses, especially in small-scale and electronic applications, e.g. Al_2O_3 to Fe–Ni–Co alloy for vacuum tube production. The Mo or Mo–Mn metallization layer is painted onto a ceramic, allowing a subsequent braze layer to wet. Brazing temperatures are high, 1580°C being typical. At this temperature the glassy phase in Al_2O_3 begins to mix with the metallizing mixture of semisintered Mo or Mo–Mn. The literature discloses several theories relating to this mechanism, including capillary action.[200–202]

Other studies of brazing[203] include an investigation into the use of an amorphous filler material between Al_2O_3 and Fe–Ni–Co alloy and the effect of brazing conditions on the resulting shear strength. This complements Naka, *et al.*[204] relative to their work in the use of amorphous fillers between two ceramic components. These results showed that, when using a Cu–Ti braze filler, the joint strength was dependent on the amount of joining between Al_2O_3 and the intermediary TiO_2.

Some excellent work was performed in the active alloy sealing of an Al_2O_3 ceramic to a Cu stud via a Ti–Ni eutectic metallization for use as a high-reliability transistor package. The Ti–Ni metallization was chosen in lieu of the more widely used Mo–Mn metallization because of batch-to-batch variations with the latter technique. The process used 0.5 μm of Ti evaporated onto the ceramic followed by 7 μm of Ni. This was followed by heating to 1000°C for 2 min in dry H_2. This heating step bonded the Ti to the ceramic by forming a liquid phase of Ti–Ni eutectic between the bulk of the Ni layer and the ceramic. The bulk of the Ni layer remained solid as the proportion of Ni was far higher than required for the 71.5Ti–28.5Ni eutectic composition with its melting point of 955°C. Close control of this post-evaporation heating was found necessary. Too short a time or temperatures below 955°C and the reaction to form the eutectic was complete; a weak joint between the metallization and the ceramic resulted. An excessively long heating time resulted in Ti diffusing through to the Ni surface, reacting with impurities in the H_2 and rendering the surface difficult to braze. After this post-evaporation heating, the joining was completed by brazing with a 72Ag–28Cu braze filler metal at >780°C.

Another unique metallizing process has been developed and applied at temperatures as low as 850°C. This makes it especially suitable for materials (such as Mg–PSZ) that cannot withstand temperatures above 1000°C. The braze filler metal consists of 95% Sn, with the balance containing carbide or carbonyl formers and other alloying elements. The process consists of the formation of a chemical bond with the surface of the ceramic that will continue to wet the surface even when it becomes molten. While the basic braze filler metal melts at relatively low temperatures, it is capable of alloying with brazing metals during joining to form ternary alloys with high melting points.

Ceramic-to-metal joining and coatings

Of course, significant improvements in the techniques of metallizing have been made and several new procedures have been developed and evaluated. Also, extensive research on the reactions that occur when a ceramic surface is metallized has contributed to the effectiveness of metallizing.

Metallizing procedures were originally developed to improve the wettability of ceramic surfaces by conventional low-temperature filler metals. Later investigators found that some active metals and their alloys or compounds (e.g. titanium and zirconium) would wet unmetallized ceramic surfaces under certain conditions. Although variations of the so-called active-metal process have been used commercially to produce ceramic-to-metal seals, they have not been accepted to the extent characterized by the metallizing-brazing concept of joining these materials.

To ensure the production of reliable ceramic-to-metal seals, most metallized surfaces are coated with Ni, Cu, or other metals. The metals are usually deposited by electroplating; in some cases, however, the coatings are produced by reducing oxides of the desired metal. These coatings perform several functions, depending on the method used to produce the ceramic-to-metal seal. If the joints are to be brazed with conventional Ag- or Cu-based filler metals, the coatings serve the following purposes:

1. A metallizing layer is composed of metals and residual oxides not completely reduced during sintering. Such a surface is not conducive to good wetting by the brazing filler metal. Plating with Ni or Cu eliminates the adverse effects of the surface on the wetting and flow characteristics of the filler metal.
2. When the metals used for metallizing are not wet readily by low-temperature filler metals, plating provides the surface with a metal easily wet by such brazing filler metals.
3. To a degree, the plated metal acts as a barrier to the penetration of the metallizing layer by the filler metal. Some filler metals react with the metals used for metallizing. If the reaction is allowed to proceed too long, the filler metal may penetrate the metallized coating and lift it away from the ceramic. Metallized coatings are usually plated with Ni to retard penetration and Cu to provide good wetting.

There are numerous methods to coat the metal/ceramic surface:

● Sintered metal powder[205–213]
● Reactive or refractory metal salt[213–215]
● Metal/glass powder[216–220]
● Vapor deposition[209]

However, with the above techniques and the early investigations, joining ceramics to metals by the active metal or active hydride process has advanced significantly. The strengths of joints made by this process are as great as those obtained with joints made by the moly-manganese process. Some difficulty has been experienced in making seals by the active metal or active hydride process in dry H_2. The dew point of H_2 must be extremely low to prevent oxidation of Ti. Producing ceramic-to-metal seals in a vacuum is advantageous in that the parts are outgassed during brazing.

The concept of fabricating ceramic-to-metal joints and seals by the active metal or active hydride process was first applied in the electronics industry. In recent years, however, these joining processes have found other uses to meet the need of high-temperature vacuum-tight seals in the nuclear and aerospace industries.

Fox and Slaughter[142, 221] reveal the use of experimentally-developed active metal alloys for producing ceramic-to-ceramic and ceramic-to-metal joints, some of which are potentially useful in nuclear reactor technology. The filler metals, 68Ti–28Ag–4Be and 49Ti–49Cu–2Be, were originally developed for joining graphite to metal. However, studies indicated that good wetting and flow occurred between 49Ti–49Cu–2Be and Al_2O_3, BeO, and UO_2.

In developing ceramic-to-metal sealing techniques for the production of output windows in high-power microwave tubes, the active metal and active hydride processes were investigated.[222] Ti, Zr, and V foils were evaluated and hydrides of Ti and Zr with foils of Ni, Cu, and 37.5Au–62.5Cu used as the brazing materials. The studies indicated that active metal seals can be developed with strengths greater than those currently obtained, if an intermediate metal with a low expansion coefficient is introduced into the joint. It was also concluded that ceramics having lower expansion coefficients than Al_2O_3 probably cannot be joined by the active metal process.

The characteristics of the moly–manganese, active metal, and active hydride processes are summarized in the following paragraphs:

1. The Mo–Mn process is a multistep sealing process in which the ceramic surface is metallized and plated with one or two metals before brazing can take place. The operations are conducted at a high temperature in a controlled atmosphere of H_2; H_2 firing may discolor some ceramics and produce conductive surfaces. Despite the number of steps required to produce a seal, the moly-manganese process can be automated quite readily, and minor deviations in the process variables can be tolerated.[194, 223–226]

2. The active hydride process is essentially a single-step process in which hydride reduction and brazing proceed simultaneously. Joining in a vacuum or in a controlled atmosphere of H_2 or an inert

gas is accomplished at relatively low temperatures, permitting a fast brazing cycle. This process is more difficult to automate than the moly–manganese process, particularly if the joints are produced in a vacuum. Careful control must be exercised in coating the ceramic with the hydride. Even though the process is considered a one-step process, the hydride process has been supplanted by the active metal process.

3. The active metal process may be a one-step operation like the active hydride process. Joining proceeds at high temperatures in a vacuum or in a controlled atmosphere; vacuum joining is not readily automated.

New ceramic/metal filler metal combinations

New active braze filler metals are constantly undergoing changes and modifications in composition to meet the ever-demanding requirements to permit metals to be joined to ceramics without the ceramic materials being metallized. Some of these Ag-based filler metals (Cusil and Incusil) are ductile and adaptable to brazing metals to such materials as Si_3N_4, PSZ, transformation-toughened Al_2O_3, and SiC, as well as many other refractory materials (Tables 4.11 and 4.12). These brazing processes discussed above are shown in Tables 4.13 and 4.14.

Yamada, *et al*.[127] recently concluded in an investigation that the hermetic seal of Kovar/Fe/Al/Al_2O_3 joints by the Al–Si interlayer method depends not only on the composition and properties of the compound layer, but also on the thermal stress induced by thermal expansion mismatch, especially on the circumferential stress parallel to the joint interface and on the radial stress.

In trying to determine the influence of Cu, Kovar, Mo, and W interlayers on the magnitude and distribution of thermal stresses and on the tensile strength of brazed Si_3N_4–steel joints, Zhou, Bao, Ren, *et al*.[228] found that joints made using low-yield strength/high-ductility interlayers, such as Cu, have lower thermal stresses and higher strengths than those made using low-thermal expansivity/high-yield strength interlayers, such as Mo or W. A composite interlayer comprising Cu and W will produce the lowest thermal stresses during brazing. Increasing the thickness of the interlayer decreases the thermal stresses produced during brazing, since the rigid restraint effect due to the high-yield strength/high-elastic modulus steel substrate is reduced.[228–232]

In other studies by Xian, Si, *et al*.[233] they found that the Ag–Cu eutectic with 5% Ti brazing filler metal for ceramic/metal joints has inherently poor oxidation resistance. It is necessary to improve this filler metal by adding other elements. Their current work has demonstrated that Al is a very effective element in enhancing the oxidation resistance of Ag–Cu–Ti-

Table 4.11 Braze filler materials for ceramics[226]

Name	Composition, (percent)	Liquidus, (^{o}F)^a	Solidus, (^{o}F)^a
Copper (BCu–1)^b	100 Cu	1981	1981
Nicoro (BAu–3)^b	62Cu, 35Au, 3Ni	1886	1832
Cu–Au(1)	65 Cu, 35 Au	1850	1814
Cu–Au(2) (BAu–1)^b	62.5 Cu, 37.5 Au	1841	1805
Cu–Au(3)	60 Cu, 40 Au	1832	1796
Cocuman	58.5 Cu, 31.5 Mn, 10 Co	1830	1645
Cu–Au(4)	50 Cu, 50 Au	1778	1751
Silver	100 Ag	1760	1760
Ticuni	70Ti, 15 Cu, 15 Ni	1760	1670
Nicuman 23	67.5 Cu, 23.5 Mn, 9 Ni	1751	1697
Nicoro 80	81.5 Au, 16.5 Cu, 2 Ni	1697	1670
Nicuman 37	52.5 Cu, 38 Mn, 9.5 Ni	1697	1616
Palcusil 15	65 Ag, 20.3 Cu, 14.7 Pd	1652	1562
Silcoro 75	75 Ag, 20 Cu, 5 Ag	1643	1625
Gapasil 9	82 Ag, 9 Ga, 9 Pd	1616	1553
Palcusil 10	58.5 Ag, 31.8 Cu, 9.7 Pd	1566	1515
Ticusil	68.8 Ag, 26.7 Cu, 4.5 Ti	1562	1526
Silicoro 60	60 Au, 20 Cu, 20 Ag	1553	1535
Palcusil 5	68.5 Ag, 26.8 Cu, 4.7 Pd	1490	1485
Nicusiltin 6^c	62.5 Ag, 29 Cu, 2.5 Ni, 6 Sn	1476	1275
Nicusil 3	71.15 Ag, 28.1 Cu, 0.75 Ni	1463	1436
Cusil (BAg–8)^b	72 Ag, 28 Cu	1436	1436
Incusil 10	63 Ag, 27 Cu, 10 In	1346	1265
Incusil 15	61.5 Ag, 24 Cu, 14.5 In	1301	1166
Georo	88 Au, 12 Ge	673	673
Au–Sn	80 Au, 20 Sn	536	536

^{a}^{o}F = 9/5°C + 32. ^{b}AWS specification. ^{c}AMS 4774A.
From R. Morrell, *Handbook of Properties of Technical and Engineering Ceramics, Part 1: An Introduction for the Engineer and Designer*, London, UK, Her Majesty's Stationery Office, 1985.

based filler metals by forming a layer of protective Cu–Al$_2$O$_3$ film. Other efforts by Moorhead[234] found that the oxidation behavior of Ti-containing brazing filler metals on PSZ and Al$_2$O$_3$ can be improved. The filler metals included Cu–41.1Ag–3.6Sn–7.2Ti, Ag–44.4Cu–8.4Sn–0.9Ti, Ag–41.6Cu–9.7Sn–5.0Ti and Ag–37.4Cu–10.8In–1.4Ti.

In investigations and evaluation of palladium-based filler metals, selected because of their oxidation resistance, ductility and relatively high melting points in brazing ceramics to metals for heat engine applications, Selverian and King[235] studied the brazed joints between Si$_3$N$_4$ and nickel. They found that the joints brazed with the low-palladium filler metals, 70Au–8Pd–22Ni, 93Au–5Pd–2Ni and 82Au–18Ni, had shear strengths of 75 to 105 MPa from 20 to 500°C while the joints brazed with the high-

Table 4.12 Selection guide to metal/ceramic braze filler metals [226]

	Carbon and low-alloy steel	Stainless steel	Tool steel	Nickel, cobalt alloys[a]	Copper[b]
Ceramic, metallized	Copper Copper-gold Cusil Nicusil Palcusil Silver	Copper Copper-gold Palcusil Silcoro	—	—	Copper-gold Cusil Georo Gold-tin Incusil Nicoro Nicusil Palcusil Silver
Carbon[b]	Ticusil	Ticuni Ticusil	Ticuni Ticusil	Ticuni Ticusil	Ticuni Ticusil
Tungsten carbide	Cocuman Copper Nicuman Nicusiltin	Cocuman Copper Nicoro Nicuman Nicusiltin	Cocuman Nicoro Nicuman Nicusiltin Palcusil	Cocuman Nicoro Nicuman Nicusiltin Palcusil	Copper-gold Nicuman Nicusiltin
Copper Copper-gold Cusil Nicoro Nicusil Palcusil Silcoro	—	—	Copper-gold Cusil Incusil Silver	—	—
Ticuni Ticusil	Ticuni Ticusil	Ticusil	—	Ticuni Ticusil	Ticusil
Cocuman Copper Copper-gold Nicoro Nicuman Nicusiltin	Gapasil Ticuni Ticusil	Cocuman Gapasil Nicuman Ticusil	—	Ticusil	Cocuman Copper Nicuman Silver

[a] Corrosion- and heat-resistant alloys.
[b] Includes alloys.
[c] Graphite and diamond.

palladium filler metals, 60Pd–40Ni, 30Au–34Pd–36Ni, and 50Au–25Pd–25Ni, all had shear strengths near zero.

Using copper filler metals, Cu–5Cr, Cu–1Nb, Cu–3V, Cu–5Ti and Cu–10Zr Nakao and his group[236] studied the reactions of Si_3N_4 and refractory metals (W, Mo, Nb, Ta).

Table 4.13 Summary of metallizing methods for ceramics

No.	Metallized layer type	Suitable ceramics	Metallizing materials in finely divided form
1.	Silver, silver/platinum	Hard glasses Most ceramics	Mixtures of $PtCl_4$, Ag_2O, Ag, and Pt
2.	Silver, silver/platinum + fluxing glass	Hard glasses Most ceramics	Mixtures of $PtCl_4$, Ag_2O, Ag, and Pt. + <20% soft glass or flux, e.g. lead borate
3.	Silver + copper oxide	Most ceramics	Ag_2O + <10% CuO, Cu_2O
4.	Molybdenum + silver	Most ceramics	Mo + ~10% Ag_2O
5.	Molybdenum + glass	Most ceramics	Mo + 10–20% glass
6.	Nickel + glass	Most ceramics	NiO + ~10% glass
7.	Copper, copper alloys	Most ceramics	CuO (+ alloying oxides)
8.	Tungsten or molybdenum (+ manganese or iron)	Oxide ceramics BeO, Al_2O_3 (debased type)	W or Mo (+Mn or Fe compounds, e.g. 80% Mo, 20% Mn)
9.	Titanium or zirconium[c] (active metal joints)	Oxide and some nonoxide ceramics	Ti and Zr, or more commonly TiH_4, ZrH_4 or other compounds
10.	Molybdenum + titanium	Pure alumina	Mo + TiN or TiC

[a]Excessive reaction between solder and metallized layer can be prevented by an additional electroplated layer of copper on the metallizing, or by using a solder with a high silver content, e.g. Cu/Ag.
[b]Nickel coating can be achieved by electroplating, or by a second coating as in (6) above but without glass.
[c]This process can be done by direct brazing in vacuum with an 'active metal' braze, using optionally TiH_4 or ZrH_4 as fluxes to wet the ceramic, i.e., a one-stage process.

Table 4.14 Metallizing and brazing conditions for ceramics

Metallizing, temperature, atmosphere	Additional layers required	Suitable solders or brazes
500–900°C in air	None	Sn, Pb, or Pb solders[a]
500–900°C in air	Ag, Ag/Pt	Sn, Pb, or Pb solders[a]
>940°C in air	Ag, At/Pt	Sn, Pb, or Pb solders[a]
1300°C in dry H_2	Ni[b]	Cu, Cu/Ag, Ag
1200–1300°C in dry H_2	Ni[b]	Cu, Cu/Ag, Ag
1300°C in dry H_2	Metallized layer needs buffing	Solders, Cu/Ag
1100°C in air + 900°C in H_2	None	Solders
1450–1650°C in wet H_2	Ni[b]	Cu/Ag, Au/Cu
>1000°C in inert atm.	None	Zr, Ti eutectic brazes
1450–1900°C in wet H_2	Ni[b]	Cu/Ag, etc.

Combinations of brazed SiC[237] and others[238, 239] and various metals for automotive and aerospace applications include two combinations of interest to, for example, the automotive industry, namely: partially-stabilized tetragonal zirconia (PSTZ) to a spheroidal graphite (SG) cast iron, and reaction-bonded silicon carbide (RB–SiC) to a 0.4%C steel.

1. Vacuum-brazed joints between PSTZ and SG cast iron were made with shear strengths of some 200 MPa.
2. As an alternative to a complex alloy filler metal, a simple Ag–Cu eutectic together with a sputtered Ti coating on the PSTZ produced joints with shear strengths up to 135 MPa.
3. Many low-stress failures resulted from excessive oxidation of Ti arising from too high an oxygen partial pressure in the vacuum furnace. The work emphasized the importance of furnace atmosphere when brazes containing Ti are used.
4. Diffusion bonding PSTZ to SG iron at 850°C temperature with Ti and Cu foil interlayers have produced an apparently crack-free joint, but others made at different temperatures or with Ti and Ag foils contained internal cracks.
5. Vacuum-brazed joints made with Ag–Cu + 5%Ti between RB–SiC and a 0.4%C steel failed by lack of wetting of the carbide surface which was attributed to an undesirable reaction between Ti and free Si in the RB–SiC.
6. Surface treatments can improve the processes.

As a result it was found by Bucklow[240] that braze filler metals containing Ti can make strong joints between PSTZ and SG iron provided that excessive oxidation of Ti is prevented. Using Ti demands standards of vacuum practice well above those for conventional vacuum brazing.

The use of sputtered coatings of Ti on oxide/ceramic surfaces prior to vacuum brazing could be commercially viable. Also the improvement of the wettability of Rb–SiC by modifying the surface or the brazing filler metal should be studied further. Finally, the treatment of cast-iron surfaces to avoid excessive diffusion of carbon into the joint and to give better bonding with the filler metal should be explored.[240]

Finally, a research program conducted by Qiu and Xia[238] on the mechanism of joining graphite to a high Al_2O_3 content ceramic (up to 95%) found that the Ti–Ag–Cu braze filler metal was very successful in brazing the combination. For the existence of C, the Ti–Ag–Cu was used to make $AlAg_3$ and Ti_3Al exist in the physical phase of the sealed region, which was an essential factor in the possible forming of a gas-tight seal between graphite and the 95% Al_2O_3 ceramic.

For the existence of $AlAg_3$ and $TiAl_3$, the use of Ti–Ag–Cu could be a significant advance to future research of sealing between 95% Al_2O_3

ceramics and oxygen-free copper,[241] Kovar, beryllium, beryllium oxide,[242] tungsten, molybdenum and stainless steel.[243]

Case histories

Joining dissimilar metals as seen in the previous section is not an easy problem to solve. Many firms have encountered problems in joining cast iron in a furnace exothermic atmosphere. The several kinds of cast irons include white, gray, malleable and ductile. There are many applications in which it is desirable to braze gray, malleable and ductile irons, either to themselves, or to dissimilar metals, however white cast irons are seldom brazed.

Brazing of ductile and malleable irons must not be performed above 760°C since the metallurgical structure may be damaged. In an exothermic atmosphere, type BAg–18 filler metal could be used.

Gray cast iron presents the most problems in brazing. Wetting the surface is difficult as there can be interference from graphite, sand, oxides, etc. The primary method of cleaning is a proprietary molten salt bath employing a reversing direct current. Other methods of cleaning include searing with an oxidizing flame, grit blasting and chemical cleaning. When heating in an oxidizing atmosphere, the silicon, if high, could oxidize, thus interfering with surface wetting of the filler metal.

It has been suggested that there is a specially formulated copper filler metal that has improved the wetting of copper on gray cast iron.

With the many formulations and mixtures of cast iron that are available, there appears to be no easy one-step method of copper furnace brazing gray cast iron (see earlier discussion under 4.2.7).

In another application a firm wished to braze aluminum bronze to naval brass in pure dry hydrogen, with a BAg–8 silver-copper eutectic filler metal. The results initially produced very poor braze joints and the filler metal would not wet and flow well.

The aluminum bronze base metal had an aluminum content of 7–10%. Since aluminum oxidizes very readily, even in the dryest of hydrogen, this was undoubtedly the major problem being encountered. Likewise, if the atmosphere is quite dry, some aluminum could gas off from the surface and aluminize the filler metal (either in wire or powder form). This also would cause the filler metal not to melt properly and leave more of a residue.

One of the simplest methods for solving this problem is to use a grade of flux specifically formulated to handle aluminum bronze base metals, AWS brazing flux type No. 4. When running the part in hydrogen, only a very small amount of this flux would be required, as the dry atmosphere protects it from oxidation; so it should be thinned down and used as a thinner liquid rather than a heavy paste.

If it is desired not to have any flux in the joint area then it is possible to plate the aluminum bronze with either nickel or copper. The thickness of the plating must be thick enough so that the aluminum does not diffuse up to the surface before the brazing filler metal melts and flows. This, of course, is dependent on time and temperature, thus the plating thickness can be varied, depending on these two variables. However, the higher temperature and the longer time to come up to heat, the thicker plating will be required. It is suggested that 0.01 mm thickness would be a good starting place. If, on brazing, the surface of the plating discolors to a grey color and flow is not adequate, a thicker plating will be required.

In brazing naval brass in a hydrogen atmosphere, the vapor pressure of zinc is increased to the point that dezincification can occur. This is true whether the atmosphere be exothermic, pure nitrogen, pure hydrogen, nitrogen/hydrogen, or any of the gas atmospheres. Since the BAg–8 silver–copper eutectic filler metal is being used, brazing would have to be around 788–816°C. To reduce the amount of dezincification, it would be better to use the BAg–18 braze filler metal, which has a lower melting point and thus will lower the amount of zinc vaporization that may occur. Brazing range on this filler metal starts at 718°C and it would be suggested that 760°C would be a suitable brazing temperature. If it is necessary to reduce the amount of dezincification further, the parts could be plated with copper or nickel. Dezincification will be noted as a change from the brass color to a white surface; or, depending on the furnace, brown fumes coming out of the furnace. Some firms have been very successful in brazing cartridge brass, which has only 30% zinc with the BAg–18 at 760°C in a pure dry hydrogen atmosphere.

How would you braze AISI type 304L stainless steel to titanium with nickel 200?

Titanium picks up oxygen quite readily, thus the parts would normally be brazed in a high vacuum of 0.13 Pa or better. In brazing titanium, it is essential that the furnace be leak free and have a clean interior prior to brazing. It is good to run a test cycle with titanium foil to assure that the furnace is clean and leak free prior to brazing. If the titanium foil, when bent exhibits brittleness, the furnace is not adequately clean. When the foil comes out essentially the same color as it goes in, and it can be bent over on itself, ironed flat and opened back up again without cracking, the furnace is adequately clean and brazing can then be accomplished. Before brazing, the 304L, nickel 200, all of the fixtures and any materials that will be in the furnace during the brazing operation (except the titanium and filler metal) should be put into the furnace and adequately outgassed. Many materials have included oxides and other elements that will gas off and be picked up by the titanium. Precleaning assures a better braze assembly. After the 304L is precleaned, it should be plated with approximately 0.01 mm of electrolytic nickel. This plating provides a better surface for the braze and improves wetting and flow. If there is a question about the bond of the nickel plate, it is desirable to test the bond by putting the part into the vacuum furnace and taking it up to 982°C for one to five minutes and then cool down and remove from the furnace. If the nickel plate is blistered, the part should not be brazed, but should be returned to the plater for stripping and replating. The blisters indicate that the cleaning was not sufficient for plating. Whereas if no blistering occurs, the nickel plating is now diffusion welded to the 304L and is ready for brazing. After all the parts, fixtures, and other materials to be put in the furnace are outgassed and cleaned, brazing can be accomplished.

There are several different filler metals that could be used but one that works well is a 68Ag–27Cu–5Pd filler metal. This filler metal has a solidus of 807°C and a liquidus of 810°C. The brazing temperature will be 816–843°C.

When vacuum brazing with filler metal at a temperature of 816–843°C, a partial pressure of argon should be used to prevent evaporation of the silver. Fifty microns partial pressure should be able to prevent the silver from vaporizing. Good, clean argon should be used as any contaminants will be picked up by the titanium. For this reason nitrogen atmosphere should not be used to provide the partial pressure as the titanium will pick up the nitrogen with deleterious effects.

As a note of caution, it is wise to remember that titanium does alloy with most other metals and if taken up in temperature too far, will alloy with and melt down the stainless steel fixturing and other metal parts in the furnace. Thus, it should not be placed directly on metal parts if the temperature is to be taken up in the 927°C range and above. Since the braze in question is to be titanium directly to stainless steel and to nickel, in this case it is essential that the brazing temperature is below the eutectic melting point of the titanium/nickel and titanium/stainless. Thus, the filler metal suggested above would accomplish this.

A manufacturing firm recently attempted to copper braze 305 to 400 series stainless steel inserts. Sometimes the copper wet the 400 series inserts and sometimes it flowed away from them and wouldn't wet, covering the surface of the 300 series stainless steel.

The problem was why wouldn't the copper wet the 400 series stainless steel, even though once in a while it brazed satisfactorily? Why did the copper flow all over the 305 steel but the 400 series stainless did not wet?

In dissecting the problem, there are several points.

Let's take the most serious problem first: the wetting of the 400 series stainless steel. In looking at the part, it is obvious that it was stamped from a sheet and thus would normally have burrs at the edges. These burrs were removed, apparently by tumbling or vibratory polishing. This part has a dull grey finish, which is quite common for parts that have been tumbled or vibratory polished. This color, and the fact that the copper will not wet the 400 series stainless steel, indicate there has been some material smeared on the surface of the part during the deburring operation. It is possible that a stone polishing media was used, or some polishing media had previously been used for deburring parts of aluminum, magnesium, or other parts that contained elements that were not cleaned up by the atmosphere or which produced an oxide in the standard brazing atmosphere. The parts that are not wet by the copper obviously have a contaminated surface. The fact that some parts braze satisfactorily and the copper wets the 400 series stainless steel indicates there isn't an element such as aluminum, or possibly selenium in the stainless steel that would also cause problems with wetting of the surface. The best method is to check the supplier of the 400 series and check the previous operations. If he used a deburring media some contaminant may be the culprit.

The problem with the copper flowing all over the surface of the 305 stainless steel is caused by the very thin 305 stainless steel heating up first and then 400 series heating up later. The brazing filler metal will flow toward the hottest surface. The top of the assembly which gets up to heat first, tends to be wet with the copper.

There are several ways to control the flow of copper. The first approach is to bring the part up to heat more uniformly. Heat shielding of the top of the parts would assist in slowing down the rapid heating of the top end of the assembly. Heat shields are very effective ways of controlling the surface heating of parts where one section heats up much faster than another.

A second approach is to lighten the fixture so the part will heat up more uniformly.

The third approach is to change the dew point of the atmosphere in the furnace. The copper will then have less tendency to flow out on the surface and will stay closer to the joint. This approach is rather difficult, and takes special equipment.

A fourth approach is to change the ratio of hydrogen to nitrogen, keeping the dew point constant.

REFERENCES

1. Funamoto T., Kokura S., Kato S., *et al.*, A study on formation of alloyed layer of low melting temperature on bonding surface of Inconel 738 LC with sputtering, *Quarterly Jrl of the Japan Welding Society*, 3(4), 1985, pp. 207–212.
2. Kato M., Funamoto T., Wachi H., *et al.*, A study of the alloyed layer of low melting temperature bonding surface of IN 738 LC with boron pack cementation, *Quarterly Jrl of the Japan Welding Society*, 5(3), 1987, pp. 84–87.
3. Funamoto T., Kato M., Wachi M., *et al.*, Diffusion weldability of IN 738 LC with alloyed layer on bonding surface by boron pack cementation, *Quarterly Jrl of the Japan Welding Society*, 5(3), 1987, pp. 87–93.
4. Liu S., Olson D. L., Martin G. P., *et al.*, Modeling of brazing processes that use coatings and interlayers, *Welding Journal*, 70(8), Aug 1991, pp. 207s–215s.
5. Tressler R. E., Moore T. L., Crane R. L., Reactivity and interface characteristics of titanium–alumina composites, *Jrl of Mater. Sci.*, 8, 1973, pp. 151–161.
6. Nicholas M. G., Valentine T. M., Waite M. I., The wetting of alumina by copper alloyed with titanium and other elements, *Jrl of Mater. Sci.*, 15, 1980, pp. 2197–2206.
7. Nicholas M. G., Active metal brazing, *Trans. and Jrl of the British Ceramic Society*, 85, 1986, pp. 144–146.
8. Nicholas M. G., Mortimer D. A., Ceramic/metal joining for structural applications, *Mater. Sci. and Tech.*, 1, 1985, pp. 657–665.
9. Schwartz M. M., *Ceramic Joining*, ASM International, Materials Park, Ohio, 1990, 200p.
10. Nicholas M. G., *Joining of Ceramics*, Chapman & Hall, New York, N. Y., 1990, 215p.
11. Wysopal R., Bangs E. R., The importance of braze alloy application in high temperature brazing, presented at *Second International AWS–WRC Brazing Conference and Colloquium*, 52nd Annual AWS Mtg., San Francisco, CA, Apr 26–30, 1971.
12. Schwartz M. M., *Metals Joining Manual*, McGraw-Hill, New York, N. Y., 1979, 565p.
13. Paulonis D. F., Duvall D. S., Owczarski W. A., U.S. Patent 3,678,570.
14. Duvall D. S., Owczarski W. A., Paulonis D. F., T.L.P.: a new method for joining heat-resisting alloy, *Welding Journal*, 53(4), Apr. 1974, pp. 203–214.
15. Schwartz M. M., *Brazing*, ASM International, Materials Park, Ohio, 1989, 439p.
16. Canonico D. A., Slaughter G. M., Brazing and the phenomenon of remelt temperature, presented at *Second International AWS-WRC Brazing Conference and Colloquium, 52nd Annual AWS Mtg.*, San Francisco, CA, Apr 26–30, 1971.
17. Hunt M., *Amorphous Metal Alloys*, ME, Nov 1990, pp. 35–38.

18. Sexton P., DeCristofaro N., *Homogeneous Ductile Brazing Foil*, U.S. Patent 4,148,973.
19. DeCristofaro N., Henschel C., Metglas brazing foil, *Welding Journal*, 57(7), July 1978, pp. 33–38.
20. Rabinkin A., New applications for rapidly solidified brazing foils, *Welding Journal*, 68(10), Oct 1989, pp. 39–46.
21. Kawase H., Takemoto T., Sano M. A., *et al.*, Study of a method for evaluating the brazeability of aluminum sheet, *Welding Journal*, 68(10), Oct 1989, pp. 396s–403s.
22. *Al–Li Alloy Bonding in the Limelight*, Connect, TWI, p6, Dec. 1989.
23. Schoer H., Aluminum brazing alloys for fluxless brazing of aluminum, VAW, *Light Metals 1989*, ed. P. G. Campbell, Minerals, Metals & Matls. Society, 1988, pp. 699–702.
24. Suganuma K., Miyamoto Y., Koizumi M., Joining of ceramics and metals, *Annual Review of Materials Science*, Vol. 18, Ed., Huggins R. A., pp. 47–73, Annual Reviews, Palo Alto, CA, 1988.
25. Dalgleish B., Trumble K., Evans A. G., Microstructural and mechanical characterization of alumina–aluminum bonds, presented at the *90th Annual Mtg. of the American Ceramic Society*, Cincinnati, OH, Paper #12-B-88, May 2, 1988.
26. Naka M., Kubo M., Okamoto I., Joining of silicon nitride with Al-Cu alloys, *J. Mater. Sci.*, 22(12), 1987, pp. 4417–21.
27. Ueki M., Naka M., Okamoto I., Joining and wetting of CaO-stabilized ZrO_2 with Al–Cu alloys, *J. Mater. Sci.*, 23(8) 1988, pp. 2983–88.
28. Ueki M., Naka M., Okamoto I., Wettability of some metals against zirconia ceramics, *J. Mater. Sci. Lett.*, 5(12), 1986, p. 1261.
29. Mahajan Y. R., Kirchhoff S., Rapidly solidified structures of an Al–Zr–Gd alloy, *Scr. Metall.*, 21(8), 1987, pp. 1125–30.
30. Rathner R. C., Green D. J., Joining of yttria–tetragonal zirconia polycrystal with an aluminum–zirconium alloy, *J. Amer. Ceram. Soc.*, 73(4), 1990, pp. 1103–1105.
31. Wigley D. A., Sandefur Jr. P. G., Lawing P. L., Preliminary results on the development of vacuum brazed joints for cryogenic wind tunnel aerofoil models, *Int. Cryogenic Materials Conf.*, San Diego, CA, 10–14 Aug 1981, 16p.
32. Takamori T., Akanuma M., Possible braze compositions for pyrolytic graphite, *Ceram. Bull*, 48 (1969), pp. 734–6.
33. Mortimer D. A., Nicholas M., The wetting of carbon and carbides by copper alloy, *J. Mater. Sci.*, 8 (1973), pp. 640–8.
34. Kim D. H., Hwang S. H., Chun S. S., The wetting and bonding of Si_3N_4 by copper–titanium alloys with other elements, *Ceram. International*, 16 (1990), pp. 333–347.
35. Datta A., Rabinkin A., Bose D., Rapidly solidified copper-phosphorus base brazing foils, *Welding Journal* 63(10), Oct 1984, pp. 14–21.
36. Precious metals boost brazing uses, *Weld. Engr*, Jan 1970, pp. 38–41.
37. Sloboda M. H., Industrial gold brazing alloys, *Gold Bulletin*, 4(1), Jan 1971, pp. 2–8.
38. Fairbanks N. P., High-temperature braze for superalloys, *Technical Rept. AFML–TR–76–155*, AFML, AFSC, Wright-Patterson AF Base, 1976.
39. Tasker A. M., Gold-containing brazing filler metals, *Gold Bull.*, 16(4), Oct 1983, pp. 111–113.

40. Ryan E. J., Doyle J. R., King D. H., Brazing tomorrow's aircraft engines, *Weld. Des. and Fab.*, 52(3), Mar 1979, pp. 108–114.

41. Morgan W. E., Bridges P. L., Development of a brazing alloy for the mechanically alloyed high temperature sheet material incoloy alloy MA 956, *Final Rept. TR–3381* (AD–A118252), Wiggin Alloys Ltd., Hereford, England, AFWAL Cont. F49620–79–C–0008, Sep 1981.

42. Bose D., Datta A., DeCristofaro N., Comparison of gold–nickel with nickel base metallic glass brazing foils, *Welding Journal* 60(10), Oct 1981, pp. 29–34.

43. Brazing with foil cuts costs, speeds assembly, strengthens jet engine parts, *Assembly Engineering*, Mar 1980.

44. Colbus J., Zimmerman K. F., Properties of gold-nickel alloy brazed joints in high temperature materials, *Gold Bull.*, 7(2), Apr 1974, pp. 42–49.

45. Nakamura M., Peteves S. D., Solid-state bonding of silicon nitride ceramics with nickel–chromium alloy interlayers, *J. Amer. Ceram. Soc.*, 73(5), 1990, pp. 1221–27.

46. Wu Y. C., Duh J. G., Eutectic bonding of nickel to yttria-stabilized zirconia, *J. Mater. Sci. Lett.*, 9, 1990, pp. 583–586.

47. Wittmer M., *Materials research society symposium proceedings*, ed. by Giess E. A., Tu K. N. and Uhlmann D. R., Vol. 40, (MRS, Pittsburgh, 1984), p. 393.

48. Wittmer M., Boer C. R., Gudmundson P., *J. Am. Ceram. Soc.*, 65 (1982), p. 149.

49. Ambrose J. C., Nicholas M. G., Young N., *et al.*, Wetting and spreading of Ni-P brazes: effects of workpiece and braze composition, *Mater. Sci. and Tech.*, 6(10), pp. 1021–1031, 1990.

50. Gale W. F., Wallach E. R., Wettability of nickel alloys by boron-containing brazes, *Welding Journal*, 70(3), Mar 1991, pp. 76s–79s.

51. Anderson J., LeHolm R. B., Meaney J. E., *et al.*, *Development of reuseable metallic thermal protection system panels for entry vehicles*, NASA CR-181783, NASI-15646, Aug 1989, 350p.

52. Heiple C. R., Christiansen S. S., Keller D. L., *et al.*, Wettability differences of braze 508 on 304 stainless steel, *Welding Journal* 69(9), Sep 1990, pp. 41–43.

53. Hardesty R., *High temperature Be panel development*, NASA Cr 18177, NASI-18613, 22p, May 1989.

54. Lashko S. V., Sokopov V. I., A brazing alloy for brazing corrosion-resisting steels operating at cryogenic temperatures, *Weldg. Prod.*, Jan 1983, pp. 37–41.

55. Grunling H. W., *High-temperature-resistant ceramic–metal joints*, Schweissen und Schneiden, Feb 1971.

56. Lugscheider E., Tillman W., Development of new active filler metals in a Ag–Cu–Hf system, *Welding Journal* 69(11), Nov 1990, pp. 416s–421s.

57. Nicholas M. G., Mortimer D. A., Jones L. M., *et al.*, Some observations on the wetting and bonding of nitride ceramics, *J. Mater. Sci.*, 25 (1990), pp. 2679–2689.

58. Xian A. P., Si Z.-Y., Joining of Si_3N_4 using $Ag_{57}Cu_{38}Ti_5$ brazing filler metal, *J. Mater. Sci.*, 25 (1990), pp. 4483–4487.

59. Ljungberg L., Warren R., Li C.-H., A novel method for identifying phases formed in brazed and soldered joints, *J. Mater. Sci. Lett.*, 9 (1990), pp. 1316–1318.

60. Peytour C., Berthet P., Barbier F., *et al.*, Interface microstructure and mechanical behaviour of brazed TA6V/zirconia joints, *J. Mater. Sci. Lett.*, 9 (1990), pp. 1129–1131.

61. Barbier F., Peytour C., Revcolevschi A., Microstructural study of the brazed joint between alumina and Ti–6Al–4V alloy, *J. Am. Ceram. Soc.*, 73(6), (1990), pp. 1582–86.

62. Johnson R., IIW Sub-Comm IA Rept. Brazing Group, 1990 Autumn Meeting SCIA-B-185/90.

63. Santella M. L., Horton J. A., Pak J. J., Microstructure of alumina brazed with a silver–copper–titanium alloy, *J. Am. Ceram. Soc.*, 73(6), (1990), pp. 1785–87.

64. Kivilahti J., Paulasto M., Turunen M., The influence of titanium content of the filler alloy and brazing parameters on the microstructures and properties of titanium/Al$_2$O$_3$ joints, *International Conf. on Evolution of Advanced Materials*, Associazione Italiana di Metallurgia, Italy, 1989, pp. 347–352.

65. Onzawa T., Suzumura A., Ko M. W., Brazing of titanium using low-melting-point Ti-based filler metals. *Welding Journal*, 69(12), Dec 1990, pp. 462s–467s.

66. Guezawa Y., Advances in joining newer structural materials, *IIW Meeting and Proceedings of International Conf.*, Montreal, Canada, 1990, pp. 223–228.

67. Chatterjee S. K., Mingxi Z., Tin-containing brazing alloys, *Welding Journal*, 69(10), Oct 1990, pp. 37–42.

68. *Silver Brazing Alloys – Cadmium Containing, Health and Safety*, The Sheffield Smelting Co. Ltd., Dec 1980.

69. Roberts P. M., Recent developments in cadmium-free silver brazing alloys, *Weld. J.*, 57(10), Oct 1978, pp. 23–30.

70. Roberts P. M., *Weld. Des. and Fab.*, Jan/Feb 1979, pp. 35–46.

71. Miller V. R., Falke W. L., *Report of Investigations 8783*, U. S. Dept. of Interior, 1983.

72. Long J. B., Evans C. J., *Tin and its Uses*, 118, pp. 12–16, 1978.

73. Rice R. W., *Joining of Ceramics, Advances in Joining Technology*, Burke J. J., Gorum A. E., Tarpinian A., eds. Boston: Brook Hill, 1976.

74. Cawley J. D., Joining of ceramic matrix composites, *Am. Ceram. Soc. Bull.*, 68(9), 1989, pp. 1619–1623.

75. Borom M. P., The mechanical and chemical aspects of glass sealing, *Glass Industry*, 59(3), pp. 12–16, 1978.

76. Kirchner H. P., Conway Jr. J. C., Segall A. E., Effect of joint thicknesses and residual stresses on the properties of ceramic adhesive joints, I, Finite element analysis of stresses in joints, *J. Am. Ceram. Soc.*, 70(2), 1987, pp. 104–109.

77. Zdaniewski W. A., Conway Jr. J. C., Kirchner H. P., Effect of joint thickness and residual stresses on the properties of ceramic adhesive joints, II, Experimental results, *J. Am. Ceram. Soc.*, 70(2), 1987, pp. 110–118.

78. Hauth III W. E., Ceramic-to-ceramic sealing of large shapes, *Am. Ceram. Soc. Bull.*, 58(6), pp. 584–586, 1979.

79. Cauley J. D., Knapp J. T., Silicate brazing of a high-purity alumina, MR9105, EWI, May 1991.

80. Lugscheider E., Partz K-D, High temperature brazing of stainless steel with nickel base filler metals BNi–2, BNi–5 and BNi–7, *Welding Journal*, 62(6), June 1983, pp. 160s–164s.

81. Lugscheider E., Partz K.-D., Kruger J., *Rechnerunterstutzte Gefugebeurteilung an Hochtemperaturlotverbindungen-Prozessoptimerung zur Qualitatssicherung und Bauteilsicherheit*, DVS-Berichte 75, Dusseldorf: Deutscher Verlag fur Schweisstechnik, 1989.

82. Lugscheider E., Knotek O., Klohn K., Development of nickel–chromium–silicon base filler metals, *Welding Journal*, 57(10), Oct 1978, pp. 319s–323s.

83. Lugscheider E., Krappitz H., The influence of brazing conditions on the impact strength of high-temperature brazed joints, *Welding Journal* 65(10), Oct 1986, pp. 261s–266s.

84. Singleton O. R., A look at the brazing of aluminum – particularly fluxless brazing, *Welding Journal*, 49(11), Nov 1970, pp. 843–849.

85. Warner J. C., Weltman W. C., The fluxless brazing of aluminum radiators, *Welding Journal*, 58(3), Mar 1979, pp. 25–32.

86. Byrnes, Jr. E. R., Vacuum fluxless brazing of aluminum, *Welding Journal*, 50(10), Oct 1971, pp. 712–716.

87. Winterbottom W. L., Process control criteria for brazing aluminium under vacuum, *Welding Journal* 63(10), Oct 1984, pp. 33–39.

88. *Aluminum Brazing Furnace*, FB Series, E3007, ULVAC Corp., 1979 (10 pages).

89. Ashburn L. L., Furnace design considerations for aluminum brazing under vacuum, *Welding Journal* 62(10), Oct 1983, pp. 45–54.

90. How Philco-Ford vacuum brazes heat exchangers, *Metal Progress*, Sep, 1972, pp. 61–68.

91. Beuyukian C. S., Fluxless brazing of Apollo coldplate development and production, *Welding Journal* 49(9), Sep 1968, pp. 710–719.

92. *Metals Handbook, 9th Ed., Vol. 6, Welding, Brazing and Soldering*, American Society for Metals, Metals Park, OH, 1983, pp. 929–995.

93. Mehrkam Q. D., Never underestimate the power of a salt bath furnace, *Weld. Des. and Fab.*, AJAX Reprint No. 184, Mar 1968.

94. *Aluminum Brazing Handbook*, 1st Ed., The Aluminum Association, New York, Jan 1971 (78 pages).

95. Dickerson P., Here's what you can do with dip brazing, *Metal Progress*, May–Jun, 1965, pp. 1–15.

96. Long J. V., Cremer G. D., *Beryllium: brazing and welding*, SAE Paper 680651, SAE Aeronautic and Space Engineering and Manufacturing Meeting, Los Angeles, CA, Oct 7–11, 1968.

97. Cremer G. D., Woodward J. R., Grant L. A., *Beryllium Brazing Technology*, SAE Paper 670805, SAE Aeronautic and Space Engineering and Manufacturing Meeting, Los Angeles, CA, Oct 2–6, 1967.

98. McFayden A. A., Kapoor R. R., Eager T. W., Effect of second phase particles on direct brazing of alumina dispersion hardened copper, *Welding Journal* 69(11), Nov 1990, pp. 399s–407s.

99. Brentnall W. D., Stetson A. R., Metcalfe A. G., *Joining of Superalloy Foils for Hypersonic Vehicles*, Tech. Rep. AFML TR-68-299, Contr. F33(615)-67-C-1211, Oct, 1968.

100. Schwartz M. M., *Metals Joining Handbook*, McGraw-Hill, New York, 1979.

101. Kenyon N., Hrubec R. J., Brazing of a dispersion strengthened nickel base alloy made by mechanical alloying, *Welding Journal* 53(4), Apr 1974, pp. 145s–151s.

102. Schwartz M. M., Brazed honeycomb structures, *WRC Interpretive Bull.* 182, Apr 1973.

103. McCown J., Wilks C., Schwartz M. M., Norton A., *Final report on development of manufacturing methods and processes for fabricating refractory metal components*, ASD Proj. 7-937, AF33(657)-7276, Sep 1963.

104. *Machine design, materials reference issue*, 57(8), Apr 18, 1985.

105. Bucklow I. A., Joining a Ni-based creep-resistant (ODS) alloy by brazing, *The Welding Inst., IIW Meeting and Proceedings on Joining*, Montreal, Canada, 1990, pp. 293–298.
106. Bucklow I. A., Diffusion bonding an Fe-based ODS alloy, *The Welding Inst., IIW Meeting and Proceedings on Joining*, Montreal, Canada, 1990, pp. 299–304.
107. Freedman A. H., Mikus E. B., High remelt temperature brazing of columbium honeycomb structures, *Welding Journal* 45(6), Jun 1966, pp. 258s–265s.
108. Cole N. C., Gunkel R. W., Koger J. W., Development of corrosion resistant filler metals for brazing molybdenum, *Welding Journal*, 52(10), Oct 1973, pp. 466s–473s.
109. A. Semeniuk, Brady G. R., Properties of TZM and nuclear behavior of TZM brazements, *Welding Journal*, 53(10), Oct 1974.
110. Gilliland R. G., Slaughter G. M., The development of brazing filler metals for high temperature service, *Welding Journal*, 48(10), Oct 1969, pp. 463s–468s.
111. McDonald M. M., Keller D. L., Heiple C. R., *et al.*, Wettability of brazing filler metals on molybdenum and TZM, *Welding Journal*, 68(10), Oct 1969, pp. 389s–395s.
112. Schwartz M. M. The fabrication of dissimilar metal joints containing reactive and refractory metals, *WRC Interpretive Bull.* 210, Oct 1975.
113. *Advanced Joining Technology*, Rept. of the committee of advanced joining technology, Publication N MAB-387, 1982.
114. Slaughter G. M., Werner W. J., Gilliland R. G., *et al.*, Recent advances in brazing, *5th Nat. SAMPE Tech. Conf.*, Vol. 5, Oct 9–11, 1973, Kiamesha Lake, N Y, pp. 115–123.
115. *Metals Handbook, 9th Ed., Vol. 4, Heat Treating*, American Society for Metals, Metals Park, OH, 1981, p. 675.
116. Zappfe C. A., *Stainless Steels*, American Society for Metals, Metals Park, OH, 1949.
117. *Metals Handbook*, 8th Ed., Vol 1, 1961, and 10th Ed., Vol 6, 1993, American Society for Metals, Metals Park, OH.
118. Burrows C. F., *Vacuum brazing gas-quenching process for the manufacturing of metal materials, components and hardware*, Department of the Army, Contr. DAAA15–68–C–0659, Final Report, Feb 1970.
119. Schwartz M., Application for gold-base brazing alloys, *Gold Bull.*, 8(4), Oct 1975.
120. Lovell D. T., Elrod S. D., *Development and evaluation of the aluminum-brazed titanium system, Vol. 1, Program summary, Final Rep.* FAA–SS–73–5–1, May 1974.
121. Bales T., *et al.*, Weld-brazing of titanium, *5th Nat. SAMPE Tech. Conf.*, Vol 5, Oct 9–11, 1973, Kiamesha Lake, NY, pp. 481–500.
122. Wells R. R., Low-temperature large-area brazing of damage-tolerant titanium structures, *Welding Journal*, 54(10), Oct 1975, pp. 348s–356s, and Tech. Rep. AFML–TR–75–50, Contr. F–33(615)–73–C–5161, May 1975.
123. Smeltzer C. E., Hammer A. N., *Titanium braze system for high temperature applications*, AFML–TR–76–145, Contr. F-33(615)–74–C–5118, Aug 1976.
124. Lugscheider E., Ruiz L. M., Brazing of titanium, paper 23, *International Conf. on Advances in Joining and Cutting Processes*, Harrogate, 30 Oct–2 Nov 1989, Abington Publish., pp. 414–423.

125. Lugscheider E., Martinez L., Brazing of Ti–Al6V4 with Al-, Ag-, Au- and Ti-based filler-metals, *AWS Annual International Brazing Conf.*, Chicago, 1987 and *DVS, brazing, high temperature brazing and diffusion welding*, Berichte 125, Dusseldorf, 19–20 Sep 1989, pp. 123–126.

126. Gamer N., Richardson J., *Investigation of ductile brazing alloy compositions for use in joining titanium and its alloys*, Wesgo Tech. Rep. 1492, Apr 1971, pp 1–32.

127. Beal R. E., Saperstein Z. P., Development of brazing filler metals for zircaloy, *Welding Journal*, 50(7), Jul. 1971, pp. 275s–291s.

128. Amato I., Ravizza M., Some developments in Zircaloy brazing technology, *Energ. Nucl.* (Milan), 16(2), 1969, pp. 35–39.

129. Thorsen K. A., Wetting of cemented carbides, Lyngby, DVS, brazing, high temperature brazing and diffusion welding, *Berichte 125*, Dusseldorf, 19–20 Sep 1989, pp. 66–71.

130. Kohl W. H., *Handbook of Materials and Techniques for Vacuum Devices*, New York, Reinhold Publishing Co., 1967.

131. Lewis C. F., Putting ceramics together, *Manf. Engineering*, Feb 1988, pp. 31–34.

132. Loehman R.E., *et al.*, Why metals adhere to Si_3N_4, *J. Am. Ceram. Soc.*, 1988.

133. Vilpas M., *Joining of ceramics for high-temperature applications*, NASA TT-20030, N87–29678, NTIS HC A03/MF A01, Oct 1987, translation of 'Korkeissa Lampotiloissa Kaytettavien Keraamisten Materiaalien Liittaminen,' Rept. UTT-TIED-481, Tech. Res. Center of Finland, Espoo, Aug 1985.

134. Tomsia A. P., Pask J. A., Chemical reactions and adherence at glass/metal interfaces: an analysis, *Dent. Master.*, 1(2), 1986, pp. 10–16.

135. Cole S. S., Sommer G., Glass-migration mechanism at ceramic-to-metal seal adherence, *J. Am. Ceram. Soc.*, 44(6), Jun 1961, pp. 265–271.

136. Rice R. W., Joining of ceramics, in *Advances in Joining Technology*, Chestnut Hill, MA, Brookhill Publishing Co., 1976, pp. 69–111.

137. Klomp J. T., Bonding of metals to ceramics and glasses, *Am. Ceram. Soc. Bull.*, 51(9), 1972, pp. 683–688.

138. Twentyman M. E., Mechanism of glass migration in the production of metal-ceramic seals, *J. Mater. Sci.*, 10, 1975, pp. 765–776.

139. Morrell R., Joining to other components, Part I, An introduction for the engineer and designer, *Handbook of Properties of Technical and Engineering Ceramics*, London, Her Majesty's Stationery Office, Section 3.5, 1985, pp. 267–278.

140. Tallman R. L., Neilson Jr. R. M., Mittl J. C., *et al.*, *Joining silicon nitride-based ceramics: a technical assessment*, EG&G Idaho, Inc., ECG-SCM-6572, DE84-011356, DOE Cont. DE-AC07-761D01570, Mar 1984.

141. Pattee H. E., Joining ceramics to metals and other materials, *Welding Research Council Interpretive Rept. 178*, Nov 1972.

142. Fox C. W., Slaughter G. M., Brazing of ceramics, *Welding Journal*, 43(7), July 1964, pp. 591–597.

143. Weymueller C. R., Braze ceramics to themselves and to metals, *Weld. Des. & Fab.*, Aug 1987, pp. 45–48.

144. Loehman R. E., Mecartney M. L., Rowcliffe D. *Silicon Nitride Joining*, AFOSR-TR-82-0304, Annual Rept. for 12/1/80 through 1/31/82, SRI International, Menlo Park, CA, Feb 1982.

145. Santella M., Brazing of titanium-vapor coated silicon nitride, *Adv. Ceram. Matls.*, 3(5), 1988, pp. 457–462.

146. Weiss S., Adams Jr. C. M., The promotion of wetting and brazing, *Welding Journal*, 46, 1967, pp. 49s–57s.

147. Brush, Jr. E. F., Adams, Jr. C. M., Vapor-coated surfaces for brazing ceramics, *Welding Journal*, 68, 1968, pp. 106s–114s.

148. Private communication with Ceramics Unit of the Matls. Technology Grp., Idaho National Engineering Laboratory, U.S. Dept. of Energy, Idaho Falls, ID.

149. Jack K. H., Review: sialons and related nitrogen ceramics, *J. Mater. Sci.*, 11, 1976.

150. Loehman R. E., Preparation and properties of yttrium–silicon–aluminum oxynitride glasses, *J. Am. Ceram. Soc.*, 62, Sep–Oct 1979, pp. 9–10.

151. Rossi G., Joining of non-oxide ceramics, *ASM 1988 World Matls. Cong.*, ASM–ACS Joining Session, Chi, IL, 1988.

152. Lugscheider E., Boretius M., Lison R., Active brazing of silicon carbide to steel using a thermal-stress reducing metallic interlayer, IIW-Doc. 1–860–88, presented at *19th AWS Braze and Solder Conf.*, Am. Weld. Soc., 1988.

153. Cassidy R. T., *et al.*, *Bonding and fracture of titanium-containing braze alloys to alumina*, Monsanto Res. Corp., MLM–3431 (OP) and MLM–3394 and DE87002197 and DE87009195, U.S. Dept. of Energy Cont DE–ACOE–76DP00053, Oct 1987 and Oct 1986.

154. Canonico D. A., Cole N. C., Slaughter G. M., Direct brazing of ceramics, graphite and refractory metals, *Welding Journal*, 56(8), Aug 1977, pp. 31–38.

155. Grunling H. W., *High temperature-resistant ceramic/metal joints*, DVS Working Subcommittee UG30.1, Schweissen und Schneiden. Feb 1973.

156. Meyer A., *The adhesion mechanism of molybdenum/manganese metallized coatings on corundum ceramics*, DKG Rept. 42, Vol. 11, pp. 405–415, and Vol 12, pp. 14–16, 1965.

157. Moorhead A. J., Keating H., Direct brazing of ceramics for advanced heavy-duty diesels, *Welding Journal*, 65(10), Oct 1986, pp. 17–31.

158. Moorhead A. J., Becher P. F., Development of a test for determining fracture toughness of brazed joints in ceramic materials, *Welding Journal*, 66(1), Jan 1987, pp. 26s–32s.

159. Dalgleish B. J., Liu M. C., Evans A. G., The strength of ceramics bonded with metals, *Acta Metallurgica*, 36(8), Aug 1988, pp. 2029–2035.

160. *Brazing Manual*, 3rd ed., Am. Weld. Soc., Miami, FL, 1976, pp. 262–263.

161. Becher P. F., Halen S. A., Joining of Si_3N_4 & SiC ceramics via solid-state brazing, *Proceedings of the DARPA/NAVSEA Ceramics Gas Turbine Demo. Engine Prog. Rev.*, Columbus, OH, MCIC, 1978, pp. 649–653.

162. Joining of silicon nitride ceramics with silicon nitride or other ceramics, *Chem. Abstracts*, 99, 1983, p. 282.
    (Ref. No. 92701z, Daido Steel Co., Ltd., Jpn., Kokai Tokkyo Koho JP 5855.381[83 55.38] [C1.CO4B37/00], Apr 1, 1983, Appl. 81/148,722, Sep 22, 1981).

163. Goodyear M. V., Ezis A., Joining of turbine engine ceramics, *Advances in Joining Tech.*, Chestnut Hill, MA, Brookhill Pub. Co., 1976, pp. 113–154.

164. Joining of silicon nitride ceramics, *Chem. Abstracts*, 99, 1983, 239 (Ref. No. 57/49z, Daido Steel Co., Ltd. Jpn, Kokai Tokkyo Koho JP 48 32, 082[83 32082] [C1, CO4B3/00], Feb 24, 1983, Appl. 81/130,055, Aug 21, 1981).

165. Becher P. F., Halen S. A., Solid-state bonding of Si_3N_4, *Ceram. Bull.*, 58, 1979, pp. 582–586.

166. Johnson S., Rowcliffe D., *Silicon nitride joining*, AFOSR-TR-83-114, F49620-81K0001, SRI International, Menlo Park, CA, 1983.

167. Ellsner G., Diem W., Wallace J. S., Microstructure and mechanical properties of metal-to-ceramic and ceramic-to-ceramic joints, *Surfaces and Interfaces in*

Ceramic and Ceramic-Metal Systems, New York, Plenum Press, 1981, pp 629–639.

168. Naka M., *et al.*, Non-oxide ceramics (Si_3N_4, SiC) joint made with amorphous $Cu_{50}Ti_{50}$ and $Ni_{24.5}$ filler metals, Technical Note, *Transactions of JWRI*, 12(2), Dec 1983, pp. 177–183.

169. Iino Y., Taguchi N., Interdiffusing metals layer technique of ceramic-metal bonding. *J. Mater. Sci. Ltrs.*, 7, 1988, pp. 981–982.

170. Moorhead A. J., Welding and brazing of film probe sensor assemblies, *Welding Journal*, 62(10), Oct. 1983, pp. 17–27.

171. Kapoor R. R., Eagar T. W., *Oxidation behavior of brazing alloys for metal/ceramic joints*, Prog. Rept. 9–41–88, Dept. of Mat. Sci. & Engr., MIT Indust. Liaison, Cambridge, MA, 1988.

172. Mizuhara H., Huebel E., Joining ceramic to metal with ductile active filler metal, *Welding Journal*, 65(10), Oct 1986, pp. 43–51.

173. Mizuhara H., Mally K., Ceramic-to-metal joining with active brazing filler metal, *Welding Journal*, 64(10), Oct 1985, pp. 27–32.

174. *Brazing Handbook*, 4th Ed., Am. Weld. Soc., Miami, FL, 1991, 493p.

175. *Designing Interfaces for Technological Applications: Ceramic-Ceramic, Ceramic-Metal Joining*, Ed. Peteves S. D., 310p, Elsevier Sci. and Publishing Co., Inc., New York, NY, 1989.

176. Nicholas M. G., Crispin R. M., The role of titanium in active metal and activated brazing of alumina, *Proceedings of 2nd International Colloquium on the Joining of Ceramics, Glass, and Metals*, Bad Nauheim, 27–29 Mar 1985, Deutsche Keramische Gesellschaft e.V., Bad Honnef, FRG.

177. Mizuhara H., Vacuum brazing ceramics to metals, *Advanced Materials and Processes*, 131(2), Feb 1987, pp. 53–55.

178. Nolte H. J., Spurck R., Metal-ceramic sealing with manganese, *Telev. Engr.*, 1(11), 1950, pp. 14–16, 18, 39.

179. Nicholas M. G., Reactive metal brazing of ceramics, *Scandinavian Journal of Metallurgy*, 20 (1991), pp. 157–164.

180. Norton M. G., Kajda J. M., Steele B. C. H., Brazing of aluminum nitride substrates, *J. Mater. Res.*, Vol. 5, No. 10, Oct 1990, pp. 2172–2176.

181. Pattee H. E., Evans R. M., Monroe R. E., *Joining ceramics and graphite to other materials*, NASA SP-5052, 1968.

182. Donnelly R. G., Slaughter G. M., The brazing of graphite, *Welding Journal*, 41(5), May 1962, pp. 461–469.

183. *Metals Handbook, 9th Ed., Vol. 6, Welding, Brazing, and Soldering*, Am. Soc. for Metals, Metals Park, OH, 1983, p. 1023.

184. Super ceramic–steel ceramic–graphite bonds, *High-Tech Matls. Alert*, 5(5), May 1988, pp. 1–2.

185. Kochetov D. V., *et al.*, Investigation of heat conditions in the brazing of graphite to steel, *Weld. Prod.*, 21(3), 1974, pp. 15–18.

186. Anikin L. T., *et al.*, The high temperature brazing of graphite, *Weld. Prod.*, 24(1), 1977, pp. 39–41.

187. Anikin L. T., *et al.*, The high temperature brazing of graphite using an aluminum brazing alloy, *Weld. Prod.*, 24(7), 1977, pp. 23–25.

188. Amato I., Cappelli P., Martinengo P., Brazing of special grade graphite to metallic substrates, *Welding Journal*, 53(10), Oct 1974, pp. 623–628, and Fiat–DCR Laboratori Centrali, SMT–25, Orbassano, Italy, Feb 1974.

189. Rabin B. H., Joining of fiber-reinforced SiC composites by in situ reaction methods, *Master. Sci. and Engr.*, A130, (1990) L1–L5.

190. Morimoto H., Tanaka T., Saito T., *et al.*, Effects of brazing temperature on SiC fiber reinforced aluminum alloy matrix composites, advances in joining newer structural materials, *Proceedings of International Conf. on IIW*, Pergamon Press, Inc., Section 1.6, p137, 1990.

191. Khorunov V. F., Kutchuk-Yatsenko V. S., Dykhno I. S., *et al.*, Brazing of sheet composite materials with aluminum matrix, advances in joining newer structural materials, *Proceedings of International Conf. on IIW*, Pergamon Press, Inc., Section 1.7, p143, 1990.

192. Young J. G., Smith A. A., Joining dissimilar metals, *Weldg. and Metal Fabrication*, July, 1959.

193. Peckner D., Joining dissimilar metals, *Materials in Design Engineering*, 56(2), 1962, pp. 115–122.

194. Schwartz M. M., The fabrication of dissimilar metal joints containing reactive and refractory metals, *WRC Interpretive Bull.* 210, Oct 1975.

195. Minegishi T., Sakurai T., Morozumi S., Electric-assisted and field-depressed segregation of reactive metals to the bonding interface in braze alloy joining, *J. Mater. Sci.*, 26(2), 1991, pp. 5473–5480.

196. Santella M. L., *Joining of ceramics for heat engine applications*, ORNL/TM-11489, DE 91-001004, Semiannual Rept. 4/89 through 9/89, pp. 227–231.

197. Kang S., Kim H., Selverian J. H., *Analytical and experimental evaluation of joining silicon nitride to metal and silicon carbide to metal for advanced heat engines application*, ORNL/TM-11489, DE 91-001004, Semiannual Rept. 4/89 through 9/89, pp. 232–247.

198. Hopper A. T., Ahmad J., Rosenfield A., *Analytical and experimental evaluation of joining ceramic oxides to ceramic oxides and ceramic oxides to metal for advanced heat engine application*, ORNL/Tm-11489, DE 91-001004, Semiannual Rept. 4/89 through 9/89, pp. 248–267.

199. Schwartz M. M., *Modern Metal Joining Techniques*, Wiley, New York, NY, 1969.

200. Pincus A. G., Mechanism of ceramic-to-metal adherence, adherence of molybdenum to alumina ceramics, *Ceram. Age*, 63(3), Mar 1954, pp. 16–32.

201. Suganama K., Okamoto T., Shimada M., New method for solid-state bonding between ceramics and metals, *Communications of the Am. Ceram. Soc.*, Jul 1983, C–117–118.

202. Burgess J. F., Neugebauer C. A., *Direct Bonding of Metals to Ceramics for Electronic Applications, Advances in Joining Technology*, Burke *et al.*, Eds., 1974.

203. Twentyman M. E., The use of metallizing paints containing glass or other inorganic bonding agents, *J. Mater. Sci.*, 10, 1975, pp. 791–798.

204. Naka M., Sampath K., Okamoto I., *et al.*, Influence of brazing conditions on shear strength of alumina-Kovar joint made with amorphous $Cu_{50} Ti_{50}$ filler metal, *Transactions of Japanese Welding Res. Inst.*, Dec 1983.

205. Nolte H. J., Metallized Ceramic, U. S. Patent 2,667,432, Jan 26, 1954.

206. Ettre K., Automatic metallizing, *Ceram. Age*, 81(6), 1965, pp. 57–60.

207. Staumanis M. E., Schlechten A. W., Titanium coatings on metals and ceramics, *Metallurgia*, 10(10), 1956, pp. 901–909.

208. Heritage R. J., Balmer J. R., Metallizing of glass, ceramic, and plastic surfaces, *Metallurgia*, 47(4), 1953, pp. 171–174.

209. LaForge L. H., Application of ceramic sections in high-powered pulsed klystrons, *Am. Ceram. Soc. Bull.*, 35(3), Mar 1956, pp. 117–122.

210. Phillips W. M., *Metal-to-ceramic seals for thermionic converters*, a literature survey, JPL/Report 32–1420, Nov 1969.

211. Spurck R. F., *et al.*, Use metallizing tape for high quality ceramic-to-metal seals, *Ceram. Ind.*, 79(3), 1962, pp. 88–91, 94.
212. Tentarelli L. A., White J. M., Buck R. W., *Low-temperature refractory metal-to-ceramic seals*, Final Rept. ECOM–03734–F, Cont. DA–36–039–AMC–03734(E) Sperry Rand Corp., Gainesville, FL, Apr 1966.
213. Varadi P. F., Dominiguez R., Tungsten metallizing of ceramics, *Am. Ceram. Soc. Bull.*, 45(9), 1966, pp. 789–791.
214. Kohl K. H., Rice P., *Electronic tubes for critical environments*, Tech. Rept. TR–57–434, Cont 33(616)–3460, Menlo Park, CA, Stanford Res. Ctr., Mar 1985.
215. Quinn R. A., Karlak R. F., *Method of coating a body with titanium and related metals*, U.S. Patent 3,022,021, Feb 20, 1962.
216. Lindquist C., Here are materials and techniques for coating ceramics, *Ceram. Ind.*, 69(1), 1957, pp. 85–88.
217. Lindquist C., How to apply conductive coatings, *Ceram. Ind.*, 69(2), 1957, pp. 101–103.
218. Lindquist C., How to dry and fire conductive coatings, *Ceram. Ind.*, 69(3), 1957, pp. 144–150.
219. Sedenka A., Studies of the effect of firing temperature on the adherence of silver to ceramics, *Am. Ceram. Soc. Bull.*, 38(4), 1959, pp. 139–141.
220. Knecht W., Application of pressed powder technique for production of metal-to-ceramic seals, *Ceram. Age*, 63(2), 1954, pp. 12–13.
221. Nogawa H., *Ceramic Processing–State of the Art of R&D in Japan*, ASM International, Metals Park, OH, 1988.
222. Grimm A. C., Strubhar P. D., *Dielectric-to-metal technology study*, Tech. Doc. Rept. TDR–63–472, Cont AF30(602)–2682, RCA, Lancaster, PA, Oct 1963.
223. Van Houten G. R., Ceramic-to-metal bonds, *Matls. Des. Engr.*, Dec 1958, pp. 112–114.
224. Kutzer L. G., Joining ceramics and glass to metals, Carborundum Co., *Matls. Des. Engr.*, Jan 1965, pp. 106–110.
225. The promise of ceramics, *AM&P Met. Prog.*, 131(1), Jan 1987, pp. 44–50.
226. White G. L., Oakley P. J., Industrial ceramics – a survey of materials, applications, and joining processes, Members Rept. Summary – Rept. 302/1986, *Welding Institute Research Bulletin, 9444.1/85/476.3*, Jun 1986, pp. 1–24.
227. Yamada T., Horino M., Yokoi K. *et al.*, Hermetic seal of ceramics and metals joints by an Al-Si interlayer, *Mater J., Sci. Lett.*, 10(1991), pp. 807–809.
228. Zhou Y., Bao F. H., Ren J. L., *et al.*, Interlayer selection and thermal stresses in brazed Si_3N_4–steel joints, *Matls. Sci. and Tech.*, Vol. 7, pp. 863–868, Sep 1991.
229. Loehman R. E., Johnson S. M., Moorhead A. J., *Structural Ceramics Joining*, Ceramic Eng. and Sci. Proc., 10(11–12), 1989.
230. Loehman R. E., Joining engineering ceramics, advances in joining newer structural materials, *Proceedings of International Conf. on IIW*, Pergamon Press, Inc., Section 1.2, 1990, 83–97.
231. Levy A., Thermal residual stresses in ceramic-to-metal brazed joints, *J. Am. Ceram. Soc.*, 74, No. 9, 1991, pp. 2141–2147.
232. Fanghan B., Jialie R., Yunhong Z., *et al.*, Properties of vacuum brazed Si_3N_4/Steel joint using active brazing filler metal, advances in joining newer structural materials, *Proceedings of International Conf. on IIW*, Pergamon Press, Inc., Section 1.3, 117–122, 1990.

233. Xian A. P., Si Z. Y., Zhou L. J. *et al.*, An improvement of the oxidation resistance of Ag-Cu eutectic-5%at Ti brazing alloy for metal/ceramic joints, *Mater. Lett.*, 12(1991), pp. 84–88.
234. Moorhead A. J., Kim H-E., Oxidation behaviour of titanium-containing brazing filler metals, *J. Mater. Sci.*, 26, No. 15, 1991, pp. 4067–4075.
235. Selverian J. H., Kang S., Ceramic-to-metal joints brazed with palladium alloys, *Welding Journal*, 71 (1), Jan 1992, pp. 25s–33s.
236. Nakao Y., Nishimoto K., Saida K., Bonding of heat-resisting fine ceramics to metals, recent trends in welding sci. and tech.; *TWR '89; Trends in Welding Research, Proceedings of 2nd International Conf. on*, ASM Intl., 1990, pp. 535–539.
237. McDermid J.R., Drew R. A., Brazing of reaction-bonded silicon carbide and Inconel 600 with an iron-based alloy, *J. Mater. Sci.*, 25, 1990, pp. 4804–4809.
238. Qiu C., Xia H., A research on the mechanism of joining graphite to ceramics of 95% Al_2O_3, advances in joining newer structural materials, *Proceedings of International Conf. on IIW*, Pergamon Press, Inc., Section 1.5, 1990, pp. 129–135.
239. Xian A-P., Si Z-Y., Hou Z-J., Oxidation of $Ag_{57}Cu_{38}Ti_5$ brazing filler metal for metal-ceramic joint, *J. Mater. Sci. Lett.*, 10, 1991, pp. 726–727.
240. Bucklow I. A., Ceramic/metal bonding: a study of joining zirconia to cast iron, and silicon carbide to steel, *Welding Inst. Bull.*, 414/1990, Sep/Oct 1990.
241. Qiu C., Mechanism research on seal between 95% alumina ceramics and activated metal of oxygen-free copper, *Proc. of 4th Ann. Meet. of PRC Elec. Soc. E-vacu. Asso.*, pp. 155–156, 1982.
242. C. Qiu, Technical Improvements on Weld of R-1 Beryllium and Beryllium Oxide Ceramics, Proc. of Exp. Exh. Mtg. on Bra. Mat. & Its Appl., I, 8, 1984.
243. C. Qiu, Advantages of Titanium and its Alloy in Welding Application, Proc. of 5th Nat. Weld Sym., 8, 1989.

5

Fluxes, atmospheres – types and forms

Fluxes, gas atmospheres, and vacuum promote the formation of brazed joints. They may be used to surround the work, exclude reactants and provide active or inert protective atmospheres, thus preventing undesirable reactions during brazing. Under some conditions, fluxes and atmospheres may also reduce oxides that are present. Caution must be observed in the use of atmospheres, because some metals are embrittled by various gases. Notable among these metals are titanium, zirconium, niobium and tantalum, which become permanently embrittled when brazed in any atmosphere containing hydrogen, oxygen or nitrogen. Also, hydrogen embrittlement of copper that has not been thoroughly deoxidized must be avoided. The use of any flux or atmosphere does not eliminate the need for thorough cleaning of parts prior to brazing.

Metals, when exposed to air, tend to react with various constituents of the atmosphere to which they are exposed. The rate of these chemical reactions is generally accelerated as the temperature is raised. The most common reaction is oxidation, but nitrides and carbides are also sometimes formed. The rate of oxide formation varies with each metal composition and the nature of the oxide. Tenacity, structure, thickness and resistance to removal and/or further oxidation must be considered. Oxide formation on some metals in air is for practical purposes instantaneous, even at or below room temperature. These reactions result in conditions, such as oxides or other compounds, which hinder the production of consistently sound brazed joints.

5.1 FLUXES

The primary purpose of brazing fluxes is to promote wetting of the base metal by the brazing filler metal. The efficiency of flux activity, which is commonly referred to as "wetting", can be expressed as a function of brazeability.[1] Flux must be capable of dissolving any oxide remaining on the base metal after it has been cleaned and any oxide films on the liquid filler metal. It is important to realize that most fluxes are not designed or intended for the primary removal of grease, oil, or dirt, and cannot take the place of proper precleaning operations. However, in some instances, fluxes may serve to suppress the volatization of high vapor pressure constituents in a brazing filler metal. Some filler metals, when in the molten stage, are self-fluxing on certain alloys.

To effectively protect the surfaces to be brazed, the flux must completely cover, be applied as an even coating and protect them until the brazing temperature is reached. It must remain active throughout the brazing cycle. Because the molten filler metal should displace the flux from the joint at the brazing temperature, the viscosity and surface tension of the flux and the interfacial energy between the flux and the surfaces of parts are important. Therefore, recommended fluxes should be used in their proper temperature ranges and on the materials for which they are designed.

Additionally, the role of flux joining is one of controlled corrosion. The corrosive attack centers on the dissolution and dispersion of oxide tarnish, but the surface layer of metal atoms may also be removed. This attack is rapid because of the elevated operating temperatures and because the oxide capacity of the flux is considerable. The interaction of the flux melt with the oxide layer is central to obtaining a good, clean, fast joint.[2]

Certain brazing filler metals contain alloy additions of deoxidizers, such as phosphorus, lithium, and other elements which have strong affinities for oxygen. For example, phosphorus–copper filler metals act as fluxes on copper and silver. In some instances, these additions make such filler metals self-fluxing without the application of prepared fluxes or controlled atmospheres. These filler metals are self-fluxing only in the molten state and will themselves oxidize during the heating cycle. In other cases, they are used in conjunction with protective atmospheres or fluxes to increase wetting tendencies. When large sections are to be brazed or where prolonged heating times are contemplated, the use of additional flux is advisable.

5.1.1 Flux constituents

Many chemical compounds[3] are used in the preparation of fluxes, and many proprietary fluxes on the market are formulated to offer specific properties. When fluxes are heated, reactions take place between the various chemical ingredients, forming new compounds at brazing temperatures which are quite different chemically and physically from the unreacted constituents. For instance, if a fluoborate is an ingredient in a flux, fluorides may be formed as the ingredients react. During brazing, the chemistry is especially transient. Reaction rates of the flux with oxygen, base metals, brazing filler metals, and any foreign materials present increase with temperature. Composition of the flux must be carefully tailored to suit all the factors of the brazing cycle, including dwell time. Attack of the flux on the metals must be limited, because the flux must react promptly with metal oxides or other tarnish to enable the joint to be satisfactorily formed. Active halides, such as chlorides and fluorides, are, for instance, necessary in fluxes for alloys containing aluminum or other highly electropositive metals.

The most common ingredients of chemical fluxes are:

- Borates (sodium, potassium, lithium, etc.)
- Fused borax
- Elemental boron
- Fluoborates (potassium, sodium, etc.)
- Fluorides (sodium, potassium, lithium, etc.)
- Chlorides (sodium, potassium, lithium)
- Acids (boric, calcined boric)
- Alkalis (potassium hydroxide, sodium hydroxide)
- Wetting agents
- Water (either as water of hydration or as an addition for paste fluxes).

Most brazing fluxes are proprietary mixtures of several of the above ingredients. The ingredients are mixed and reacted in ways that give satisfactory results for specific purposes. Their functions are as follows:

1. *Borates* are useful in formulating the fluxes that melt at higher temperatures. They have good oxide-dissolving power and provide protection against oxidation for long periods. Most borates melt and are effective at temperatures around 760°C or higher. They have a relatively high viscosity in their molten condition and therefore must be mixed with other salts to increase fluidity.
2. *Fused borax* is another high-temperature melting material that is active at high temperatures. It is little used in lower-temperature brazing processes.
3. *Elemental boron powder* is added to increase overall fluxing action. Silver brazing fluxes that contain elemental boron offer improved protection on carbides and on materials that form refractory oxides such as chromium, nickel, and cobalt.
4. *Fluoborates* react similarly to other borates in many respects. Although they do not provide protection from oxidation to the same extent as other borates, they flow better in the molten state and have greater oxide-dissolving properties. Fluoborates are used with other borates or with alkaline compounds, such as carbonates.

5. *Fluosilicaborates.* Another class of compound is fluosilicaborates, which have somewhat higher melting points than fluoborates and provide good coverage and surface adherence. Their high melting points limit their use.

6. *Fluorides* react readily with most metallic oxides at elevated temperatures and, therefore, are used extensively in fluxes as cleaning agents. They are particularly useful when refractory oxides, such as those of chromium and aluminum, are encountered. Fluorides are often added to increase the fluidity of molten borates, thereby facilitating their displacement and improving the capillary flow of the molten brazing filler metal. Fluorides can generate dangerous fumes, however, and so their use warrants strict attention to good safety practices. Fluorides, up to 40% in flux content, give silver brazing fluxes their characteristically low melting points, 560°C, and high activity for dissolving metal oxides.

7. *Chlorides* function in a manner similar to fluorides but have a lower effective temperature range. Chlorides must be used with caution because, at lower temperatures, they are used to depress the melting points of fluoride-based fluxes.

    As seen in Table 5.1, aluminum and magnesium brazing fluxes contain alkaline chlorides or fluorides. Lithium salts give these fluxes low melting points, 540–615°C, and high chemical activity, enabling the fluxes to dissolve stubborn aluminum oxide.

8. *Boric acid* is a principal constituent used in brazing fluxes because it facilitates the removal of the glass-like flux residue left after brazing. Its melting point is below that of borates but higher than that of fluorides. Silver brazing fluxes contain boric acid and potassium borates, combined with complex potassium fluoborate and fluoride compounds.

    High-temperature fluxes, based on boric acid and alkaline borates, sometimes contain small additions of elemental boron or silicon dioxide to increase activity and protection, good up to 1204°C. Fluoride content of these fluxes is usually low, at most 2–3%. These braze ferrous and high-temperature alloys and carbides.

9. *Alkalis,* such as potassium and sodium hydroxides, are used sparingly, if at all, to elevate the useful working temperature of the flux. Their drawback is that they are deliquescent; even small amounts in other flux agents can cause problems in humid weather and can severely limit the storage life of the flux. Alkalis elevate the useful working temperature of the flux.

10. *Wetting agents* are used in paste and liquid fluxes to facilitate the flow and spreading of the flux onto the workpiece prior to brazing.

11. *Water is present* in brazing fluxes either as water of hydration in the chemicals used in formulating the flux or as a separate addition for making a paste or liquid. Water used in forming a paste must be evaluated for suitability, and hard waters should be avoided.

Table 5.1 List of brazing fillers[3]

AWS spec.	Flux category	Form	Base materials	Filler metals	Application method	Heat source	Joining methods	Typical ingredients(b)	Active temp. range (°C)	Notes	Flux Discription
FB1A	Aluminum brazing	Powder	Aluminum alloys	BAlSi	Manual	Torch, furnace	4,5,6,7	Chlorides, fluorides	580–615	For torch or furnace brazing; water or alcohol may be added as the flux is used.	Powder flux for torch or furnace brazing
FB1B	Aluminum brazing	Powder	Aluminum alloys	BAlSi	Manual	Furnace	4,5,6,7	Chlorides, fluorides	560–615	For furnace brazing; water or alcohol may be added as the flux is used.	Powder flux for furnace brazing
FB1C	Aluminum brazing	Powder	Aluminum alloys	BAlSi	Dip brazing	Salt bath	4,5,6,7	Chlorides fluorides	540–615	For chemical bath dip brazing.	Powder flux for dip brazing
FB4A	Aluminum bronze	Paste	Brazeable base metals containing aluminum (aluminum brass, aluminum bronze, Monel K500); may also have application when minor amounts of titanium or other metals are present which form refractory oxides	BAg. BCuP[1]	Manual	Torch, furnace, induction	4,5,6	Chlorides, fluorides borates	595–870	Water, 35% max; paste may be thinned with water if desired; usually applied by brushing.	Paste flux

Table 5.1 Continued

AWS spec.	Flux category	Form	Base materials	Filler metals	Application method	Heat source	Joining methods	Typical ingredients(b)	Active temp. range (°C)	Notes	Flux Discription
FB3D	High-temperature brazing	Paste[2]	Copper, ferrous & nickel alloys, carbides	BAg, BCu, BNi, BAu, RBCuZn	Manual automatic	Torch, furnace, induction	4,5,6	Borates, fluorides	760–1,205	—	Paste flux used for controlled-atmosphere furnace brazing
FB3I	High-temperature brazing	Slurry[2]	Copper, ferrous & nickel alloys, carbides	BAg, BCu, BNi, BAu, RBCuZn	Automatic	Torch			760–1,205	—	—
FB3J	High-temperature brazing	Powder[2]	Copper, ferrous & nickel alloys, carbides	BAg, BCu, BNi, BAu, RBCuZn	Manual	Torch, furnace			760–1,205	—	—
FB3K	High-temperature brazing	Flammable liquid	Copper, ferrous & nickel alloys, carbides	BAg, RBCuZn	Manual automatic	Torch			760–1,205	—	—
FB2A	Magnesium brazing	Powder	Magnesium alloys	BMg	Dip brazing	Salt bath	6,7	Chlorides, fluorides	480–620	—	—
FB3A	Silver brazing	Paste	Copper, ferrous & nickel alloys, carbides	BAg, BCuP[1]	Manual, automatic	Torch, induction	4,5,6	Borates, fluorides boron	565–870	Water, 35% max; paste may be thinned with water if desired; usually applied by brushing or dipping the work into the flux.	General-purpose paste flux for most ferrous and non-ferrous alloys
FB3C	Silver brazing	Paste[3]	Copper, ferrous & nickel alloys, carbides	BAg, BCuP[3]	Manual, automatic	Torch, induction			565–925	—	

Table 5.1 Continued

AWS spec.	Flux category	Form	Base materials	Filler metals	Application method	Heat source	Joining methods	Typical ingre-dients(b)	Active temp. range (°C)	Notes	Flux Discription
B3E	Silver brazing	Water-based liquid	Copper, ferrous & nickel alloys, carbides	BAg, BCuP[1]	Manual, automatic	Torch, furnace			565–870	All brazeable ferrous and non ferrous metals except those containing aluminum or magnesium	
B3F	Silver brazing	Powder	Copper, ferrous & nickel alloys	BAg, BCuP[1]	Manual	Torch, furnace	4,5,6	Borates, fluorides	650–870	All brazeable ferrous and nonferrous metals except those containing aluminum or magnesium	Paste flux
B3G	Silver	Slurry brazing	Copper, ferrous & nickel alloys, carbides	BAg, BCup[1]	Automatic	Torch			565–870		
B3H	Silver brazing	Slurry[3]	Copper, ferrous & nickel alloys, carbides	BAg	Automatic	Torch			565–925		

1. Used with copper and copper–alloy base metals only. 2. May contain elemental boron or silicon dioxide. 3. Boron-modified.
Note: Pastes have high viscosities and are typically applied by brushing. Slurries have low viscosities and can be sprayed or automatically dispensed.4. Apply dry powder to joint. 5. Dip heated filler-metal rod in powder or paste. 6. Mix flux to paste consistency with water, alcohol, or other carrier. 7. Dip or immerse in molten flux bath. (b) Fluxes in form of aqueous paste may also contain wetting agents.

5.1.2 Groups of fluxes

There is no single flux which is best for all brazing applications. Fluxes are classified (Table 5.1) according to their performance on certain groups of base metals in rather specific temperature ranges. The five categories of brazing fluxes are: aluminum, aluminum–bronze, silver, magnesium and high-temperature flux. Within each type and class, there are numerous commercial and proprietary fluxes available, and selection of an appropriate flux must be done by careful analysis of the properties or features required for a particular application. Reference to Table 5.1 is not a substitute for thorough evaluation in selecting an optimum flux for a specific high-production joint. For successful use, a flux must be chemically compatible with all the base metals and filler metals involved in the brazement. It must be active across the entire brazing temperature range and throughout the time at brazing temperature. If the brazing cycle is long, a less active but more protective flux should be selected. Conversely, if the cycle is short, a more active flux which will promote quick filler-metal flow at the minimum temperature may be used. Where more than one flux is suitable for the application, other considerations, such as safety and cost, should be evaluated.[3]

5.1.3 Flux selection criteria

Base material type determines flux selection more than any other factor. To braze aluminum alloys, coat parts with aluminum brazing fluxes. Similarly, aluminum-bronze and magnesium fluxes braze only with their respective base metals. To braze ferrous alloys and nickel alloys, two flux types can be used: silver-brazing or high-temperature fluxes. Which of the two is better depends on base and filler material type, brazing conditions, and cost. Fabricators call on silver-brazing fluxes, more expensive than high-temperature fluxes, to minimize heat input and distortion to the work. These also braze copper alloys. To braze carbides – tungsten carbide infiltrated with cobalt to impart high strength with toughness – coat with boron-modified fluxes and fill the joint with silver brazing filler metals containing nickel. High-temperature fluxes and filler also braze carbides, when the carbide–steel combination can tolerate the high brazing temperatures, 1093°C.

Within a particular flux type, there are several criteria for choosing a specific flux for maximum efficiency:

- For dip brazing, water (including water of hydration) must be removed, usually by preheating prior to immersion in the salt bath.
- For resistance brazing, the flux must permit the passage of current. This usually requires a wet, dilute flux.

- The effective temperature range of the flux must include the brazing temperature for the specific filler metal being used.
- Controlled atmospheres may modify flux requirements.
- Ease of flux residue removal should be considered.
- Corrosive action on the base metal or filler metal should be minimized.

5.1.4 Flux/temperature range specification

To be effective, flux must be molten and active before the filler metal melts, and it must remain active until the filler metal flows through the joint and solidifies upon cooling.

Therefore, filler metal solidus determines minimum working temperature of the flux and filler metal liquidus dictates maximum brazing temperature that the flux must withstand. Generally, select a flux that is active approximately 30°C below the solidus of the brazing filler metal and that remains active at least 90°C above the filler metal liquidus.

If overheating is likely to occur during brazing, as when torch brazing, select a flux active at 120–175°C above the filler metal liquidus. This gives the flux a wide temperature range to remove surface oxides before the filler metal melts and will keep it effective at brazing temperatures.

Brazing time affects flux performance. Molten flux forms a semi-protective blanket that prevents oxidation only for a finite period – oxygen will eventually diffuse through the flux to the base materials. Flux must continue to remove newly-formed oxide until the end of the heating cycle. Because flux can dissolve only a limited amount of oxide, the longer the heating cycle the greater the likelihood that the flux will become saturated with oxide, a condition called flux exhaustion.

Rated temperature range of a flux, which depends on brazing temperature, flux type and volume, and base-material type, assumes a brazing cycle of 15–20 seconds. With a longer heating cycle, flux exhaustion may occur even when brazing below the maximum operating temperature, because over time the flux becomes saturated with metal oxide. To avoid flux exhaustion over prolonged heating cycles, switch to a flux with a higher working temperature range. When the heating cycle is short, a fabricator can braze with a flux above its maximum rated working temperature. Using a low-temperature flux above the maximum working temperature eases flux removal, since these fluxes are more soluble in water than are high-temperature fluxes.

5.1.5 Flux application

Ideally, apply flux to both joint surfaces; for some applications, coating only one surface suffices–the flux will transfer to the mating surface on assembly.

Application methods depend on joint design, production volume, and joint-heating technique. Operators brush to apply paste flux to the joint and to surrounding surfaces, or they may dip parts into a container of flux. Flux for dipping is of a thinner consistency than that used for brushing. In some cases, parts are dipped in boiling flux solutions in which the solids are completely dissolved. Automatic application of flux can be carried out by spraying, pumping, blotting, or dipping.

Fluxes for brazing are generally available in the form of powder, paste, slurry, or liquid. The form selected depends on the individual work requirements, the brazing process, and the brazing procedure used. Fluxes are most commonly applied in paste form because of the ease with which pastes can be applied to small parts and their adherence in any position. The particle size of paste or dry flux should be uniform and small for the most effective application. It is frequently helpful to heat the paste slightly before application. A low-viscosity slurry or diluted paste flux is used when the flux is to be sprayed on a joint. Certain fluxes (types FB3A and B) will completely dissolve in water to produce a liquid solution called "**liquid flux**". Automatic torch brazing has been made possible by the development of face-feeding machines. One of these is the paste feeder, which applies a mixture of flux and filler metal powder. Methods of applying fluxes, and techniques employed in the use of fluxes, are as follows.

Dipping

This is the most popular production method for applying flux. Preformed rings should be in position prior to dipping to ensure a thorough and uniform coating of flux on the ring and to avoid having the operator touch the flux with their fingers. Continual contact with flux may cause a skin disease known as dermatitis. Flux may be thinned for dipping by dilution, by heating, or both. Flux heated to 60 to 70°C will adhere to the metal much better than will cold flux. Heating also reduces spattering when the water is boiling out during the brazing operation. Flux pots are available with thermostatic controls that maintain desired temperatures.

Spraying

Thin layers of flux may be applied with a standard paint sprayer. The container should be an integral part of the spray gun to simplify cleaning and to prevent flux from caking in the lines. Air pressure keeps the ports clear. Between applications, immerse the container in hot water. Following the operation, the container should be filled with hot water and the ports blown free of all flux. The spraying station should be well ventilated, and all other precautions should be taken to keep the operator from inhaling the spray mist.

Brushing

Brushing of flux on parts has several advantages. For example, the scrubbing action of the bristles helps wet the metal with flux – a very important feature when working with dense materials such as tungsten carbide and high-chromium stainless steels. Brushing does not lend itself particularly well to automatic applications. However, it has been applied successfully by using indexing turntables, with the parts rotating. The brushes dip into the flux, rise, and advance to the parts. Rotation of the parts under the brushes ensures complete coverage.

Pressure oil can

On parts with chamfers or concave surfaces, flux can be applied quite easily with a simple pressure-type oil can. The flux must be thinly diluted to ensure good coverage around the entire well. This method is especially good for small joints.

Thin coatings

Thin coatings may be applied with a sponge set in a bath of flux. The method is particularly helpful where flux is desired on only one surface or on projections from the surface.

Improving filler metal low

To obtain an effective brazed joint, the braze filler metal must displace all of the flux in the joint. Generally this is not difficult since the braze filler metal melts at one point in the assembly and, by capillary action between closely fitted members, flows through the joint, flushing the flux ahead of it. On sharp shoulders and close fittings emanating both directions from the shoulder, the braze filler metal may have some difficulty in flowing. In this case, the sharp corner mating to the shoulder should be broken to assist flux displacement. Unless this is done, the flux will boil up through the braze filler metal causing pinholes, or the braze filler metal may separate, wetting both mating surfaces but leaving a layer of flux between.

Fluxing of large flat surfaces

On large flat surfaces, a very thin coating of flux should be applied. Often it may be necessary to wipe the flux from the surface, leaving only the pores of the metal filled with flux. On flat surfaces, the oxygen is usually excluded and not much flux is required. Also, since shim stock is generally used, the same flushing action does not occur as with a preformed ring, and a heavy coating of flux may result in too many entrapped flux islands. A heavier shim or washer of smaller area will flush out the flux with fewer voids. A heavy coating of flux can be applied around the perimeter to prevent the entrance of oxygen to the joint.

Flux removal

Flux subjected to atmospheric oxygen is difficult to remove. To aid removal, flux should be used with less dilution. Various methods for reducing scaling should be tried to find the most effective. Flux that has not reacted generally comes off easily in hot water. Quenching of parts after the filler metal has set, but while they are still warm, aids flux removal considerably. Parts that cannot be quenched must be permitted to cool slowly. If flux removal is difficult, a warm bath consisting of 10% sulfuric acid should be used.

Dilution Of Fluxes

Generally, paste fluxes are supplied so thick that they must be diluted with water to a suitable consistency. When oxidation is light, as on copper and silver, the flux may be thinned considerably. When the flux is spent and the filler metal does not flow properly, the paste should be heavier. When a heavier flux does not help, brazing conditions must be changed or another flux must be used. Copper and copper alloys require a considerably diluted flux to remove their light oxides. A heavier flux and general rather than locally applied heat are required for carbon steels. A fluoride flux should be used for alloy steels containing chromium, vanadium, or manganese; stainless steels require an active, almost undiluted flux.

Particle size

The particle size of dry flux or paste flux is important because better fluxing action will result when all constituent particles of a flux are small and thoroughly mixed. Stirring, ball milling, or grinding of a flux mixture is helpful if the flux has become lumpy. Preheating of the paste or liquid flux may facilitate application.

Powdered fluxes

Powdered flux can be applied to the joint in four different ways:

1. dry;
2. mixed with water and alcohol to form a paste;

3. in torch brazing, by dipping the heated filler-metal rod into the flux as needed;
4. sprinkled on the joint.

Mixtures of powdered filler metal and flux are sometimes used where it is desirable for both flux and filler metal to be preplaced.

Spraying

Liquid fluxes (type FB3D), in which fluxing ingredients are completely in solution, may be sprayed on the joint or entrained in the fuel gas.

Gas/flux mixture

Liquid fluxes are sometimes used in torch brazing. The fuel gas is passed through the liquid flux container, thus entraining the flux in the fuel gas, and the flame and flux are applied where needed. Usually a small amount of additional preplaced flux is used for the joint surroundings.

Slurries

Most slurries are water-based; some organic-based fluxes – petroleum, or polyethylene-glycol-based for example – suit precision dispensing due to lower evaporation rates and better viscosity control.

Hot rodding

Hot rodding, used to braze-weld, plunges a hot brazing rod into powder flux. Heat from the rod causes a small amount of flux to adhere to the rod surface. This method is best suited to brazing of shallow joints, up to 6.4 mm in steel, as it results in poor capillary penetration in deep joint areas.

General

The paste and liquid flux should adhere to clean metal surfaces. If the metal surfaces are not clean, the flux will ball up and leave bare spots. Thick paste fluxes can be applied by brushing. The proper consistency depends upon the types of oxides present, as well as the heating cycle. For example, ferrous oxides formed during fast heating of the base metal are soft and easy to remove, and only limited fluxing action is required. However, when joining copper or stainless steel or when the heating cycle is long, a concentrated flux is required. Flux reacts with oxygen, and once it becomes saturated, it loses all its effectiveness. The viscosity of the flux may be reduced without dilution by heating it to 50 to 60°C, preferably in a ceramic-lined flux or glue pot with a thermostat control. Warm flux has low surface tension and adheres to the metal more readily.

5.1.6 Fluxes and specific processes

Flux is required for induction brazing. The flux used should decompose oxides without corroding the base metal or the brazing filler metal, should be extremely active because of the short brazing times employed, and should be easy to remove after brazing.

Type FB3A flux is used for an estimated 95% of the induction brazing applications that involve steel. Paste and liquid fluxes are most often applied to the joint by brushing.

A flux is used in almost all resistance brazing. It serves the same purposes in resistance brazing as in other brazing processes:

- providing a coating to prevent or minimize oxidation of the work metal during heating;
- dissolving oxides that are present or that may form during heating;
- assisting the molten filler metal in wetting the work metal to promote capillary flow.

The flux in resistance brazing, however, has the additional function of serving as an electrical conductor to permit passage of the brazing current through the joint; most dry fluxes are nonconductors and must be mixed with water in order to conduct current.

The flux is usually applied as a dilute water-base paste shortly before the parts and filler metal are assembled for brazing. If the filler metal is in powder form, flux can be combined with it in a fine particle paste.

The same fluxes are used for resistance brazing as for other brazing processes on the same work metal. Type FB3C fluxes are general-purpose fluxes suitable for most metals that are commonly resistance brazed (although type FB4A flux is needed for copper alloys that contain tin, aluminum or silicon); type FB1A, B, and C fluxes are used on aluminum alloy work metals.

The two general situations in which a flux is not used in resistance brazing are brazing in a vacuum or protective reducing gas or inert atmosphere and brazing of copper with a BCuP filler metal. A flux is not ordinarily needed in resistance brazing of copper when a BCuP filler metal is used, because these filler metals are self-fluxing on copper by virtue of their phosphorus content.

A non-corrosive flux brazing process has been recently introduced, "Nocolok"*, and has established itself as an accepted process for brazing aluminum heat exchangers of all types. Although brazing with Nocolok can be successfully achieved using flame and induction heating the process is at its most effective using a continuous furnace.

In common with all fluxes used for brazing aluminum the action of Nocolok flux is to remove the tough persistent oxide film from the metal surface and promote filler metal wetting and flow. The potassium fluor-aluminate flux operates at brazing temperatures by melting and dissolving the oxide film. The flux melts at 562°C just below the eutectic temperature of the Al–Si filler metal (577°C). In contrast, chloride salt-based fluxes work by melting, followed by penetration and separation of the oxide film from the metal substrate.

The benefits of using Nocolok flux include in essence that Nocolok flux and its flux residue are both non-corrosive, unreactive with aluminum and almost insoluble in water. The flux residue is known to improve the corrosion resistance of brazed components, both in the as-brazed and painted conditions. In commercial terms the non-corrosive nature of the flux together with its tolerance to brazing assembly fit-up and flexible process control ensure that Nocolok flux brazing is one of the lowest cost methods for the joining of aluminum heat exchangers. The best braze quality and lowest flux loadings (3–5 g/m^2) result from continuous furnace brazing with a dry gas atmosphere having a dew point below –40°C and an oxygen content of less than 1000 ppm.

Torches usually apply one of the high-temperature fluxes, FB3K. This flux is a flammable liquid containing trimethyl borate. A dispenser installed in the fuel gas line feeds flux vapor into the flame.

5.1.7 Application quantity

One must apply enough flux to coat the joint faces and adjacent surfaces with a thin layer. Excess flux will not compromise joint quality and may even assist flux removal since residues will be less loaded with metal oxide and more soluble in water. Also, applying flux to surfaces adjacent to the joint helps to prevent oxidation of the workpiece and may act as a flux reservoir, draining flux into the joint. Using too little flux, however, can lead to premature flux exhaustion and inadequate coverage, producing unsound or unsightly brazed joints. Better to err on the side of too much rather than too little flux. The choice of heat source has little effect on flux selection. Exceptions include salt bath heating, which requires dip brazing fluxes; specialized high-temperature torches using a flammable-liquid flux; and furnace brazing, which often calls for a

*Nocolok is a registered trademark of Alcan Aluminium Ltd.

powder flux to minimize the amount of vapor. Boron-modified fluxes are often preferred for induction heating.

5.1.8 Base metal/filler metal/flux combinations

The filler metal-flux combination can be either a brazing paste or flux-coated rod. Pastes, mixtures of filler-metal powder and flux, and sometimes an organic binder to ease dispensing, work well for automated processes; aluminum, silver and high-temperature brazing pastes are most popular. Flux-coated rods perform brazing and braze welding. The most common flux-coated filler-metal rods are silver-brazing and low-fuming bronze, used primarily to braze weld.

The variety of base metals and alloys that are joined by brazing has prompted the development of many different fluxes in addition to those listed in Table 5.1. Two fluxes have been used successfully for furnace or induction brazing of beryllium with good results: a 60%LiF–40%LiCl flux and a tin chloride flux. It is important from a safety standpoint that beryllium and beryllium compounds as flux residues are toxic. Only approved installations should consider brazing of beryllium regardless of the methods used.

The FB2–type flux has been used to clean the surface of magnesium, which permits capillary flow. Because of the corrosive nature of this flux, complete removal is of utmost importance if good corrosion resistance is to be obtained in brazed joints.

In the refractory metal family, tantalum may be brazed in air using fluxes normally used for brazing with aluminum filler metals or fluxes that are suited to the particular filler metal being used. However, tantalum and niobium require protective coatings, such as nickel or copper electroplate, to induce wetting during brazing. Conventional low-temperature fluxes have been used in brazing tungsten for electrical contact applications when silver- and copper- based filler metals are used.

In brazing molybdenum with an oxyacetylene torch, fair protection may be obtained by using a combination of fluxes – a commercial borate-based or silver-based brazing flux, plus, a high-temperature flux containing calcium fluoride. The temperature range over which these fluxes are active is the range from 565 to 1425°C. The molybdenum is first coated with the commercial silver brazing flux and then the high-temperature flux is applied. The silver flux is active at the lower end of the active temperature range; the high-temperature flux then takes over and is active up to 1425°C.

When BAg filler metals are used for brazing nickel-based alloys, FB3A- and FB3B-type fluxes are suitable for most alloys not containing aluminum, whereas FB4 flux may be used with aluminum-containing nickel alloys.

Because most cast irons are brazed at relatively low temperatures, the filler metals used are almost exclusively silver-based filler metals. Of these, BAg–1 is most often used for brazing of cast iron, principally because it has the lowest brazing temperature range. A fluoride-type FB3A flux is usually used with the BAg–1 filler metal whereas FB3B flux is used with other filler metals in the BAg series.

The selection of flux for brazing low-alloy and carbon steels depends on the brazing filler metal.[4] Fluxes FB3A, FB3B and FB4A are suitable for BAg filler metals; type FB4A normally is used with RBCuZn filler metals.

Fluxes and atmospheres may also be used together. The flux can be in either paste or powder form or can be combined with the brazing filler metal. For example, in a face-fed operation, the hand-held brazing filler metal can be coated with the appropriate flux. Typical fluxes employed for prefluxing of low-alloy and carbon steels that are brazed in a neutral chloride salt bath are FB3A and FB3B. Generally, the application of flux to an assembly is not necessary when a cyanide bath is used.

In brazing of copper–aluminum alloys (aluminum bronzes), the formation of refractory aluminum compounds creates difficulty in wetting, and as a result strong fluxes are required. Aluminum bronzes can be brazed with silver-bearing filler metals and type FB4A flux.

The RBCuZn brazing filler metals may be used for brazing of coppers and of copper–nickel, copper–silicon, and copper–tin alloys. However, they are not useful for brazing aluminum bronzes because the required brazing temperatures destroy the effectiveness of the fluxes required for these base metals (see Table 5.1). With the copper–zinc filler metals, care should be taken not to overheat the metal, because volatization of the zinc causes voids in the joint. In torch brazing, an oxidizing flame will reduce zinc fuming, and FB3D brazing flux should be used.

The BCuP brazing filler metals are useful for brazing high-leaded cast brass pipe fittings if precautions are taken to flux properly and avoid overheating. Brasses containing aluminum or silicon require treatment similar to aluminum or silicon bronzes. Lead added to brass to improve machinability may alloy with the filler metal and cause brittleness. Major brazing difficulties occur when the lead content is over 2 or 3%. To maintain good flow and wetting during brazing, leaded brasses require complete flux coverage to prevent the formation of lead oxide or dross.

Additionally, the FB3A and FB3B fluxes are suitable for use with BCuP and BAg filler metals in brazing all the copper base metals except aluminum bronzes.

Refractory oxides form easily on aluminum bronzes, and the more active FB4-type fluxes are needed to cope with them. The effectiveness of type FB3A flux may be reduced rapidly at the temperatures needed for brazing with RBCuZn filler metals, and is completely destroyed in brazing with BCu. Type FB5 flux may be used with these filler metals except in brazing of aluminum bronze or beryllium copper. More active fluxes are needed for these base metals, and mixtures of FB4 and FB5 fluxes may be found satisfactory for the few applications of this kind.

In brazing of copper, the copper–phosphorus and copper–silver–phosphorus filler metals are self-fluxing. Flux is beneficial, however, for heavy assemblies where prolonged heating would otherwise cause excessive oxidation.

The special coppers that contain small additions of silver, lead, tellurium, selenium, or sulfur (generally no more than 1%) are brazed readily with the self-fluxing BCuP filler metals.

Wetting action is improved when a flux is used and when there is a small amount of shearing motion between the components while the filler metal is molten.

Finally, FB3-type flux is suitable for most applications in which copper–nickel alloys are brazed.

5.1.9 Post-braze cleaning and flux removal

There are seven major reasons for removing residual flux after brazing:

1. The joint cannot be inspected for soundness until the cover of flux residue has been removed;
2. The joint may be bound together by the flux in the semblance of a brazed joint, only to break apart later in service;
3. In fluid or pressure service, the flux may block pinholes that might withstand a pressure test but would leak soon after being placed in service;
4. If left on the joint, the flux attracts available water, resulting in oxidation and corrosion;
5. Painting, coating, or plating cannot be done satisfactorily on areas covered with flux residue;
6. If a joint is overheated, excess heat impairs flux removal; spent flux residues, saturated with metal oxides, are most difficult to remove;
7. To avoid flux exhaustion, apply excess flux to ease removal of residues from the base material.

If parts have been well cleaned before brazing and not overheated during brazing, flux residue can usually be removed by a hot water rinse followed by thorough drying. To avoid corrosion, flux removal should be delayed no more than 48 h.

A quick method of removing glass-like residues is to quench the joint in cold water after brazing and thus crack off the deposit by thermal shock. However, in some applications, such treatment may cause distortion of the brazed assembly.

Scrubbing, applying a stream jet, and most of the standard abrasive techniques, such as wire brushing and abrasive blasting, are also used to dislodge stubborn flux residues, provided the operation does not impair the function of the assembly.

When the flux cannot be removed from steel assemblies by rinsing in cold or hot water, a cold 5% solution of sulfuric acid will prove more effective. This works for nonferrous metals and mild steels. More aggressive solutions are needed for stainless steels and high-temperature alloys. The solution may be warmed to accelerate the action, provided that care is taken to prevent excessive attack on the assembly. A small addition of sodium dichromate to the solution makes the action even faster, but the time of immersion must be carefully controlled to avoid the greater risk of etching the steel.

Phosphate solutions similar to those used for cleaning steel are effective flux removers and have the added advantage of giving carbon steel assemblies a temporary protective coating. However, the coating will hamper subsequent brazing operations.

Boric acid, as applied in gas fluxing, can be removed by washing in clean water heated to at least 65°C. (Boric acid is only slightly soluble in cold water.)

Mixed borax and boric acid fluxes are more difficult to remove than other types. Fortunately, moisture absorption and corrosion are minimal with borax fluxes. In fact, rather than risk damage to delicate assemblies, such as electronic components, when mixed borax and boric acid fluxes are used, the flux is sometimes allowed to remain after brazing.

These fluxes can be removed by quenching, shot blasting, sand blasting, chipping, filing, scraping, and wire brushing. The rate of solution in water is slow, and even if dilute sulfuric acid is used, the necessary period of immersion may be inconveniently long for production work.

Fluoride fluxes are soluble in water and are much easier to remove than borax fluxes. Holding under running cold water while brushing with a wire or bristle brush will usually suffice.

Alternatively, the assemblies may be boiled in water for a few minutes, and then rinsed in cold water. Brazed parts should be quenched while still hot, if at all possible, but the brazing filler metal must be completely solidified before any water quenching begins. If the flux has not been too heavily oxidized, quenching and agitating the part in hot water should be sufficient to remove all flux residue. Dilute sulfuric acid solutions and phosphate solutions can also be used for quicker results. The residue of fluoride fluxes is hygroscopic, and if the assembly is not quenched after brazing, it is often advantageous to postpone post-brazing cleaning for 24 h. Under normal atmospheric conditions, the residue will absorb moisture during this period and will become more readily soluble in any of the solvents previously mentioned.

It is helpful to think of flux as performing like an ink blotter. The blotter can absorb only so much ink, after which it becomes saturated and unable to absorb any more ink. Similarly, flux will absorb oxides which are continuously generated during the brazing cycle (oxides are not generally formed in brazing methods which exclude oxygen-bearing atmospheres), but only to its point of saturation. Once that point has been reached, the flux becomes useless, and further heating will only oxidize the joint. Fully-saturated fluxes are generally hard to remove. In such cases, acidifying the water will accelerate removal of the flux residue.

Flux residue should be removed completely to avoid corrosion by the residual active chemicals. The residue obtained from a flux, particularly when considerable oxide removal has occurred, is a form of glass. The less flux required to clean the metal, the less will be the formation of glass and the easier the task of removing the flux residue. The thermal shock cracks off the residue. By the same token, the use of an adequate amount of the proper flux will usually make residue removal easier because spent flux has a high glass content.

Flux removal from properly cleaned, brazed parts usually can be accomplished by rinsing in hot water accompanied by light brushing. Preferably, this rinse should be done immediately after the brazing operation, and thorough drying following rinsing is recommended.

Tenaciously adhering flux particles may be removed either by dipping the parts in one of several proprietary chemical dips or by mechanical means such as fiber brushing, wire brushing, shot blasting, or chipping. The method used should be compatible with the required properties of the base metals. If, for example, stainless steel parts are cleaned with a wire brush, a stainless steel brush should be used. Soft metals, such as aluminum and copper, should be cleaned by methods which will not damage or roughen their surfaces. Cleaning with nitric acid is not recommended where brazing filler metals containing silver or copper have been used. Parts should be thoroughly washed and dried after chemical dipping.

As brazing technology becomes more sophisticated, methods of detecting flux residues not removed by traditional chemical methods are being developed. Ion scattering spectrometry (ISS) and secondary ion mass spectrometry (SIMS) have successfully been adapted for detecting residual flux in brazed joints in service. Further, ISS and SIMS can differentiate between chloride and fluoride fluxes. These new methods offer an opportunity to literally "fingerprint" flux residuals. ISS and SIMS analysis also can detect residuals and contaminants solely on the surface. Then, by depth profiling, the braze deposit can be analyzed independently of the surface and/or the base metal, and finally the braze/base metal joint can be analyzed.

Fluxes used in brazing of aluminum alloys can cause corrosion if allowed to remain on the parts. It is therefore essential to clean joints after brazing. A thorough water rinse followed by a chemical treatment is the most effective means of complete flux removal.

As much flux as possible should be removed by immersing the parts in an overflowing bath of boiling water just after the filler metal has solidified.[5] If such a quench produces distortion, the parts should be allowed to cool in air before immersion, to decrease the thermal shock. When both sides of a brazed joint are accessible, scrubbing with a fiber brush in boiling water will remove most of the flux. For parts too large for water baths, the joints should be scrubbed with hot water and rinsed with cold water. A pressure spray washer may be the best first step. A steam jet is also effective in opening passages plugged by flux.

Any of several acid solutions (Table 5.2) will remove any flux after washing. The choice depends largely on the thickness of the brazed parts, accessibility of fluxed areas and the adequacy of flux removal in the initial water treatment.

Agitation and turbulence improve the efficacy of any flux-removal treatment. Ultrasonic cleaning is effective for cleaning inaccessible areas, decreases the immersion time, and reduces the possibility of attack on the aluminum.

Checking for complete flux removal should be a routine inspection procedure. To detect the presence of flux, a few drops of distilled water are put on the surface to be tested and left there for a few seconds. The water is then picked off with an eyedropper and placed in an acidified

Table 5.2 Solutions for removing brazing flux from aluminum parts[3]

Type of solution	Composition	Operating temperature (°C)	Procedure (a)
Nitric acid	58 to 62% HNO_3 H_2O	20	Immerse for 10 to 20 min; rinse in hot or cold water.
Nitric–hydro–fluoride acid	58 to 62% HNO_3 48% HF H_2O	20	Immerse for 10 to 15 min; rinse in cold water; rinse in hot water; dry.
Hydrofluoric acid	48% HF H_2O	20	Immerse for 5 to 10 min; rinse in cold water; immerse in nitric acid solution (first entry in table); rinse in hot or cold water.
Phosphoric acid–chromium trioxide	85% H_3PO_4 CrO_3 H_2O	82	Immerse for 10 to 15 min; rinse in hot or cold water.
Nitric acid–sodium dichromate	58 to 62% HNO_3 $Na_2Cr_2O_7.2H_2O$ H_2O	60	Immerse for 5 to 30 min; rinse in hot water.

(a) Before using any of these solutions, it is recommended that the assembly first be immersed in boiling water to remove the major portion of the flux.

Fluxes, atmospheres

solution of 5% silver nitrate. If the solution stays clear, the metal is clean. If a white precipitate clouds the solution, flux is still present on the surface. Flux-removal procedures must then be repeated until the brazed assembly tests clean. Complete removal of the flux is essential, because it is corrosive to aluminum in the presence of moisture.[6]

5.1.10 Self-fluxing filler metals

When discussing fluxes, it is well to mention the self-fluxing brazing filler metals, even though many of them cannot be used for high-temperature applications. These filler metals contain small additions of phosphorus, lithium, and/or other elements that have high affinities for oxygen. The best known of these materials are the copper–phosphorus and silver–copper–lithium filler metals.

A self-fluxing filler metal designed specifically for high-temperature brazing of stainless steels and other heat-resistant alloys was developed in the USSR. The composition of this filler metal is Cu–28Ni–28Mn–5Co–1Si–1Fe, plus small additions of lithium, phosphorus, and sodium.[7]

Gaseous fluxes have been developed for brazing, but they are not used widely. Such fluxes are usually derived from fluoride compounds and are used as additives to neutral atmospheres. The activity of gaseous fluxes is low, and precautionary measures must be taken because flux decomposition products are often toxic. For example, an ammonium fluoride–argon atmosphere has been used to braze a stainless steel heat exchanger.[8] The decomposition products of ammonium fluoride are nitrogen, hydrogen, and hydrogen fluoride; hydrogen fluoride is very toxic. Other gaseous fluxes suggested for high-temperature brazing are the volatile halides of vanadium, boron, and tungsten for iron-based alloys, and zirconium chloride for nickel–chromium alloys.[9]

5.1.11 Safety precautions with fluxes

Whenever one uses fluxes it is good practice to try to prevent brazing fluxes from contacting skin. Occasional contact is not dangerous, but all flux should be thoroughly washed off before eating or taking anything into your mouth. Cuts or breaks in the skin must be properly covered by a dressing. Flux, especially if it contains fluorides or chlorides, can delay the healing of wounds.

Fluxes produce fumes when heated, especially above the temperatures given as their maximum. It is suggested to braze in work stations with large air space into which fumes can escape. Ventilate with fans or exhaust hoods to carry fumes away from workers, or equip operators with air-supplied respirators. More on Safety in Chapter 6.

5.2 ATMOSPHERES

The second way to control the formation of oxides during brazing and also reduce oxides present after precleaning is to surround the braze area with an appropriate controlled atmosphere. Like fluxes, controlled atmospheres are not intended to perform primary cleaning for removal of oxides, coatings, grease, oil, dirt, or other foreign materials. All parts for brazing must be subjected to appropriate pre-braze cleaning operations as dictated by the particular metals. When flux is used, a controlled atmosphere may be desirable to extend the useful life of a flux and to minimize post-braze cleaning. In controlled-atmosphere applications, post-braze cleaning is generally not necessary.

Controlled atmospheres are used extensively for high-temperature brazing. While performing the same basic function as fluxes (i.e., the prevention of oxidation during the brazing cycle), they have several advantages over fluxes:

1. The joint members are maintained in a clean, oxide-free condition when brazing is done in a controlled atmosphere. After brazing, the brazement can often be used in the as-brazed condition or finish machined without cleaning.
2. Controlled atmosphere brazing is particularly useful in joining complex assemblies such as heat exchangers, thrust chambers, and honeycomb sandwich structures. Complete removal of fluxes from such assemblies after brazing is difficult or impossible.
3. Problems associated with flux entrapment in the brazed joint can be avoided if controlled atmospheres are used.

Although many controlled atmospheres are available, those used primarily for brazing fall into three broad categories:

- reducing atmospheres;
- inert atmospheres;
- vacuum.

The reactions resulting from the use of reducing, inert gas, and vacuum atmospheres are diverse. Certain conditions, however, apply to all three. The general techniques of atmosphere brazing can involve:

- gaseous atmospheres alone;
- gaseous atmospheres together with a solid or liquid flux preplaced at the interfaces;
- high vacuum;
- combinations of vacuum and gas atmospheres.

5.2.1 Atmosphere application

One type of controlled atmosphere is the product of combustion of a torch flame: a neutral or reducing flame is normally used. Separately supplied controlled atmospheres may also be used with induction or resistance brazing. But controlled atmospheres are most commonly used in furnace or retort brazing operations. In fact, furnace brazing requires the use of a suitable atmosphere to protect assemblies against oxidation and, in the case of steels, against decarburization during brazing and during cooling, which is accomplished in chambers adjacent to the brazing, especially where titanium, zirconium, and refractory metals are concerned.

The principle followed in the use of controlled gas atmospheres involves the preparation of a special protective gas and its introduction into the furnace or brazing retort at pressures above atmospheric. As the gas is continuously supplied to the furnace and circulated through it, the furnace becomes purged of air. The protective gas atmosphere is maintained at a slight pressure, which prevents air from seeping into the brazing retort or furnace. In some operations, work is placed in a cold retort or furnace prior to purging, and the retort or furnace is not opened until the brazing cycle is completed. Where parts must be fed continuously or periodically into a furnace which is at brazing temperature, gas curtains or intermediate chambers are provided to avoid contamination of the furnace atmosphere.

The ability to control the composition and therefore the effectiveness of a furnace atmosphere depends not only on the condition and proper operation of the atmosphere-producing equipment but also on the proper set-up and operation of the furnace being used.

When certain types of controlled atmospheres, such as those containing hydrogen, are employed, extreme care must be taken to prevent the formation of explosive mixtures of gas. Mixtures of hydrogen with air ranging from 4 to 75% hydrogen are explosive.

As a safety precaution when potentially explosive gas atmospheres are used, the furnace or retort should be thoroughly purged with the gas to ensure the removal of all air before heat is applied. Waste gases from the furnace can be either continuously burned or directed into the open air outside the building.

Some atmospheres, such as those containing carbon monoxide, are toxic. Proper burning off or disposal of the waste gases from these atmospheres is especially important for safety. In brazing of toxic metals such as beryllium, waste gases should be carefully filtered or piped to an outside area.

Many brazing atmospheres are generated by passing metered mixtures of hydrocarbon fuel gas and air into a retort for reaction. Most of these atmospheres are rich exothermic mixtures in which the heat liberated from the reaction is sufficient to continue it. A rich exothermic atmosphere is the least expensive of the generated atmospheres, is adequately reducing for many applications, has relatively low sooting potential, and requires a minimum of generator maintenance. About 70 to 80% of all brazing atmospheres are exothermic, and they are generally used to braze mild steel or low-carbon steel.

Recently a firm attempted unsuccessfully to furnace-braze carbon steel components in an exothermic atmosphere using BAg–1 braze filler metal. They also found that the following filler metals were not acceptable since they contained cadmium and/or zinc and neither any atmosphere or even vacuum would allow for a good braze joint: BAg–1a; BAg–2; BAg–2a; BAg–3; BAg–27.

The cadmium in a braze filler metal will normally vaporize in an atmosphere and will be dropped out as a fine dust in the furnace, to be stirred up later, or it will be carried out into the room if there is an inadequate exhaust system. In a vacuum furnace however, the cadmium and zinc will be evaporated from the filler metal and deposited in the heat shielding and on the colder electrical insulators. This can cause considerable problems. Furthermore, this will occur even with the use of a partial pressure in the vacuum furnace. Thus, cadmium should not be used in a vacuum/inert atmosphere furnace.

Filler metals of silver-copper that contain zinc (but no cadmium) will also vaporize in an atmosphere, whether it has a dew point of +26.7°or down to –62°C. Sometimes flux is used to reduce vaporization and, while this works, it is not completely satisfactory. Furthermore, flux also introduces a cleaning operation.

It is recommended that, if needed, silver filler metals containing elements other than cadmium and zinc be used in atmosphere furnaces. In a vacuum furnace, it is necessary to use a partial pressure, particularly at higher temperatures, to prevent vaporization of silver and copper.

Exothermic or endothermic gases are chiefly made by the controlled combustion of natural or synthetic gases with air to form a mixture essentially composed of nitrogen, hydrogen, methane or ethane, carbon dioxide, carbon monoxide, and water vapor.

As the ratio of fuel gas to air is increased, a mixture becomes endothermic: it requires the addition of heat and a catalyst for combustion to occur. Endothermic gas mixtures are used in brazing medium and high-carbon steels and sometimes mild steels.

5.2.2 Atmosphere composition

The compositions of controlled atmospheres recommended for brazing cover a wide range, some of which are presented in the Appendix Table

A.16. These data are not intended as a comprehensive tabulation of atmosphere/metal combinations but rather as a general outline of some of the more widely used combinations.

Dew-point control

The combustion of gas mixtures results in a controlled atmosphere containing entrained moisture, which is largely undesirable in brazing. The moisture can sometimes be removed by condensation. The use of certain brazing filler metals, however, requires cooling in conjunction with absorption-type driers to reduce the dew point to satisfactory levels. Accurate dew-point control is especially important when dry hydrogen atmospheres are required because of the sensitivity to moisture of the metals usually brazed in this type of atmosphere. Dissociated ammonia atmospheres do not always require such accurate control.

The ability of pure hydrogen to reduce metal oxides is determined by the temperature, the oxygen content (measured as dew-point), and the pressure of the gas. Because furnaces typically operate at atmospheric pressure, only temperature and dew-point play a part.[3]

The diagram presented in Figure 5.1 is a plot of the dew point at which the oxide and the metal are in equilibrium at various temperatures. The 20 curves shown in this diagram define the equilibrium conditions for 20 pure metal/metal oxide systems. The positions of 13 additional elements whose curves fall outside the chart are also indicated. The oxides chosen for the calculations of this diagram represent the most difficult-to-reduce oxide of each metal.

The metal/metal oxide equilibrium curves slope upward and to the right for each metal. The region above and to the left of each curve represents conditions that are oxidizing for that metal. All points below and to the right of each curve cover the conditions required for reducing the oxides. The diagram therefore illustrates that the higher the processing temperature, the higher the dew point (or oxygen content) that can be used for any particular metal. In other words, a given purity of hydrogen becomes progressively more reducing at progressively higher temperatures. Or, to put it another way, the higher the brazing temperature, the lower the $H_2:H_2O$ ratio can be for any given metal.[3]

Use of the diagram in Figure 5.1 for practical purposes requires, first, that the correct curve be selected. For processing of any alloy, the element having the most stable oxide (farthest to the right) is the governing curve. If copper is to be brazed to stainless steel, for example, then a ratio suitable for reducing the chromium oxide must be selected. The chromium oxide curve applies, because chromium oxides are more stable than those of iron or nickel. Generally, it has been found that, when the

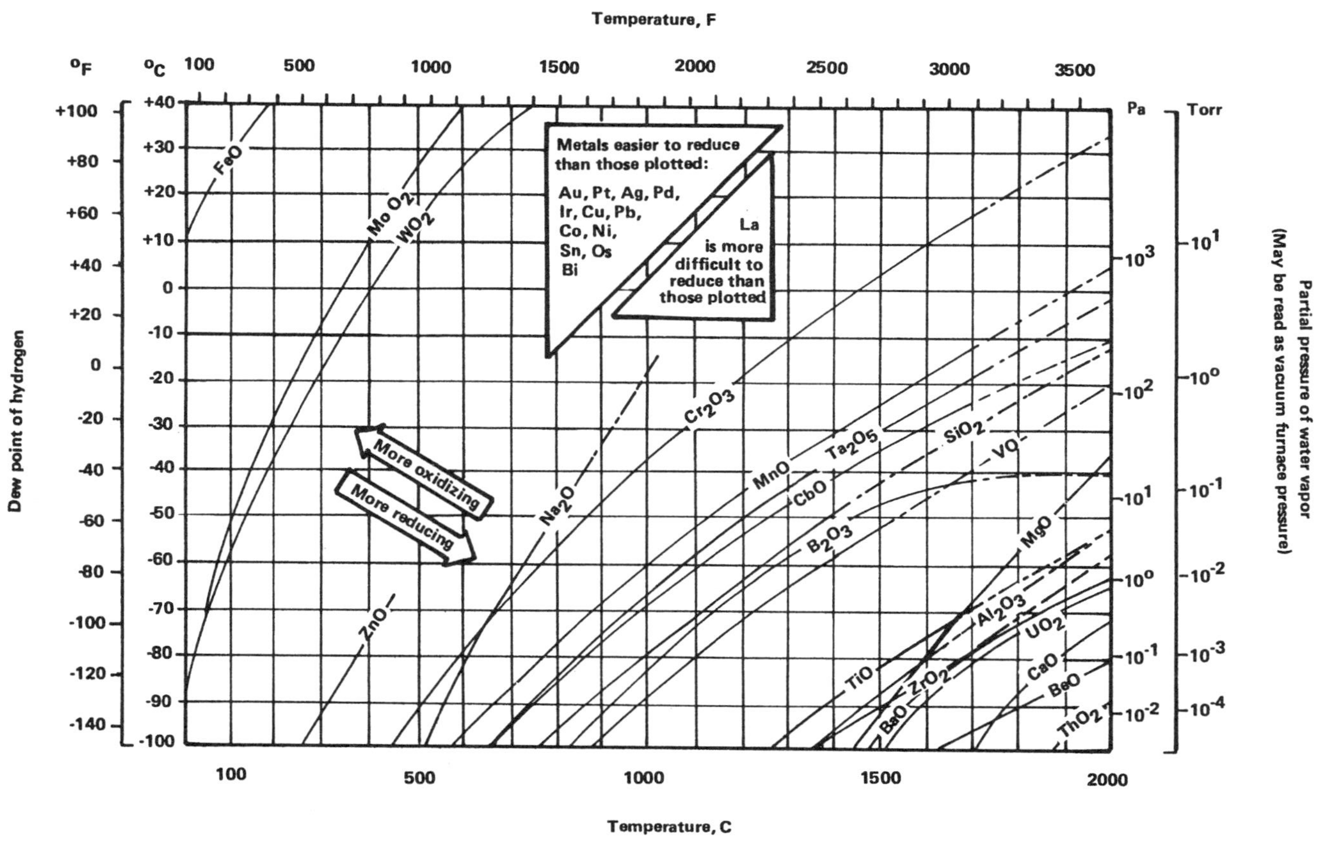

Fig. 5.1 Metal/metal oxide equilibria in hydrogen atmospheres.[3]

most difficult-to-reduce constituent of an alloy is present in more than about 1 at.%, a continuous film of its oxide is formed and its curve therefore is applicable. Alloys having a concentration progressively lower than 1 at.% of the most stable oxide-former appear to lie progressively closer to the curve of the next most stable oxide-former.

From Figure 5.1 it can be seen that chromium oxide can be reduced at 815°C if the dew point of hydrogen is lower than –56°C; at 1095°C, oxide reduction will occur in a hydrogen atmosphere with a dew point lower than –29°C. When a pure metal is brazed, the curve representing that metal is used to determine the temperature and hydrogen dew point at which oxide reduction will occur. In the case of an alloy, the curve for the element that forms the most stable oxide is used to determine the conditions for oxide reduction.

The need to control the dew point of the hydrogen atmosphere is evident from these considerations, and many devices and systems have been developed to accomplish this objective. To ensure accuracy, dew-point measurements should be made at the furnace outlet, rather than at the inlet.

In practice it is necessary to use hydrogen that has a somewhat lower dew point than that indicated by the curve for any given metal, partly because the surface becomes oxidized during heating until the temperature is reached that corresponds to the equilibrium temperature for that dew point. The reduction of these oxides formed during heating requires that sufficient time be allowed at conditions sufficiently below, or to the right of, the equilibrium curve. It is also necessary in practice to provide a continuous flow of hydrogen into the work zone during processing to sweep out the outgassed contaminants and thus maintain the necessary atmosphere purity at the metal surface.

A practical example of dew point is one involving an automotive fluid coupling which is a bowl-shaped part with a number of vanes brazed to the bowl and a stamping on the top part of the vane. The copper filler metal is only right at the brazed joint and in the fillet, but does not flash out over the rest of the low-alloy steel. The brazing takes place in a continuous furnace using a copper oxide paste, BCu–2.

In the past engineers considered that the flowing out of the copper from the joint, which is sometimes called flashing or blushing, indicated a better braze. In reality, both can have equal strength and, on occasion, if the filler metal is retained at the braze joint, there may be a somewhat larger fillet on a horizontal. However, not much difference can be obtained on a vertical joint, which is quite small due to gravity.

The primary controlling factor that determines the wetting of the filler metal across the alloy steel surface is the dew point of the atmosphere, everything else being equal. Brazing specialists found with a 7-to-1-ratio exothermic atmosphere that a standard dew point of around 16 to 21°C would cause copper to flow from the joint across the surface for some distance. To reduce the flow and keep it in a large clearance gap, it was necessary to increase the atmosphere dew point in the furnace to 32° and above, so this would hold the copper in the braze joint and still produce good, high-strength copper brazed joints in the press fit joint. In most furnace application experiences, the higher dew point was obtained by reducing the flow of atmosphere into the furnace, thus allowing the incoming parts to bring in the oxygen and moisture on their surfaces and internally, when the part was so designed. With less atmosphere flow, as the parts outgassed on heating, less of the partial pressure of oxygen was removed from the furnace with the flow of the atmosphere; thus increasing the dew point.

With everything else being held constant, the flow can also be altered by reducing the percentage of hydrogen, which means changing the ratio from seven parts of air to one part of gas to eight or nine parts of air to one part of gas.

The same copper flow effect would also be noted in a nitrogen/hydrogen atmosphere, in which there is normally 2% hydrogen in the nitrogen carrier gas. In this atmosphere, the change in dew point in the furnace will also change the Cu flashing characteristics of the copper on the carbon steel.

Many years ago, it was noted that in a hydrogen atmosphere, as the dew point increased to the (27°C to 32°C) dew point, that the copper stayed close to the joint and did not flash out onto the carbon steel. Conversely, as the dew point was lowered, more flashing occurred across the carbon steel. When dew points reached the −51 to 57°C range, it was found that you could flow the copper in a 0.025-mm thick joint in carbon steel out of the joint and across the surface of the carbon steel. Thus, while you needed very low dew points to braze chromium-containing steels, to prevent oxidation of chromium one found it impractical to run carbon steel parts brazed with copper in the same load, as you encountered difficulties keeping the copper in the joint area. When running carbon steel in hydrogen, you reduced the flow of drier hydrogen, thus allowing the buildup of the partial pressure of oxygen in the retort system so the copper would stay in the joint.

The copper brazing of carbon steel has been used for many years to fabricate large production quantities of parts and has been a very good tool for the brazing industry. However, a better understanding of the atmosphere, particularly on the shop floor, would be helpful in obtaining the desired results. Some people use a 21°C dew point atmosphere with very good success, and others feel that it is necessary to have a −1.5°C dew point to get proper results. In certain cases, these lower dew points are required, particularly if there is a higher chromium content in the low-alloy steel.

One last point on dew point control. A 300-series stainless steel was being brazed with a brazing atmosphere of nitrogen with 20% hydrogen. When dew point readings were taken, a different dew point was found when the flow of atmosphere into the furnace was increased. Therefore, the problem was why the dew point changed with different flow rates of atmosphere through the piping before it even reached the furnace?

In working with high-purity gases with low dew points, as would be coming from a liquid nitrogen tank and a hydrogen trailer, which originally came from a liquid system, these atmospheres work much different than you would normally expect gases to behave. The change in dew point with a change in flow rate through the piping system is a classic 'footprint', which is stating that there is a leak in the system. In directly stating the problem, any leak-out of a system of high-purity gases is also a leak into the system.

Since the potential across the leak is essentially with the same dry gas on the inside and same humid, oxygen-containing air on the outside of the leak, the diffusion potential is the same all the time. As the flow is changed in the piping system, we are passing more gas through the pipe, with the same amount of oxygen and moisture coming through the leak. Thus we dilute the amount of oxygen and moisture per cubic foot of gas flowing through the pipe. This is recognized at the dew point sensor as a lower dew point. Thus, when you will have given dew point at a specific flow rate and you increase the flow rate, the dew point would normally get better.

Noting this change in dew point with flow rate allows one to have a good indicator of leaks in their system, by just measuring the dew point at two widely different flow rates. If there is no difference in dew point, then the system does not have any leakage. However, if there is a difference in the dew point, this is an indicator that there are leaks in the system and maintenance of the piping system is required.

The dew point instrument can be a useful tool in detecting leaks in the dry gas supply system.

Practice has shown that certain metal oxides can be reduced to an extent in a high-purity inert-gas environment due to the low partial pressure of oxygen in the atmosphere. A continuously pumped vacuum roughly equivalent to the partial pressure of water vapor, as presented on the right hand ordinate of Figure 5.1, also gives similar results in practice. Because the equilibrium diagram is presented only for H_2/H_2O atmospheres, all the oxygen (O_2) in the hydrogen atmospheres must be converted to H_2O before the dew point is determined.

Such curves as those shown in Figure 5.1, and others which have been published,[10,11] aid in comprehending the actions of hydrogen and water vapor in respectively reducing and oxidizing metal oxides and metals, but they do not portray the complete story involved in the use of controlled atmospheres. They indicate neither the rate at which reduction will occur nor the physical form of the oxide.

Oxides of aluminum, titanium, beryllium, and magnesium cannot be reduced by hydrogen at ordinary brazing temperatures. If these elements are present in small amounts, satisfactory brazing can be done in gas atmospheres. When these elements are present in quantities exceeding 1 or 2%, the metal surface should be plated with a pure metal that is easily cleaned by hydrogen, or a flux should be used in addition to the hydrogen.

Accurate control of dew point cannot be overemphasized in gas atmosphere brazing if sound and completely bonded joints are to be produced in metals which form oxides of high stability.[12]

5.2.3 Atmosphere components (gases)

The components of brazing atmospheres have individual characteristics which affect their suitability for brazing various metals and alloys.

Carbon monoxide (CO)

CO is an active agent for the reduction of some metal oxides (e.g., those of iron, nickel, cobalt, and copper) at elevated temperatures. Carbon monoxide can serve as a source of carbon, which may be desirable in brazing some carbon steels but is undesirable in other applications. When decomposed, it may release oxygen, which is undesirable in many controlled atmospheres for brazing. Carbon monoxide can be generated from oil on the parts at brazing temperatures. Carbon monoxide is toxic, and adequate ventilation must be provided unless waste gas is trapped and burned.

Carbon dioxide (CO_2)

CO_2 is neutral to most metals and is an inert constituent of some brazing atmospheres, except when it is decomposed to carbon monoxide or carbon and oxygen, all of which are reactive with metals. It will oxidize iron, however, and some alloying elements, such as chromium, manganese, and vanadium. At high temperatures, CO is more stable; at low temperatures, CO_2 forms preferentially. Its presence may be undesirable as a source of oxygen, carbon, and carbon monoxide when decomposed.

In CO–CO_2 atmospheres, the carbon dioxide content of a furnace atmosphere can be undesirably increased by air leakage. In applications such as the brazing of carbon steels, it must be removed from the atmosphere to avoid oxidation and decarburization of the metal surfaces.

Methane (CH$_4$)

Methane may come from the atmosphere gas as generated from organic materials left on the part by inadequate cleaning. It can serve as a source of carbon and hydrogen. Methane is sometimes added to certain atmospheres to balance decarburizing gases present.

Oxygen (O$_2$)

Free oxygen in the brazing atmosphere is always undesirable. In addition to the sources already mentioned, oxygen may come from gases adsorbed on surfaces in the heating chamber.

Nitrogen (N$_2$)

Nitrogen is used in a controlled atmosphere to displace air from the furnace and to act as a carrier gas for the other atmosphere constituents. The typical high purity of nitrogen allows low levels of reducing gases to be used. A nitrogen atmosphere is applicable whenever exothermic gas or dissociated ammonia is used as the reducing agent. Nitrogen is inert to most metals, but high levels of nitrogen should be used cautiously when working with metals that are susceptible to nitriding, such as chromium and molybdenum.[13] Proper use of nitride-inhibiting atmosphere constituents can minimize nitrogen pickup where it is a concern. Nitrogen is noncombustible and nonexplosive and therefore is desirable from a safety standpoint.[14] There are several other advantages of nitrogen-based atmosphere.

Cryogenic nitrogen has a very low dew point, is a very dry gas, and so, when hydrogen (from exothermic reaction or from dissociated ammonia) is added, the resulting H$_2$:H$_2$O ratio is relatively high, which makes for a high reducing capacity, or good fluxing. In fact, a nitrogen-based atmosphere usually permits the required amount of hydrogen to be reduced to below the explosive level of the mixture.

Another advantage of nitrogen-based atmospheres stems from the elimination of chemical fluxes when they are used primarily to reduce oxides. The use of fluxes requires larger joint clearances to allow flux to escape and be displaced by the brazing filler metal. This may produce a weaker joint.

A particular advantage of a nitrogen-based atmosphere is that it can be tailored to provide just the right level of reduction, depending on the material being processed or the stage within the brazing cycle. For example, it may be desirable to have a slightly oxidizing atmosphere in the preheating section of a furnace to help burn off organic compounds used in paste-type brazing filler metals.

As a result of the above, provision for adjustments in furnace atmosphere composition can be made by introducing different compositions at different points in the cycle or, in the case of continuous furnaces, in different zones.

Finally, it should be noted that a nitrogen-based atmosphere containing methanol has been developed which results in a very dry atmosphere and subsequently in better wetting by brazing filler metals.[15]

A firm was brazing fuel manifold assemblies and encountered inconsistent results. Some parts brazed well while in others the braze filler metal balled up, wouldn't wet the surface, and would fall off. The materials being joined were AISI 347 stainless steel and N155 stainless steel, in a vacuum of approximately $1 \mu m$ at 1010 °C. The braze filler metal, 82Au–18Ni successfully wetted the 347 stainless steel but did not wet the N155. This result tended to indicate that there was no problem in satisfactorily wetting the 347 stainless steel, however the 'footprint' left by the N155 gave a clue relative to where to attack the problem.

The main difficulty in using N155 is the nitrogen content (0.10 to 0.12%) in the steel. This is a lot of nitrogen and, depending on the pumping system and atmosphere quality, a low nitrogen content may or may not braze adequately. Usually a high nitrogen content of 0.20% could be expected to cause problems. Since this part was brazed in vacuum furnace, some outgassing of the nitrogen will occur, however, enough nitrogen remains at the surface to prevent wetting and flow of the filler metal.

One difficulty in brazing this type of material is the consistently varying results, depending on the number of pieces in the furnace, quality of the atmosphere, heating rate, and nitrogen content in the base metal. When production quantities are put into the furnace, they would not braze at all. This inconsistency occurs when the production quantity dumps out a lot of nitrogen into the furnace and causes a nitrogen buildup on the surface of the N155. Whereas, in the case of a single part in the furnace, assuming that the atmosphere and other variables are good, the nitrogen layer will be sufficiently removed from the surface and the part can achieve adequate brazing.

If the vacuum furnace is set up with a nitrogen backfill and there happens to be a small leak, there may be a good enough vacuum for normal brazing. However, with extra nitrogen in the atmosphere, this could cause some variation in the existing problem.

If the nitrogen-containing base metal has a fairly high content of nitrogen on the surface, it will appear to have an iridescent bluish grey color. Under these circumstances, you should not expect adequate wetting and flow on the surface of this base metal. In some base metals with a low nitrogen content, a pretreatment, consisting of a bake-out of the detail parts in a separate vacuum furnace load, at or above the brazing temperature, can outgas sufficient amounts of nitrogen to allow adequate brazing. Unfortunately, on some base metals, and furnace equipment, the precleaning bake-out cycle has not proven adequate to remove enough nitrogen to allow suitable brazing. Experiments to ensure that the proper thickness of nickel plating is used with various nitrogen contents in the base metal should be conducted. If there is nitrogen in the atmosphere, from leakage or backfilling for partial pressure brazing, the thickness of electrolytic nickel may have to be greater. A good starting point would be with 0.01 mm thickness of the electrolytic nickel plating. Finally it is also good practice to run a prototype set of parts to determine if adequate vacuum coverage and braze filler metal selection are proper and suitable.

Inorganic vapors

In equipment designed for their use, vapors such as those of zinc, cadmium, lithium and fluorine compounds can serve to reduce metal oxides and scavenge the atmosphere of oxygen. They are useful for replacement of constituents of alloys evolved during brazing. Such vapors are toxic, and proper safety precautions should be used.[16]

Hydrogen (H_2)

Reducing atmospheres not only prevent the formation of surface oxides on the base metal at the brazing temperature, but also reduce residual surface oxides and the oxides that form during the low-temperature stages of the heating process. Although some reducing atmospheres can be used to braze metals whose oxides are easily reduced, hydrogen and atmospheres containing large amounts of hydrogen (e.g., cracked ammonia) are most suitable for high-temperature brazing.

Hydrogen is one of the most active agents for reducing the oxides of many metals during brazing. If an oxidized metal is heated to a sufficiently high temperature in a dry hydrogen atmosphere, the oxide will be reduced and water vapor will form. Oxide reduction continues until the amount of water vapor increases to the point where the ratio of H_2O to H_2 reaches equilibrium for the metal oxide at that particular temperature; further oxide reduction will not occur unless the moist hydrogen is replaced by dry hydrogen. Although all metal oxides can be reduced, some are more difficult to reduce than others, for example, hydrogen with a much lower dew point at a given temperature is required before oxide reduction can occur. Figure 5.1 is the most useful diagram in determining the condition in pure dry hydrogen (temperature and hydrogen dew point) under which a particular metal oxide will be reduced.

Heat-resistant base metals that contain appreciable amounts of aluminum and/or titanium are difficult or impossible to braze in a hydrogen atmosphere, because the oxides of these metals cannot be reduced at the temperatures used for brazing. From Figure 5.1 it can be seen that a hydrogen dew-point of –90°C or lower is required to reduce titanium oxide at 1370°C; even lower dew-points are required for reduction of aluminum oxide. Although hydrogen dew-points below –73°C can be obtained under laboratory conditions, they are difficult to obtain and maintain in production. However, several approaches to the problem of brazing metals alloyed with titanium and/or aluminum can be considered:

1. The joint can be brazed in a vacuum.
2. The surfaces of the joint members can be plated with nickel, copper or another metal that does not form a refractory oxide and is easily wetted by the brazing filler metal.
3. The joint members can be oxidized in a wet hydrogen atmosphere, and then the oxides of titanium or aluminum can be leached from the joint surfaces in a nitric acid–hydrofluoric acid solution. Surfaces relatively free from titanium and aluminum oxides can be obtained in this manner. Long brazing cycles should be avoided, because aluminum and titanium may diffuse to the surface and reoxidize.
4. A high-temperature flux can be used to prevent the formation of oxides and promote wetting.

Of the techniques described above, plating is the most straightforward method for production brazing of heat-resistant alloys containing aluminum and/or titanium.

Several grades of hydrogen are available for brazing, and the grade that is appropriate to the application should be selected. Dissociated ammonia has a high hydrogen content and is, therefore, a very reducing atmosphere. It is used mostly in brazing of stainless steels or mild nickel alloys.

The use of boron (B)-containing filler metals is not suitable in N_2-containing furnace atmosphere. B combines with N_2 to produce a black boron nitride. Depending on the percent nitrogen, also time and temperature, there may be only a small reaction or a reaction sufficient to put a heavy black layer on the top of the filler metal and to prevent all filler metal flowing. Taking the work to 760°C, with N_2 as an atmosphere, certainly would not be recommended for B-containing filler metals. Purging at room temperature may also be objectionable when the flow rate of the atmosphere is not sufficient to remove all of the N_2 by 538°C or before (preferably 204°C).

The B–N$_2$ reaction, like so many others, is a time-rate reaction. Experience has shown the following reactions:

1. Dissociated NH$_3$ (25%N$_2$–75%H$_2$) is suitable for a fast induction brazing operation, yet is not suitable for furnace brazing with a heating rate of 15 min from room temperature (RT) to brazing temperature.
2. 0.4% N$_2$ in pure dry H$_2$ is suitable for furnace brazing with a heating rate of 15 min from RT to brazing temperature, but is not suitable for furnace brazing with a heating rate of 60 min from RT to brazing temperature.
3. 0.2% N$_2$ in pure dry H$_2$ is suitable for furnace brazing with a heating rate of 60 min from RT to brazing temperature, and is not suitable for furnace brazing with a heating rate of 7 h from RT to brazing temperature.
4. 0.01% N$_2$ in H$_2$ was needed for furnace brazing with a heating rate of 7 h from RT to brazing temperature, yet is not suitable with a heating rate of 7 days from RT to brazing temperature.

The above observed examples are based on filler metals containing from 2–3.5% B and it is not known how this reaction decreases with decreasing B content.

When Ni brazing in N$_2$-based atmospheres, filler metals containing Si or P will be suitable as long as the surface is not nitrided or a light blue-gray N$_2$ film is not present. This film will occur from N$_2$-containing base metals or base metals heat treated in a N$_2$-based atmosphere where interstitial N$_2$ is picked up in the base metal.

B-containing nickel filler metals should be used in pure dry H$_2$, A–H$_2$ or vacuum atmospheres.

Water vapor

Water vapor is objectionable because its presence can promote oxidation or cause decarburization. Water vapor may be added intentionally by controlled humidification or unintentionally due to air leakage, air carried into the furnace with the work, reduction of metal oxides, leakage from water jackets, contaminated gas lines, diffusion of oxygen through inadequate flame curtains, and other less obvious sources. However, a carefully controlled amount of water vapor will aid in cleaning of carbonaceous material from brazed surfaces and in removal of binder left behind by brazing filler metals. In addition, water vapor can be used to inhibit filler metal flow where filler metal containment is desirable. This is particularly beneficial in brazing of wide-gap joints. The amount of water vapor required in the latter instance depends on the amount of hydrogen that is present in the atmosphere.

Table 5.3 Relationship between dew-point temperature and moisture content[3]

Dew-point temperature		Moisture content	
(°C)	(°F)	(vol %)	(ppm)
−18	0 .... 0.150		1500
−34	−30 .... 0.0329		329
−51	−60 .... 0.0055		55
−62	−80 .... 0.0014		14
−73	−100 .... 0.0002		2

The reducing ability of a hydrogen-based atmosphere depends primarily on the $H_2:H_2O$ ratio, which must be higher than 10:1 if the atmosphere is to be effective. The amount of water in an atmosphere is specified by the dewpoint, the temperature at which moisture in the gas will condense. The relationship between dew-point temperature and moisture content of gases is shown in Table 5.3.

Argon and helium (A and He).

Inert gas atmospheres can be used to braze most metals. However, they are most useful in brazing base metals whose properties are adversely affected by exposure to hydrogen. For example, alloys of titanium, zirconium, niobium, and tantalum are extremely sensitive to the presence of minute quantities of hydrogen and become embrittled. Such metals can be brazed satisfactorily in a controlled atmosphere of argon or helium; they can be brazed equally well in a vacuum. The parts to be brazed must be cleaned and handled carefully, because the primary purpose of the inert gas atmosphere is to prevent the formation of oxides during brazing.

Inert gases such as helium and argon form no compounds with metals. In equipment designed for their use, they inhibit evaporation of volatile components during brazing and, when used at ambient or partial pressure, reduce the evaporation rate of volatile elements relative to vacuum.

Although both argon and helium can be used for controlled-atmosphere brazing, argon is most frequently used. Commercial argon is available at a guaranteed as-delivered purity of 99.996%, or no more than 40 ppm total impurities. Assuming that water vapor is the only impurity, this impurity level corresponds to a dew point of about -50°C. The actual dew point may be considerably lower, because impurities other than water vapor may be present in the gas. As a result, argon can often be used as a protective atmosphere without further purification.

Vacuum

Vacuum brazing has made great strides in the past decade, and there is every reason to believe that the usefulness of this process will continue to grow in the future. The growth of vacuum brazing is due in part to improvements in equipment design and performance. Vacuum brazing is particularly well suited for joining:

 (a) heat-resistant nickel- and iron-based alloys that contain aluminum and/or titanium;
 (b) reactive metals;
 (c) refractory metals;
 (d) ceramics.

Vacuum conditions are especially well suited for brazing very large, continuous areas where solid or liquid fluxes cannot be removed adequately from the interfaces during brazing, and where gaseous atmospheres are not completely efficient because of their inability to purge occluded gases evolved at close-fitting brazing interfaces. Vacuum is also suitable for brazing many similar and dissimilar base metals, including titanium, zirconium, niobium, molybdenum, and tantalum. The characteristics of these metals are such that even very small quantities of atmospheric gases may result in embrittlement and sometimes disintegration at brazing temperatures. These metals and their alloys may also be brazed in inert-gas atmospheres if the gases are of sufficiently high purity to avoid contamination and the resultant loss in properties of the metals.

 Compared with other types of brazing, vacuum brazing has the following advantages and disadvantages:

1. Vacuum removes essentially all gases from the brazing area, thereby eliminating the necessity of purifying a supplied atmosphere. During brazing, the pressure within the furnace is maintained at a level such that oxidation of the workpieces does not occur. Commercial vacuum brazing facilities operate in the range from 0.013 to 0.00013 Pa; the gas impurity level corresponding to a pressure of 0.013 Pa is about 0.3 parts per million (ppm). The actual pressures used depend on the materials being brazed, the brazing filler metals being used, the area of the brazing interfaces, and the degree to which gases are expelled from the base metals during the brazing cycle.
2. Certain oxides of base metals will dissociate in vacuum at brazing temperatures. Vacuum is used widely to braze stainless steel,[17,18] superalloys, aluminum alloys, and refractory materials by special techniques. The mechanism of oxide removal in a vacuum is not clearly understood. Oxide films can be removed by evaporation,

dissociation, diffusion, or a combination of diffusion and chemical reaction.

3. The low pressure existing around the base and filler metals at elevated temperature removes volatile impurities and gases from the metals. Frequently the properties of the base metals themselves are improved. This characteristic is nevertheless a disadvantage where elements of the filler metal or base metals volatilize at brazing temperatures because of the low surrounding pressure. This tendency can be corrected by adherence to proper vacuum brazing techniques.

Many vacuum furnaces have the ability to operate under a partial pressure of inert gas. There are two general types of vacuum brazing then: brazing in a high vacuum, and brazing in a partial vacuum. High vacuum is particularly well suited for brazing of base metals containing hard-to-dissociate oxides.[12]

Partial vacuums are used where the base metal or filler metal, or both, volatilize at brazing temperatures under high-vacuum conditions. The lowest pressure at which the metals will remain in the solid or liquid phase at the brazing temperature is determined by calculation or experimentation. The brazing chamber is evacuated to high-vacuum conditions. The heating cycle proceeds under high vacuum until just below the temperature where vaporization would begin. High-purity argon, helium or, in some instances, hydrogen is gradually introduced in sufficient amounts to overcome the vapor pressure of the volatile metals at brazing temperature. This technique appreciably widens the range of materials for which vacuum brazing is effective.

Vacuum purging prior to high-purity dry-hydrogen brazing is frequently employed where extra precautions must be taken to ensure optimum freedom from foreign or contaminating gases. Similarly, dry-hydrogen or inert-gas purging prior to evacuation is sometimes helpful in obtaining improved brazing results in a high-vacuum atmosphere.

Zirconium, titanium, and other elements with high affinities for oxygen and other gases are sometimes strategically placed close to, but not in contact with, the part being brazed in a high-vacuum atmosphere. These so-called "getters" rapidly absorb very small quantities of oxygen, nitrogen and other occluded gases which may be evolved from the metals being brazed and thus improve the quality of the brazing atmosphere.

Another method of reducing contamination in vacuum is by another gettering technique. Lithium, magnesium, sodium, potassium, calcium, titanium and barium can all be vaporized in the chamber to reduce the volume of oxides and nitrides present in the vacuum atmosphere. These materials may condense on the chamber walls; therefore, care must be taken. Their disadvantage is that most of them will either react with the workload or form a coating on the wall when exposed to atmospheric moisture.

Successful brazing in vacuum depends on the presence of a promoter –either a metal or a reactive gas. The key action of metal promoters in vacuum brazing is to reduce chemically the oxide films to permit wetting by the filler metal. In addition, they must also scavenge remaining oxygen and moisture in the vacuum, but these are not the key mechanisms. Many metals can fulfill the function of a braze promoter, but magnesium is the best. Magnesium contained in the brazing filler metal does double duty. In the case of aluminum, as it vaporizes, magnesium tends to disrupt the aluminum oxide at the aluminum boundary layer, and later, as a vapor in the vacuum chamber, it reacts with oxygen and oxides (mainly water) to reduce or eliminate the formation of additional aluminum oxide surface films that could inhibit good wetting and capillary flow. Table 5.4 shows the metal activators and their actions in vacuum brazing.

5.2.4 Atmospheres, base metals and filler metals

The protective atmospheres most commonly used in furnace brazing with silver alloy brazing filler metals are rich exothermic gas, endothermic gas, dissociated ammonia, dry hydrogen, and commercial nitrogen-based atmosphere blends, principally with hydrogen. Even when a flux is used, an atmosphere is usually employed to minimize or prevent oxidation and discoloration of the base metals and to ensure that the flux performs its functions.

Exothermic and endothermic atmospheres are less expensive than dissociated ammonia or dry hydrogen; they are used in furnace brazing of steel-to-steel or to oxygen-free copper or copper alloys, using a flux and BAg–1a or BAg–5 filler metal.

Table 5.4 Possible promoters (activating metals) in vacuum brazing[3]

Activating metal	Removes oxygen?	Removes water vapor?	Vaporizers?	Reacts with aluminum oxide?	Promotes vacuum brazing?
Rare earths, beryllium, scandium, yttrium	Yes	Yes	No	Yes	Yes
Magnesium, calcium, strontium, lithium	Yes	Yes	Yes	Yes	Yes
Barium, sodium, zinc	Yes	Yes	Yes	No	No
Antimony, bismuth	Yes	No	Yes	No	No

Provided that dissociation is complete (100%), dissociated ammonia can be used in nearly all furnace brazing applications involving the use of silver alloy brazing filler metals.

Case History

A problem was encountered by a brazing firm whereby they were obtaining a high leak rate on a copper and brass valve assembly. The main body and fitting of the assembly were brass and the tube was copper. The brazing filler metal was BAg–1 and the parts were brazed in a continuous furnace at 704–732°C. The atmosphere used was dissociated NH_3 and a liquid flux to assist brazing. The parts were previously torch brazed successfully with a minimum number of rejections, however production demands shifted the brazing method to furnace. Why then was there a problem?

Zinc- and cadmium-containing filler metals should not be used in a controlled-atmosphere furnace, thus BAg–1 filler metal is not suitable for this furnace atmosphere. The first problem with the filler metal is that it contains approximately 24% cadmium, which is considered to be a poisonous material. It is readily dissociated from the filler metal, and either deposited in the furnace as a powder or carried out in the atmosphere; in both cases leading to a hazardous condition. This is not the proper use of this type of filler metal. Secondly, the 16% zinc also dissociates, thus leading up to 40% loss in the volume of filler metal applied to the joint. A 1.59 mm diameter wire was used in torch brazing and was reduced to a 1.0 mm diameter wire for furnace brazing; thus the filler metal has been decreased by 59%. Up to an additional 40% of filler metal can be lost by evaporation of the zinc and cadmium from the BAg–1 filler metal, leading to an overall reduction of 75.4%, which is an insufficient quantity of filler metal to fill the large joint properly. An additional problem could be caused by continual vaporization of the zinc and cadmium, adding to some porosity in the joint area.

The liquid flux used in this instance ran down on the fitting part in one area and appeared to keep the base metal bright and clean in that area, indicating the flux was doing its job. It also left a residue on the filler metal fillet, indicated by black spots on the surface of the fillet, meaning additional cleaning was required. Flux used in an atmosphere furnace will vaporize partially and build up a flux residue in the cooler areas of the furnace. Since the fluxes are usually hygroscopic materials, they will pick up moisture when the furnace is shut down, causing considerable difficulty when a dry atmosphere is required.

Therefore, in summary, the filler metal should be changed to a silver-copper or silver-copper-tin composition, which would not have any elements that would vaporize in a dissociated NH_3 atmosphere. A better-quality part and joint would be obtained if the fitting part and casting (main body part) were brass plated, particularly in the joint area. Plating the ID of a hole, even though it is approximately 19 mm in diameter is difficult, as the throwing power is not too good and internal electrodes may have to be used. With more filler metal in the joint, no bubbling from vaporization, and covering up the brass with a copper-tin plating including in the joint area, the quality of the brazement should improve. However, more filler metal is the most important variable.

5.2.5 Other considerations

Protective atmospheres are used in furnace brazing of copper and copper alloys, although numerous exceptions exist. Exothermic-based, endothermic-based, dissociated ammonia, and other suitable prepared atmospheres are widely used to protect copper and copper alloys that are not adversely affected by hydrogen at elevated temperatures.

Depending on the base metal/filler metal combination, the use of a brazing flux may be avoided by brazing in either a dry-hydrogen or a prepared nitrogen-based atmosphere. These are among the more expensive furnace atmospheres; however, at low dew points, they have the advantage of being highly reducing. Prepared exothermic-based atmospheres are considerably less expensive and are effective in preventing oxidation at elevated temperatures, but the reducing potential of these atmospheres is limited because hydrogen content does not exceed about 13%, and consequently they cannot be used as substitutes for chemical fluxes.

Vacuum atmospheres are suitable for brazing of copper and copper alloys when the alloys contain few or no elements having high vapor pressures at the brazing temperature (lead, zinc, etc.). The brazing filler metal should also be restricted to vacuum-grade materials containing low contents of high vapor pressure elements, such as zinc and cadmium, see Appendix Table A.3.

All of the refractory metals have been successfully brazed in inert, reducing or vacuum atmospheres. The environment in which the refractory metals are brazed is determined by the reactivities of these metals with oxygen, hydrogen and nitrogen as well as the effects of these elements on the mechanical properties of the refractory metals.

All of the refractory metals react with oxygen at moderately elevated temperatures, but they form different types of oxides. Niobium and tantalum form hard, adherent oxides at temperatures above 205 and 400°C, respectively. On the other hand, molybdenum and tungsten form volatile oxides at temperatures above 400 and 510°C, respectively. In either case, the surfaces of the refractory metals must be protected during brazing to ensure wetting by the brazing filler metal. Also, these metals must be coated with an oxidation-resistant material if they are exposed in air at elevated temperatures. For such service conditions, the filler metal must be compatible with both the base metal and the coating.

Niobium and tantalum are embrittled by the presence of hydrogen at relatively low temperatures. In contrast, molybdenum and tungsten can be brazed in a hydrogen atmosphere.

Niobium, molybdenum, and tantalum are embrittled by nitrogen at high temperatures; however, reactions with nitrogen begin at relatively low temperatures. For example, nitrogen is dissolved in molybdenum at temperatures as low as 595°C, but severe embrittlement does not occur until the temperature exceeds about 1095°C. Tantalum behaves in much the same manner.

Tungsten can be brazed in an inert-gas atmosphere (helium or argon), a reducing atmosphere (hydrogen) or vacuum. Two precautions should be observed for vacuum brazing. The vapor pressures of the compositional elements of the brazing filler metals should be compatible with the soundness of the base metals, and the deposited filler metal should be evaluated.

Purified, dry hydrogen and inert gas (helium and argon) atmospheres have been found suitable for brazing molybdenum. For brazing pure molybdenum, the purity of the hydrogen atmosphere is not critical. A dew point of 26.7°C can be tolerated in hydrogen when reducing molybdenum oxide. A low dew point (–46°C or lower) at 1205°C is required for brazing of titanium-bearing molybdenum alloys. Vacuum furnaces have been used for brazing of molybdenum; however, precautions should be taken in selecting a proper filler metal that will not volatilize during the brazing cycle. Pressures of less than 0.13 Pa are desirable for brazing of molybdenum alloys containing titanium.

Inert and vacuum atmospheres are satisfactory for brazing of titanium, zirconium, and beryllium.

A vacuum atmosphere of 0.13 Pa or better is required for brazing of titanium, whereas the dew point of an argon or helium inert atmosphere should be –57°C or less to prevent discoloration of the titanium and produce satisfactory brazed joints at temperatures from 760 to 925°C.

Fabricators of titanium brazements generally favor liquid argon from cryogenic storage vessels because of its inherent purity.

Although information is limited on brazing of zirconium as compared with brazing of titanium, these two metals are very similar chemically. In general, the brazing techniques used are applicable to both systems.

Beryllium is most commonly brazed in purified argon, although helium and vacuums of less than 0.13 Pa are also suitable.

Finally, where premetallized ceramics are to be brazed, the operation can be carried out in high-purity inert gas, hydrogen or vacuum atmospheres. If the ceramic is to be brazed directly – i.e., without being premetallized – a vacuum atmosphere is preferred.

The current increased interest in ceramics[19-33] and cemented carbides[34] has created new and expanded work in dew-point curves, gas mixtures, and reducing atmospheres to allow wetting by brazing filler metals to take place.

REFERENCES

1. Blanc G. M. A., Colbus J., Keel C. G., Notes on the assessment of filler metals and fluxes, *Weld. J.*, 42(5), May 1961, pp. 210s–222s.
2. *Research Studies of Brazing Fluxes*, Bulletin F-37, United Wire and Supply Corp., Providence, R1, 1952 (11 pages).
3. Schwartz M. M., *Brazing*, ASM International, Metals Park, OH, 1987, 440p.
4. Schillinger D. E., Addison, Jr, H. J., Relationships among brazing defects and brazing conditions, *Weld. J.*, 56(10), Oct. 1975, pp. 321s–329s.
5. No. 554 Classifying brazing fluxes, *AWS Brazing Handbook* and AS 3–92 Spec. for fluxes for brazing and braze welding.
6. Kay W. D., Postbraze cleaning of silver brazed joints, *Weld. J.*, 57(10), Oct. 1976, pp. 872–873.
7. Gubin A. J., Bobkina E. N., Self-fluxing filler metals for the brazing of stainless steels and heat-resistant alloys, *Weld Prod.*, 13(8), 1966, pp. 48–52.
8. Savichev R. V., Kulik S. G., Brazing heat exchangers made from Kh18N10T steel, *Weld. Prod.*, 15(1), 1968, pp. 66–67.
9. Cibula A., The soundness of high-temperature brazed joints in heat resisting alloys, *Brit. Weld. J.*, 5(5), 1958, pp. 185–201.
10. Sibley A. T., Ahuja R. K., Buck D. M., Choosing a nitrogen-based atmosphere, *Metal Powder Report*, 36(1), 1981.
11. *Brazing Handbook 4th Ed.*, AWS C3 Comm. on brazing and soldering, 493p, 1991.
12. Sakamoto A., Wetting in vacuum-inert gas partial pressure atmosphere brazing, *Weld. J.*, 64(10), Oct. 1983, pp. 272s–281s.
13. Nitrogen furnace atmosphere particularly favorable in commercial brazing of stainless steel tubing, *Industrial Heating*, Aug. 1982, pp. 46–47.
14. Bannos T. S., *The effect of atmosphere composition on braze flow*, Air Products and Chemicals, Inc., 1983 (6 pages)
15. Whitman W., Nelson T. A., Solomon J., Furnace brazing with nitrogen atmospheres, *Weld. J.*, 61(10), Oct. 1980, pp. 21–25.
16. American National Standard Z49.1, *Safety in welding and cutting*, AWS, 550 N.W. LeJeune Road, P.O. Box 351040, Miami, FL 33135.
17. Hooven W. T., Vacuum furnace brazing of missile guidance component, *Industrial Heating*, Aug. 1982, pp. 43–44.
18. Pattee H. E., High-temperature brazing, *WRC Bull.* #187, Sept. 1973 (47 pages).
19. Chang W. H., A dew-point/temperature diagram for the metal/metal oxide equilibria in hydrogen atmosphere, *Weld. J.*, 35(12), Dec. 1956, pp. 622s–624s.
20. Rey M. C., Kramer D. P., *et al.*, Dew-point/temperature curves for selected metal/metal oxide systems in hydrogen atmosphere, *Weld. J.*, 65(5), May 1984, pp. 162s–166s.
21. Williams R. O., *Thermodynamics of copper–nickel alloys containing aluminum, silicon, titanium and chromium relative to their use in ceramic brazing*, ORNL–6072, Metals and Ceramics Division, Oak Ridge National Laboratory, Contract No. DE–ACO5–840R21400, Nov. 1984 (16 pages).

22. Nicholas M. G., Mortimer D. A., Ceramic/metal joining for structural applications, *Mater. Sci. Technol.*, 1(9), 1985, pp. 657–665.
23. Pattee H. E., Joining ceramics to metals and other materials, *WRC Bull.* #178, WRC, New York, 1972.
24. Mizuhara H., Huebel E., Joining ceramic to metal with ductile active filler metal, *Weld. J.*, 65(10), Oct. 1984, pp. 43–51.
25. Vilpas M., *Joining of ceramics for high-temperature applications*, NASA TT–20030, N87–29678, NTIS HC A03/MF A01, Oct. 1987; translation of *Korkeissa lampotiloissa kaytettavien keraamisten materiaalien liittaminen*, Rept. UTT–TIED–481, Tech. Res. Center of Finland, Espoo, Aug. 1985.
26. Hammond J. P., David S. A., Santella M. L., Brazing ceramic oxides to metals at low temperatures, *Weld. J.*, 67(10), Oct. 1988, pp. 227s–232s.
27. Mizuhara H. Vacuum brazing ceramics to metals, *Advanced Metals and Processes*, Feb. 1987.
28. Rice R. W., Joining of ceramics, in *Advances in Joining Technology*, Chestnut Hill, MA, Brookhill Publishing Co., 1976, 69–111.
29. Morrell R., *Ceramics in Modern Engineering*, Institute of Physics, London, UK, 0305–4624/84/050252, 1984, 252–261.
30. Weymueller C. R., Braze ceramics to themselves and to metals, *Weldg. Des. & Fab.*, Aug. 1987, 45–48.
31. Becher P. F., Halen S. A., Joining of Si_3N_4 & SiC ceramics via solid-state brazing, *Proceedings of the DARPA/NAVSEA Ceramics Gas Turbine Demo. Engine Prog. Rev.*, Columbus, OH, MCIC, 1978, 649–653.
32. Loehman R. E., Transient liquid phase bonding of silicon nitride ceramics, in *surfaces and interfaces in ceramics and ceramic-metal systems*, Matls. Sci. Res., 14, New York, Plenum Press, 1981, 701–711.
33. Tallman R. L., Neilson, Jr. R. M., Mittl J. C., *et al.*, Joining silicon nitride-based ceramics: a technical assessment, EG&G Idaho, Inc., ECG–SCM–6572, DE84–011356, DOE Cont. DE–AC07–761D01570, March 1984.
34. Aa. Thorsen K., Fordsmand H., Praestgaard P. L., An explanation of wettability problems when brazing cemented carbides, *Weld. J.*, 65(10), Oct. 1984, pp. 308s–315s.

6 Joint Design, Tooling/Fixturing, Surface Preparation, Inspection and Safety

6.1 JOINT DESIGN

If a joint design works reliably in the application for which it is intended, it is a 'good' joint design whether it is a very simple or a very sophisticated design. However, the design of a brazed joint requires some special considerations dictated by the nature of the joining process:

1. *Composition and strength of the brazing filler metal.* Generally, the bulk strength of the brazing filler metal is lower than that of the base metals, and so a correctly designed joint is required to obtain adequate mechanical strength.
2. *Capillary attraction.* Because brazing depends on the principle of capillary attraction for distribution of the molten brazing filler metal, joint clearance is a critical factor affecting the brazing process.
3. *Flux and air displacement.* Not only must the filler metal be drawn into the joint, but flux and air must be displaced from it. This requirement influences joint design and clearances.
4. *Type of stress.* In general, it is preferred that any load on a brazed joint be transmitted as shear stress rather than tensile stress. Different style of joints subject the filler metal film to different stresses and therefore alter its behavior under stress. These stresses usually are tensile stresses in butt joints, shear stresses in lap joints and tensile and/or shear stresses in scarf joints.
5. *Composition and strength of the base metals.* In a joint made according to recommendations between high-strength members, the filler metal film in the joint may actually be stronger than the base metal itself. The rigidity and freedom from yielding of the members of the joint confine the filler metal film between them, causing the film to have properties different from those of the base metal. Therefore, it is

possible for joints in high-strength base metals to be stronger than joints of the same type in lower-strength base metals.

6.1.1 Types of joints

One design factor that can be altered for best brazing results is the type and bonding area of the joint. This is an important factor as well as clearance between members of the joint. Both of these affect not only the strength of the completed joint but also the ease of brazing. Several factors influence the selection of the type of joint to be used. These factors include the brazing process to be used, fabrication techniques prior to brazing, the number of items to be brazed, the method of applying the brazing filler metal, and the ultimate service requirements of the joint.

There are basically only two types of brazed joints: butt and lap. All other joints are really only modifications of these two basic types. Common types of brazed joints are shown in Figure 6.1.

Butt and lap joints[2]

The butt joint has the advantage of a single thickness at the joint. Preparation is relatively simple, and the joint has sufficient strength for

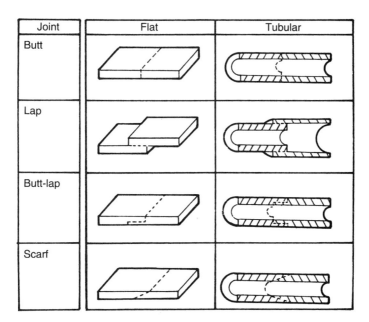

Fig. 6.1 Types of brazed joints.[1]

many applications. However, the strength of any joint depends in part on the bonding area available, and in a butt joint this area is determined by the thinnest member of the joint. The thinnest member, therefore, dictates the maximum strength of the joint.

Another drawback of the butt joint is that almost all of the load is transmitted as tensile stress, which is not very desirable. For these reasons a butt joint should be chosen only when the thickness of the joint is a critical consideration and strength requirements are secondary.

Butt joints are used where the thickness of lap joints would be objectionable, and where the strength of the brazed joint will satisfactorily meet the service requirements. The strength of a properly brazed butt joint may be sufficiently high so that failure will occur in the base metal, or it may be below the strength of the base metal so that failure will occur in the braze. Joint strength will depend on the strength of the filler metal and on the filler metal/base metal interactions that take place during the brazing cycle. High efficiency will not be obtained with the butt joint when the filler metal in the joint is much weaker than the base metal. To obtain the best efficiency, the brazed joint should be free of defects (no flux inclusions, voids, unbrazed areas, pores, or porosity).

Another means of obtaining high butt-joint strength is to utilize minimum joint clearances compatible with the base and filler metals involved, as well as with the brazing process to be used. Although the minimum joint clearance produces optimum joint strength, producing such clearances is sometimes considered economically impractical. However, with the current sophisticated metalworking techniques, maintaining proper clearances is not ordinarily a major problem. It is important to point out that if high-quality, high-reliability brazements are to be manufactured, it is imperative that clearances be controlled.

The bonding area of a lap joint can be made larger than that of a butt joint. In fact, the area of overlap may be varied so that the joint is as strong as the weaker member, even when a lower-strength filler metal is used or when small defects are present in the final braze. The lap joint has a double thickness at the joint, but the load is transmitted primarily as shear stress, which is desirable. As a rule of thumb, an overlap of at least three times the thickness of the thinner member usually yields maximum joint efficiency. Longer overlaps waste preparation time and filler metal, and do not increase joint strength. For an exact determination of the length of a lap for maximum strength (Figure 6.2), use one of these formulas:

$$\text{Flat}: \ L = F(T \times \frac{t}{S})$$

$$\text{Tubular}: \ L = F(T \times \frac{t(D-t)}{S})$$

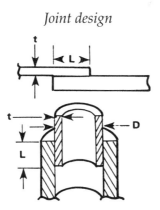

Fig. 6.2 Flat and tubular lap joints.[1]

where L is length of lap, F is factor of safety, T is tensile strength of thinner member, t is wall thickness of thinner member, S is shear strength of filler metal, and D is diameter of lap.

Strength is only one reason that the majority of brazed joints are of lap design. Lap joints can be readily designed to be self-jigging or, as in the case of tubing, self-aligning. Also, pre-placed filler metal can be held in position better with such joints. Lap joints do, however, have disadvantages. They can interfere with fit or function in some applications, they result in increased metal thickness at the joint, and they create stress concentration at the edges of the lap where there is an abrupt change in cross section.

Butt-lap and scarf joints

Variations of the two basic joint designs include the butt-lap joint and the scarf joint. The butt-lap joint is an attempt to combine the advantage of a single thickness with maximum bonding area and strength. It requires more preparation than straight lap or butt joints and may not be applicable to thin members. The butt-lap joint is generally easier and less costly to prepare than the scarf joint, which requires more fixturing. However, both find use with flat and tubular parts.

The scarf joint represents another attempt to increase the cross-sectional area of the joint without increasing its thickness. Angling of the butting surfaces (see Figure 6.1) increases the effective bonding area. Such joints are relatively difficult to prepare properly and are even more difficult to align, particularly with thin members. Because the scarf joint is at an angle to the axis of tensile loading, its load-carrying capacity is similar to that of the lap joint and greater than that of the butt joint.

Finally, the designer wishing to distribute the stresses in a joint, and thereby maximize the mechanical strength of the joint by reducing stress

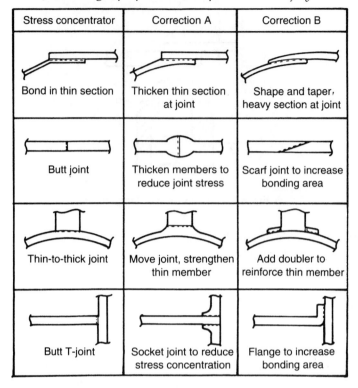

Stress concentrator	Correction A	Correction B
Bond in thin section	Thicken thin section at joint	Shape and taper, heavy section at joint
Butt joint	Thicken members to reduce joint stress	Scarf joint to increase bonding area
Thin-to-thick joint	Move joint, strengthen thin member	Add doubler to reinforce thin member
Butt T-joint	Socket joint to reduce stress concentration	Flange to increase bonding area

Fig. 6.3 Design changes that reduce stress concentrations.[1]

concentration, can determine where the greater stress falls and then counter such stress concentration (Figure 6.3) by:

- Thickening the thinner members at the point of stress concentration;
- Reshaping the thicker member to spread the transfer of loads;
- Thickening both parts at the joint to enlarge the joint area and reduce stress;
- Changing the joint type;
- Moving the joint location;
- Adding reinforcement.

In assemblies of small or thin parts, a fillet should be used to distribute stresses and strengthen the joint. To produce a fillet, a little more than the minimum amount of brazing filler metal is applied, or a more sluggish filler metal is used, or both.

6.1.2 Joint clearance

The single most important design consideration in achieving good brazements is joint clearance – the distance between the faying surfaces to be joined. Joint clearance affects the mechanical performance of a brazed joint in several ways. It affects:

1. The purely mechanical effect of restraint of plastic flow of the filler metal offered by the greater strength of the base metal.
2. The possibility of voids.
3. The relationship between joint thickness and capillary force, which accounts for filler metal distribution.

The actual proper clearance for a brazed joint depends on the type of flux, the surface finish of the mating parts, the base metal/filler metal interaction, the base metal, the filler metal and the type of brazing process to be used.

In general, the smallest possible clearance will result in the strongest joint. Why should small clearances be used? The smaller the clearance, the easier it is for capillarity to distribute the filler metal throughout the joint area, and the less will be the likelihood that voids or shrinkage cavities will form as the filler metal solidifies. Small clearances and correspondingly thin filler metal films make sound joints.

According to some experts, the ideal clearance for production work (see Figure 6.4) is 0.05 to 0.13 mm. Joint clearances up to 0.13 to 0.20 mm are good when silver base filler metals are used. However, some metals actually require interference fits, whereas others require clearances as great as 0.25 mm.[3,4] As a rule, mineral fluxes, strong interactions between filler metal and base metal, long joints, wide-melting-range metals, and easily-oxidized base metals call for increased joint clearances.

A factor that complicates determination of the proper design of a brazed joint is that the clearance must be proper at the brazing temperature. Although this is easily attained when both joint members are made of the same base metal, brazing of dissimilar metals with different coefficients of expansion can lead to complications. Figure 6.5 may be used as a guide for determining the optimum range at brazing temperatures when designing brazed joints for maximum strength.

For example, if a brass bushing is to be brazed into a steel sleeve, the fact that brass expands much more than steel when heated to the same temperature must be considered. A clearance of, say, 0.08 mm at room temperature may disappear completely at a brazing temperature of 725°C. With similar metals of about equal mass, the room temperature clearance is a satisfactory guide.

The joint clearances indicated in Figure 6.5 are radial clearances for tube-type lap joints. The clearances should be used as diametral clearances for some applications if there is no provision in the design to ensure alignment and concentricity of the parts. Excessive joint clearance will result in voids in the joint, particularly when a gas flux is used.

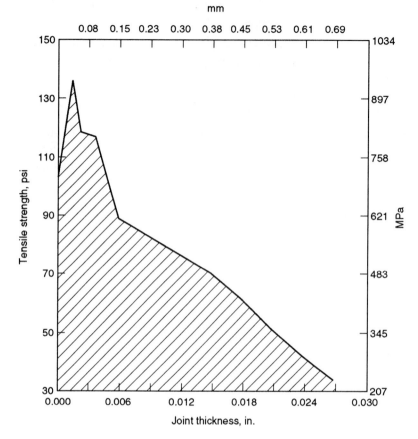

Fig. 6.4 Tensile strength versus joint clearance.[1]

Joint strength is related to test specimen design and testing method. Thus tests must be conducted in accordance with the proposed production joint design and brazing procedures to obtain specific design strength data. Variations in brazing procedures and joint design will alter the effect of joint clearance on strength properties. For example, a free-flowing filler metal used for brazing in a high-quality atmosphere will adequately flow through a joint having a very small clearance. However, when the atmosphere deteriorates, it is often necessary to use a larger clearance to obtain adequate flow.

When designing a brazed joint, the brazing process to be used and the manner in which the filler metal will be placed in the joint should be established. In most manually-brazed joints, the filler metal is simply fed

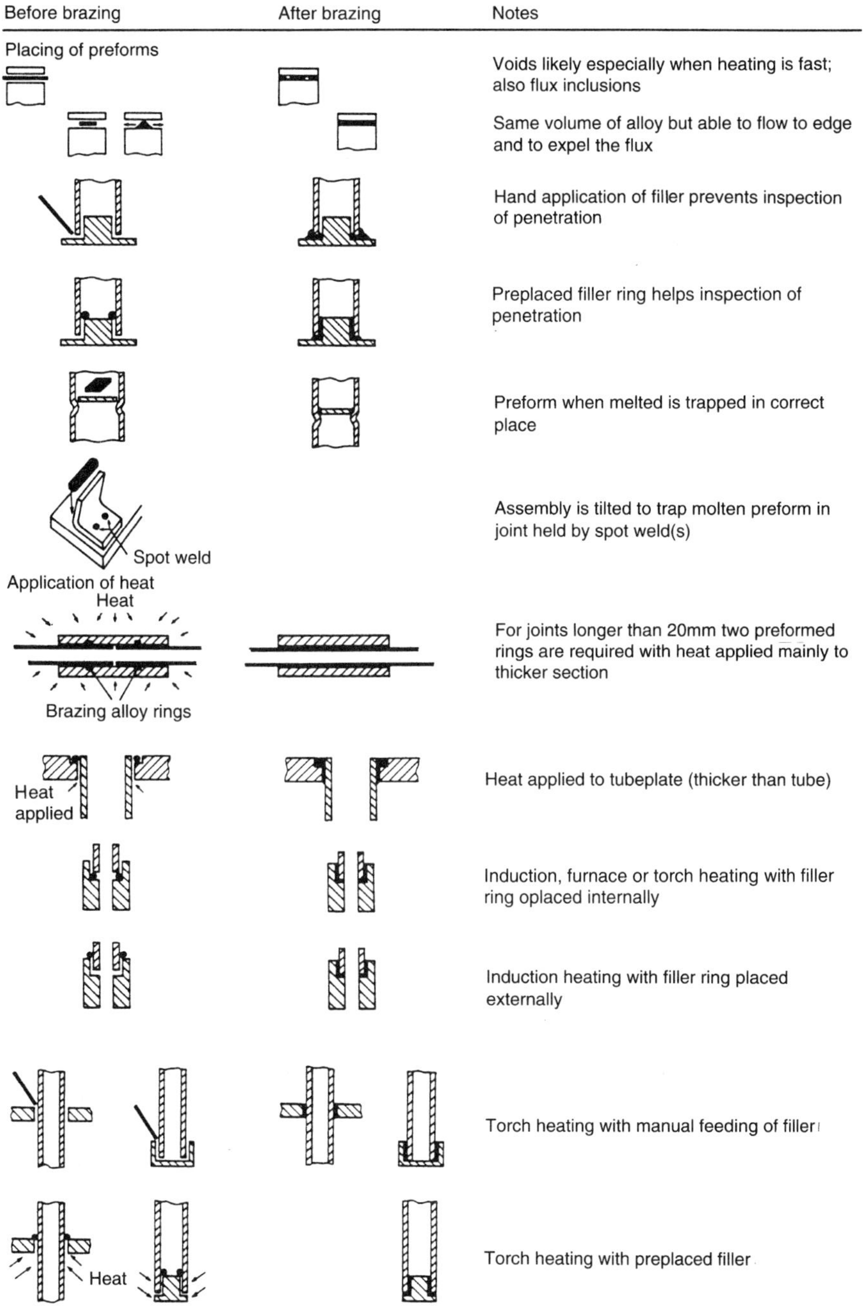

Before brazing	After brazing	Notes
Placing of preforms		Voids likely especially when heating is fast; also flux inclusions
		Same volume of alloy but able to flow to edge and to expel the flux
		Hand application of filler prevents inspection of penetration
		Preplaced filler ring helps inspection of penetration
		Preform when melted is trapped in correct place
Spot weld		Assembly is tilted to trap molten preform in joint held by spot weld(s)
Application of heat Heat Brazing alloy rings		For joints longer than 20mm two preformed rings are required with heat applied mainly to thicker section
Heat applied		Heat applied to tubeplate (thicker than tube)
		Induction, furnace or torch heating with filler ring oplaced internally
		Induction heating with filler ring placed externally
		Torch heating with manual feeding of filler
Heat		Torch heating with preplaced filler

Fig. 6.5 Recommendations for assembly of brazed joints.[5]

Before brazing	After brazing	Notes

Not properly filled

When torch brazing is used heat must not be applied to preforms

Reducing stresses in service

Flexure loading   Stress-induced tear

Flexure loading   Joint integrity maintained

When a thin flexibile component brazed to a thicker item is subject to stress, it may fracture at the joint unless the thickness of the thicker component is reduced to avoid stress concentration

Fig. 6.5 *cont'd*

from the face side of the joint. For furnace brazing and high production brazing, the filler metal is pre-placed at the joint. Automatic dispensing equipment may perform this operation.

Figures 6.6 and 6.7 illustrate methods of preplacing brazing filler metal in wire and sheet forms. When the base metal is grooved to accept preplaced filler metal, the groove should be cut in the heavier section. When computing the strength of the intended joint, the groove area should be subtracted from the joint area, since the brazing filler metal will flow out of the groove and into the joint interfaces, as shown in Figure 6.8.

Powdered filler metal can be applied in any of the locations indicated in Figure 6.6. It can be applied dry to the joint area and then wet down with binder, or it can be premixed with the binder and applied to the joint. The density of powder is usually only 50 to 70% of a solid metal, so the groove volume must be larger for powder. Where pre-placed shims are used, the sections being brazed should be free to move together when the shims melt. Some type of loading may be necessary to move them together and force excess filler metal and flux out of the joint. Many assemblies are simple and require only a push fit.

It is essential to have well-fitted joints with square corners on the female and male parts so that capillarity is continuous throughout the joint. Furnace brazing and other high-production brazing methods, however, frequently involve preplacement of the brazing filler metal and may also incorporate some sort of automatic dispensing equipment. Provisions must be made to allow the components to accept the preform or paste and to secure it properly in position. Examples of preplacement

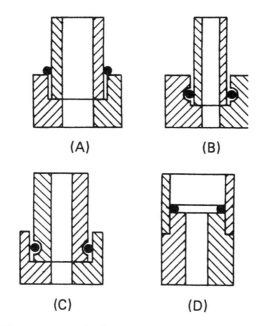

Fig. 6.6 Methods of preplacing brazing filler wire.

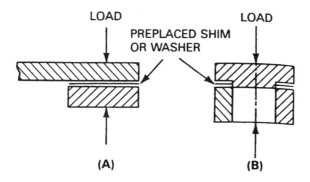

Fig. 6.7 Preplacement of brazing filler shims.

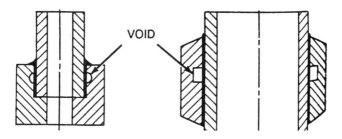

Fig. 6.8 Brazed joints with grooves for preplacement of filler metal; note that after the brazing cycle the grooves are void of filler metal.

of filler metal wire and shims are shown in Figure 6.1. In some applications, additional filler metal is added by extending the filler metal shim beyond the joint.

In dip brazing, joint clearances should be from 0.10 to 0.25 mm, and the preplaced filler metal is subject to very uniform preheating and subsequent immersion in the salt bath. Preheating is necessary to drive off any moisture on the assembly as well as to minimize thermal shock and flux entrapment.

Very close attention must be given to joint clearances for induction brazing, because in this process the heat may be induced in only one component of the joint, causing it to heat more quickly and thereby expand more rapidly than the other. This uneven expansion could produce a very undesirable uneven heating condition and cause a change in joint clearance during the brazing cycle, which should be compensated for in the initial joint clearance.

Finally, as a rule of thumb for torch, furnace, or mechanized torch or induction brazing with laps of 6.4 mm or less, clearances of 0.10 to 0.25 mm can be used. Clearances up to 0.6 mm are used for longer laps, because the brazing filler metal changes composition by dissolving base metal and becomes sluggish as it flows through long lap joints. The correct clearance for any given joint is best determined by trial.

It should be kept in mind that, in most cases, a properly designed and produced high-strength brazement will ultimately fail in the base metal; the braze is by no means the weakest link. In general, loading of a brazed joint demands the same design considerations given to any joint or change in cross section. In that regard, any joint design that moves the high stress concentration from the joint to the base metal is a good design. Whenever possible, joints should be designed to be stressed in shear or compression, rather than in tension.

6.1.3 Design for assembly

For a sound brazed joint, parts must be positioned securely during brazing and cooling cycles. If possible, parts should be designed so that gravity keeps them in position during brazing. If their shapes and weights permit, parts should be positioned so that they will stay put or move into position when the brazing filler metal melts.

The parts to be brazed should be assembled immediately after fluxing, before the flux has time to dry and flake off. Assemblies designed to be self-locating and self-suppporting are the most economical.

When fixtures are needed to maintain alignment or dimensions, the mass of a fixture should be minimized. It should have pin-point or knife-edge contact with the parts, away from the joint area. Sharp contacts minimize heat loss throughout conduction to the fixture. The fixture material must have adequate strength at brazing temperature to support the brazement. It must not readily alloy at elevated temperatures with the work at the points of contact. In torch brazing, extra clearance will be needed to access the joint with the torch flame as well as the brazing filler metal. In induction brazing, fixtures are generally made of ceramic materials to avoid putting extraneous metal in the field of the induction coil. Ceramic fixtures may be designed to serve as a heat shield or a heat absorber.

When gravity or gravity-aided fixturing cannot do the job, other methods of fixing parts in position must be considered. Such methods include press fitting; expanding, crimping, peening, or swaging; staking or pinning; and tack welding (Figure 6.9).

Self-jigging

This is the method of assembly in which the component parts incorporate design features that will ensure that the parts, when assembled, will remain in proper relationship throughout the brazing cycle without the aid of auxiliary fixtures. This is the preferred method of assembly, because it eliminates the initial and replacement costs of auxiliary fixtures and the cost of heating them during brazing, and it usually is a more effective method of holding the components. Self-jigging can be accomplished by the various methods mentioned previously.[3,6]

Gravity locating

Perhaps the simplest method of assembling two components is to rest one on top of the other with the brazing filler metal either wrapped around one component near the joint or placed between the components. The principal disadvantage of gravity locating is the lack of a dependable means of orienting the components or keeping them from moving in relation to one another. Nevertheless, some production components are assembled in this manner, especially those in which the upper component is relatively heavy.

Interference or press fitting

This assembly method requires expansion or contraction of mating component surfaces and provides a very tight fit – sometimes called a 'tight press fit'. Most interference fits require considerable force to achieve assembly, a force generally provided by an arbor press or similar tool. Thus, an interference fit is a press fit.

Lighter interference fits may provide zero clearance or a very slight gap between the mating surfaces of the components. These fits also

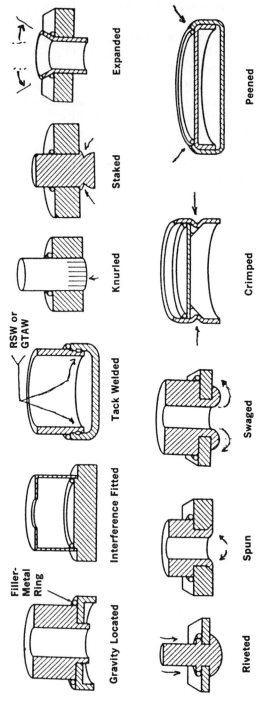

Fig. 6.9 Typical self-fixturing methods for brazed assemblies.

require some external force, such as that provided by an arbor press, to achieve assembly. Fits with zero clearance are referred to as 'size-to-size' fits. Some method is used to prevent slippage when the components are heated in the furnace, particularly if the joint has a vertical axis. A shoulder on one of the components can be used to ensure stability.

Knurling

In high-production manufacturing, there is considerable variation in joint clearance among the assemblies being brazed. Typical brazed assemblies in which a round male member is fitted into a female member are subject to either of two conditions: (a) the male part is off-center, or (b) the male member is out-of-round. Knurling of the male member is sometimes used to correct these conditions and obtain uniformity among brazed joints.

Staking

Figure 6.9 shows how staking will effectively lock two components in position. Burrs are turned in on the shaft by driving a punch into it. This method, of which there are a number of modifications, is commonly used to retain the orientation of such assemblies as cams, levers, and gears on shafts or on common hubs. It is sometimes a substitute for tack welding, knurling, or interference fitting.

Expanding

This method is commonly used in assemblies of tubes to tube sheets. A tubular component is pressed into a header sheet and expanded in the hole to lock the assembly, as shown in Figure 6.9.

Spinning

When the diameter of a hole in an assembly may not be altered during assembly, as when a hub is fastened to a lever, the assembly can be locked together by spinning in a riveting machine. The spinning method of assembly is used for parts for various types of business machines, many of which were formerly assembled by cross drilling and pinning of hubs.

Swaging

An inexpensive and effective method of assembling a stud in a hole in a hollow body is to swage it in place on the detail parts or in the punched hole, because the swaging operation forces the components into intimate contact.

Crimping

Figure 6.9 shows the assembly of a disk, a shell, and a filler-metal ring in which the disk and ring are held in place by crimping the end of the shell. In general, it is preferable to set an assembly of this type on end in the furnace so that the filler metal will flow downward through the joints.

Tack welding

The tack welding method of assembly usually requires careful investigation to determine the most strategic point(s) for placing the weld(s). For economy, the number of tack welds per assembly should be held to a minimum.

Peening

Assembly of two hollow shells by the peening method is shown in Figure 6.9. Components are pressed together and the outer shell is peened. For application of filler metal to an assembly of this type, spraying on the joint interfaces before assembly is preferred.

Riveting and folding or interlocking

Riveting could be considered to be a modification of the spinning and swaging methods in which a rivet is used as part of the assembly. It is widely used to assemble the vanes to the outer disks of fan wheels before furnace copper brazing. Several methods and designs of folding or interlocking can be used to secure joints. These methods are widely used in the manufacture of brazed tubing or tubular assemblies.

6.1.4 Effects of brazing variables on clearance

Mineral and gas-phase fluxes

Use of mineral-type or gas/atmosphere-type fluxes, or a combination of both, will have an important bearing on joint clearance. When the clearance is too small, the mineral flux will be held in the joint, and displacement by the liquid brazing filler metal may be difficult or impossible. Thus, joint defects may be produced. When the clearance is too large, the liquid filler metal will travel around pockets of flux, thus giving rise to excessive flux inclusions.

Gas-phase fluxes are atmospheres which also effect optimum clearances for specific brazements. Gas fluxes permit lower clearances for optimum strength, and so the load-carrying capacity of the joint can be higher. In atmosphere furnace brazing of joints in the vertical position, free-flowing filler metals will flow out of joints having clearances in excess of 0.08 mm.

Surface finish

In general, the brazing filler metal is drawn into the joint by capillary attraction. Thus, if the surface finish of the base metal is too smooth, the filler metal may not distribute itself throughout the entire joint and may leave voids. To ensure adequate filler-metal flow throughout the joint, particularly when the clearance is zero or is a press fit, the faying surfaces of the joint should be roughened, preferably with a clean metallic grit compatible with the base metal.

Surfaces that are too rough also result in lower joint strength, because only the points may be brazed or because the average clearance is too large. A surface roughness of 0.8 to 2 μm RMS is generally acceptable but is not to be considered optimum for all base metal/filler metal combinations. Tests must be conducted to ensure optimum conditions for a specific brazement.

Base metal/filler metal interaction

The mutual solubility of many base metals and filler metals causes interaction to take place through solution of the base metal by the liquid filler metal and diffusion in the liquid and solid states. Such interaction affects the permissible clearance for a specific brazement. If the interaction is low, the clearance may, in general, be small; however, when the interaction is high, which usually occurs in the liquid phase, then the clearance will have to be larger. When the joint is long and interaction is large, the clearance should be increased. For example, when aluminum base metal is brazed with a lower-melting aluminum-based filler metal at a temperature close to the base metal melting point, interaction can be expected.

Base metals

These often contain one or more elements whose oxides are not easily dissociated in a specific atmosphere or by a specific mineral flux. Because the metal/metal oxide dissociation of a given base metal by a specific mineral flux or atmosphere is dependent on many factors, it is important to match the proper clearance with the brazing process and flux or atmosphere. For example, for small aluminum additions, the clearance may have to be increased. With larger aluminum additions, brazing may not take place unless the atmosphere is made more active (lower dew point or lower vacuum pressure) or the surface is protected with a barrier coating such as an electrolytic nickel coating. The clearances would then require reappraisal to obtain optimum joint properties.

Coefficients of thermal expansion will, of course, have an effect on joint clearance at brazing temperature when the base metals are dissimilar.

Brazing filler metals

Free-flowing filler metals generally require smaller clearances than sluggish filler metals. Filler metals that have single melting points, such as copper, silver, eutectic filler metals, and self-fluxing filler metals, will usually be free flowing, particularly when there is very little interaction with the base metal.

Variations in the quality of fluxes and atmospheres (low-quality or oxidized fluxes; low-purity atmospheres) can enhance the free-flowing characteristic of the filler metal or can result in no flow at all. Thus, clearances may appear to be improper for a given set of brazing conditions when in fact the flux or atmosphere requires improved control.

Joint length and configuration

Joint length affects clearance, particularly when there is interaction between the brazing filler metal and the base metal. As the filler metal is drawn into a long joint, it may pick up enough base metal to freeze before it reaches the other end. The more interaction that exists with a specific base metal/filler metal combination under given brazing conditions, the larger the clearance must be as the joint becomes longer. This is only one of a number of reasons why it is important to make the joint length as short as possible, consistent with optimum joint strength.

Dissimilar base metals

In designing joints where dissimilar base metals are involved, the joint clearance at the brazing temperature must be calculated from thermal expansion data. The brazing temperature must be also taken into consideration.[7, 8] When there is high differential thermal expansion between two details, the filler metal must be strong enough to resist fracture, and the base metal must yield during cooling. Some residual stress will remain in the final brazement as a result of joining at the brazing temperature and subsequent cooling to room temperature. Thermal cycling of such a brazement during service will also stress the joint area, which may or may not shorten service life. Whenever possible, the brazement should be designed so that residual stresses do not add to the stress imposed during service.

In a few specific cases, base materials (such as carbides) do not possess sufficient ductility, and it is necessary to use a soft ductile spacer, such as nickel or copper, which will yield during the cooling cycle to prevent high stress between two high-strength base metals e.g., tungsten carbide brazed to a nickel-based or cobalt-based stainless steel that is heat and corrosion resistant. Problems can also be expected in brazing of low-ductility materials, such as carbides and ceramics, to heat-treatable base metals requiring high to moderate quenching rate or to base metals that undergo volume increases as a result of transformation.[9]

6.1.5 Joint design and ceramics

The mechanical behavior of ceramics is often a key factor in their technological application. For structural applications, strength, thermal shock resistance, fracture toughness, reliability and lifetime are invariably critical issues. Even in nonstructural applications, mechanical behavior is still of importance in producing an optimum design and in degradation mechanisms. In the last twenty years there have been substantial advances in understanding the mechanical behavior of ceramics, especially the way this behavior is influenced by processing and microstructure. At the same time, the scientific framework for describing mechanical properties has become more sophisticated, and new testing techniques have been developed for the measurement of critical parameters.

In order to understand fracture, flaw, size, shape, and so on, various developments have occurred. Detailed information on fracture mechanisms, joint design and associated joining processes can be found in References 10 through 17 as well as Figure 6.10.

Electroforming

Electroforming is a process by which joining is obtained by electroplating a metallic layer on the surface of a joint or a layer of metal is deposited on a form. Ceramic joints joined in this manner require that the ceramic first be metallized and plated with Cu and Ni or Au. After masking off conductive areas on which plating is not desired, the joint is plated with Cu in an electroplating bath. After plating, the form is removed to leave a shell of metal whose inside configuration matches that of the form. The technique has been used to produce ceramic-to-metal seals in a traveling wave tube.[18]

Typical joint designs are shown in Figure 6.11. Work performed by Hare *et al.*[18] resulted in the fabrication of forty-one electroformed seals

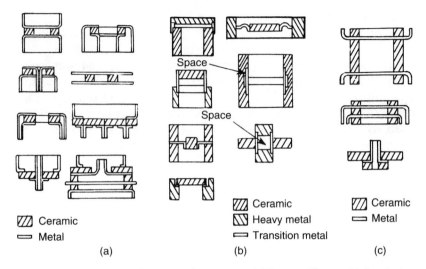

Fig. 6.10 Ceramic-to-metal joint configurations. (a) Butt and lap seal joint designs. (b) Joint designs for transition to thick all-metal members. (c) Backup of ductile metal seal with blank ceramic.

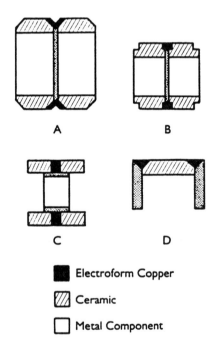

Fig. 6.11 Joint designs for electroformed seals.[18]
(A) Vee design of metal-to-ceramic electroform seal;
(B) Step design of metal-to-ceramic electroform seal;
(C) Step design using plain ceramic cylinders and metal sleeves;
(D) Vee design using ceramic disc and metal cylinder.

joined by electroplating. The materials used included Mo, Cu, and Kovar; the ceramic was a high-Al_2O_3 body.

Braze filler metals and coatings

Most braze filler metals do not wet ceramics easily unless their surfaces are treated in a manner to promote wetting. Such difficulties are to be anticipated when one recalls that oxide ceramics (Al_2O_3, BeO, ZrO_2, and so on) comprise the largest group of these structural materials.

Although the metallizing of ceramic surfaces is costly and time-consuming, the brazing of metals to such surfaces is a relatively straightforward process because the metallized layer ensures wettability of the ceramic by the filler metal. However, certain metals and hydrides possess the ability to wet ceramic surfaces that have not been metallized, and 'active metal' and 'active hydride' processes based on this characteristic have been developed for producing ceramic-to-metal joints and seals.

The joining of ceramics to metals with the active metal or active hydride processes dates back to the middle 1940s, when titanium hydride was used for this purpose.[19] Fine Ti hydride powders (300 mesh) suspended in a suitable binder were painted on the area to be joined, dried, and then the ceramic and metal parts were assembled with a Ag-based filler metal in contact with the hydride area. The assembly was heated to 900 to 1000°C in a vacuum or an H_2 atmosphere. Pure Ti remained on the ceramic surface after the Ti hydride dissociated. When the Ag braze filler metal melted, it alloyed with the Ti to form a Ag–Ti alloy that bonded strongly with the metal and the areas of the ceramic that were coated with Ti hydride.

Pearsall and Zingeser[20] reported that in studying the bonding of ceramics with active metals and their hydrides and extending the work of Bondley,[19] the hydrides of Zr, Ta, and Cb were just as effective as Ti hydride in ceramic-to-metal joints. The effectiveness of various filler metals in making bonds with Al_2O_3, synthetic sapphire, BeO, and ThO was evaluated. In addition, it was found that Ti and Zr, produced in reducing Ti and Zr hydride, could also be used in powder form for ceramic-to-metal joints, thus marking the beginning of the 'active metal' joining process. In developing experimental filler metals, it was noted that excellent bonds to ceramics, diamonds, sapphires and other materials were made with a filler metal containing 85%Ag–15%Zr; Al–Zr, Al–Ag–Zr, and Ag–Ti were also evaluated as filler metals. The effects of various brazing environments including vacuum and gases (H_2, A) were also investigated.

Since these early investigations,[20-22] joining ceramics to metals by the active metal or active hydride process has advanced significantly. The strengths of joints made by this process are as great as those obtained with joints made by the moly-manganese process. Some difficulty has been experienced in making seals by the active metal or active hydride process in dry H_2. The dew point of H_2 must be extremely low to prevent oxidation of Ti. Producing ceramic-to-metal seals in a vacuum is advantageous in that the parts are outgassed during brazing.

The concept of fabricating ceramic-to-metal joints and seals by the active metal or active hydride process was first applied in the electronics industry. In recent years, however, these joining processes have found other uses to meet the need of high-temperature vacuum-tight seals in the nuclear and aerospace industries.

The characteristics of the moly-manganese, active metal, and active hydride processes are summarized in the following paragraphs.

1. The moly-manganese process is a multistep sealing process in which the ceramic surface is metallized and plated with one or two metals before brazing can take place. The operations are conducted at a high temperature in a controlled atmosphere of H_2. Hydrogen firing may discolor some ceramics and produce conductive surfaces. Despite the number of steps required to produce a seal, the moly-manganese process can be automated quite readily, and minor deviations in the process variables can be tolerated.[23-27]

2. The active hydride process is essentially a single-step process in which hydride reduction and brazing proceed simultaneously. Joining in a vacuum or in a controlled atmosphere of H_2 or an inert gas is accomplished at relatively low temperatures, permitting a fast brazing cycle. This process is more difficult to automate than the moly-manganese process, particularly if the joints are produced in a vacuum. Careful control must be exercised in coating the ceramic with the hydride. Even though the process is considered a one-step process, the hydride process has been supplanted by the active metal process.

3. The active metal process may be a one-step operation like the active hydride process. Joining proceeds at high temperatures in a vacuum or in a controlled atmosphere; vacuum joining is not readily automated.

New active braze filler metals are constantly undergoing changes and modifications in composition to meet the ever-demanding requirements to permit metals to be joined to ceramics without the ceramic materials being metallized. Some of these Ag-based filler metals (Cusil and Incusil) are ductile and adaptable to brazing metals to such materials as Si_3N_4,

PSZ, transformation-toughened Al_2O_3, and SiC, as well as many other refractory materials.[25, 27]

Geometrical considerations

Since most metals and their alloys have a coefficient of thermal expansion greater than most ceramics, rigid joints between ceramics and metals are stressed on cooling. The magnitude of residual stresses depend on:

- joint geometry
- relative thickness of ceramic and metal
- ability of metal and braze to relax stresses.

Consideration of thermal expansion mismatch is often not necessary when using soft solders or soft metal components. As the temperature capability of the joint is raised by using more refractory metals and higher brazing temperatures, however, not only is there a greater mismatch on cooling to ambient temperature but there is also a lesser ability to relax stresses. It is therefore desirable to try to match the thermal expansions of ceramic and metal over the temperature range from ambient to the brazing temperature and to ensure that the ceramic components are thick-walled compared with the metal attached (as in Figure 6.12A).

It is desirable to place the ceramic in slight compression by allowing the metal to clamp down on it by relative thermal contraction from the brazing temperature, as in a disc seal (Figure 6.12E). If the metal component is made too stout, the clamping stress produces excessive axial tension in the ceramic and weakens the product.

In the tube seal (Figure 6.12), the stresses are predominantly in shear at the interface between ceramic and braze. Here a successful joint requires a strong metalized layer and a strong braze not subject to embrittlement. A balancing ring of ceramic may be required.[25]

With internal seals involving, for example, tubes or rods (Figures 6.12C and 6.12D), the metal ideally should be closely matched to the ceramic in thermal expansion coefficient. Thin-walled tubes of soft metal can be sealed without difficulty, since the metal yields on cooling.[25] With solid rods, the stresses developed are rather higher. Soft pins of Cu or Pt can be brazed into metallized holes in a ceramic disc such as Al_2O_3, provided they do not exceed about 0.5 mm in diameter; otherwise the metallized layer splits. For larger pins, a low-expansion metal such as W or Mo can be used. These contract less than most ceramics on cooling from the brazing temperature and place the interface in compression. They also, however, put the ceramic into hoop tension, which has to be balanced by an outer compression seal on the ceramic disc. This presents no great difficulty since the outer seal can be used to fix the component to a structure.

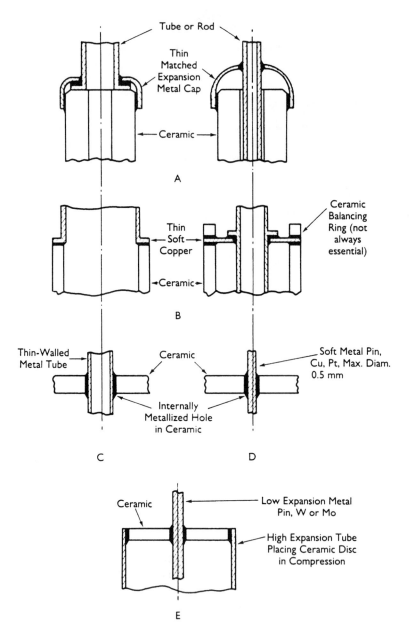

Fig. 6.12 Various types of seal[25] (braze shown in dark in exaggerated thicknesses):
(A) Leadthrough joints for heavy section conductors;
(B) Large bore tube and disc joints
(C) Tube through a disc
(D) Soft pin in a disc
(E) Hard pin in disc seal

Tubular bore seals consisting of a thin ceramic section hermetically sealed to metal members at each end for space power alternators delivering 300 W electrical power for the system have utilized designs discussed above and in Figure 6.12 as well as active metal filler metals.[9,28,29]

Ceramic/ceramic and ceramic/metal applications

The applicability of the aforementioned brazing methods to marine diesel engine composites is shown in Figures 6.13 and 6.14. Figure 6.13 shows a basic outline of a ceramic bonded exhaust valve in which a sintered ceramic component is solid-state diffusion-bonded to the valve face. From the result of an exhaust valve damage simulator test, the exhaust valve in which Si_3N_4 is used as the face is expected to have a burn-out resistance ten to fifty times higher than the conventional value (stellite/stainless steel).[30,31]

Figure 6.14 shows the ceramic bonded fuel nozzle being developed. When SiC is solid-state diffusion-bonded to the atomizer seat of the nozzle, its wear resistance is several tens of times higher than that of the conventional fuel nozzle. In view of the fact that the working temperature of the seat zone is relatively low, a method of bonding with Al as the insert metal is under investigation.[31]

The application of the ceramic/metal bonding technology to marine diesel engines described above may also be applied to the wheel shafts of superchargers and gas turbines and to high-temperature machine components in gas turbine combustion chambers and elsewhere.

Titanium-containing braze filler metals were used to join both 94% and 99+% Al_2O_3 compositions. Resulting tensile strengths of 76.8 to 109.8 MPa

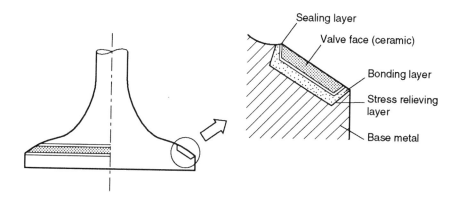

Fig. 6.13 Ceramic bonded exhaust valve.[32]

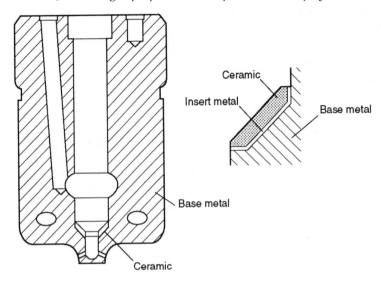

Fig. 6.14 Ceramic bonded fuel nozzle.[32]

compared favorably with conventional Mo–Mn metallizing.[33] The two different Ti-containing braze filler metals which were used were essentially Cu–Ag eutectic compositions which contain small (1 to 3) percentages by weight of Ti. The basic difference between the two filler metals was that one contained 10%wt. In whereas the second filler metal contained no In.

In order to verify the braze filler metal as an active braze alloy (ABA), a component was selected (Figure 4.15) to see how well the component (94%Al_2O_3) performed when brazed with the In-containing filler metal. The component chosen was a 94%wt. Al_2O_3 ceramic header with two $MoAl_2O_3$ cermet electrical feedthroughs. Attached to the connector end of this header are two Cu contacts, which are subsequently brazed to the Mo/Al_2O_3 cermet surface.

The results of the shear tests are contained in Table 4.7. The braze filler metal preforms with 10% In yielded the highest shear strength. As a result it was found that ABAs provide a simplified method of joining Al_2O_3 ceramics. Second, comparable tensile strengths can be obtained from ABAs as from conventional Mo–Mn metallizing techniques. ABA sealing results in the migration of Ti from the bulk braze to the ceramic (Al_2O_3) surface.

Braze filler metal systems have been developed for joining partially stabilized ZrO_2 to nodular cast Fe (NCI). The process was developed for

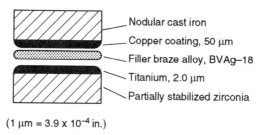

Nodular cast iron
Copper coating, 50 μm
Filler braze alloy, BVAg–18
Titanium, 2.0 μm
Partially stabilized zirconia

(1 μm = 3.9 x 10⁻⁴ in.)

Fig. 6.15 Iron–zirconia braze joint.[32]

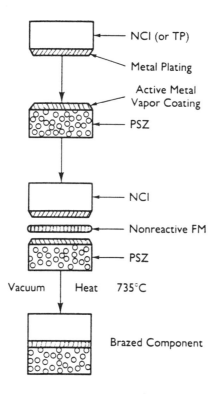

NCI (or TP)

Metal Plating

Active Metal
Vapor Coating

PSZ

NCI

Nonreactive FM

PSZ

Vacuum   Heat   735°C

Brazed Component

Fig. 6.16 Schematic drawing of principal approach of active substrate process.[34] NCI = nodular cast iron; PSZ = partially stabilized zirconia; FM = filler metal; TP = transition piece. (°F = 9/5°C + 32).

advanced design diesel engines and termed active substrate process.[34] These joints call for low brazing temperatures, to avoid the loss of properties of the ZrO_2 and the NCI, both heat-treated. The process consists of vapor depositing a layer of Ti onto the ceramic, then braze at 735°C with a 60Ag–30Cu–10Sn braze filler metal (Figure 6.15). Subsequent work[34] has shown that by electroplating NCI with Cu, enhanced wetting occurs. Adding a transition piece minimizes strain on the ceramic (Figure 6.16). Tolerances to thermal cycling and shock (argon quench from braze) resulted in no failures after 24 thermal cycles of 375 to 600°C at average heating and cooling rates of 56°C min^{-1}.[34] Shear strengths of 136 MPa were attained for the NCI/ZrO_2 joints.

In regard to the transition piece to minimize thermal strain on the ceramic in the joints between the NCI and the ZrO_2, a Ti metal transition piece was introduced. The results showed that the thermal expansion coefficient differential at the ceramic surface was held below 1×10^{-6}°C, and differential strain during braze cool-down was minimized. Testing of the ZrO_2–NCI joints showed that the various braze interfaces had excellent shear strength and resistance to thermal shock and cycling.

In addition, the process can be applied for joining a number of other ceramics and metallic materials for structural applications, including Al_2O_3 and dispersion-toughened Al_2O_3 for the ceramic component and cast Fe, Ti, Cb–1 Zr alloy, and TiC–Ni, Mo cermet as the nonceramic component.

Other applications illustrating the use of transition pieces are shown in Figures 6.17 and 6.18. If the thermal expansion difference between metal and ceramic is large, brazing with a graded seal can be used. For instance, this process has been used to join Si_3N_4, Mo, and stainless steel rings together simultaneously (Figure 6.17). A two-step (step braze) joining technique can also be used, with a stainless steel–Mo braze first followed by a lower-temperature braze to a ceramic ring with a lower-temperature

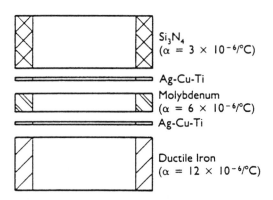

Fig. 6.17 Graded seal assembly.[35]

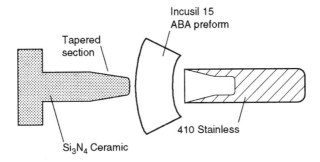

Fig. 6.18 Tapered Si_3N_4 joined to stainless steel shaft.[35]

brazing filler metal. This technique minimizes the joint stresses on cooling to room temperature.

Figure 6.18 shows a tapered joint for joining a Si_3N_4 turbocharger blade to a 410 stainless steel shaft. The turbocharger brazed joint should withstand up to 400°C and the tapered joint allows the alignment of both ceramic and metal shaft axes.

Recent work in the development of a high-temperature ceramic-to-metal seal was disclosed whereby procedures[36] were developed for fabricating vacuum-tight metal-to-ceramic ring seals between Inconel 625 and MgO–3%(wt.)Y_2O_3 tubes metallized with CaO–29% Al_2O_3–35% SiO_2(wt.) glass containing 50% (volume) Mo filler. A braze filler metal of Au–25% Pd–25% Ni (wt.) was found to be the most reliable braze for joining Inconel to metallized MgO–3%(wt.)Y_2O_3 bodies. Another braze filler metal Au–34% Pd–36% Ni (wt.) has also been used successfully. The temperature program for brazing was significant:

1. Heat to 928.6°C in 4.5 hr.
2. Increase temperature by 26.6°C over 5-min interval.
3. Cool rapidly to 782°C and hold 1 hr.
4. Cool to 582°C at 66.6°C per minute.
5. Furnace-cool from 582°C to room temperature.

The reliabilities of joint design and processing procedures like those above, the material systems evaluated and proven to be successful, and leak test procedures for 3000 hr without failures in vacuum or air resulted in a prototype electrical feedthrough.

Adiabatic piston joining

In order to evaluate brazed joints and ceramic-to-ceramic and metal-to-ceramic brazing, a reliable data base which includes testing and design

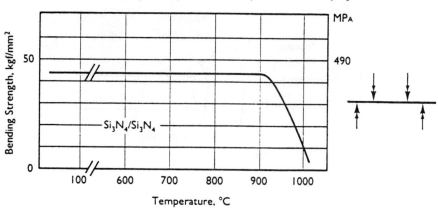

Fig. 6.19 Bending strength of ceramic/ceramic bonding.

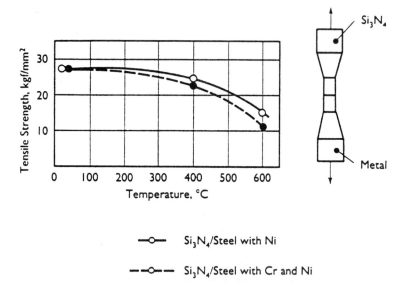

Fig. 6.20 Tensile strength of steel and ceramic bonding with nickel and nickel–chromium interlayer.

criteria must be put in place for advanced ceramic materials. An example is the technology required for the joining of an adiabatic piston. Initially bending strengths of Si_3N_4 (Figure 6.19) and tensile strengths of joined steel and Si_3N_4 (Figure 6.20) were developed in order to join the piston head (Si_3N_4) to the steel ring (Incoloy) (Figure 6.21). The process being developed to join the ceramic to the metal is

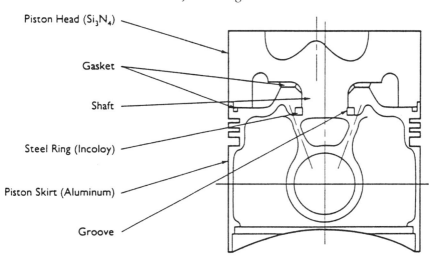

Fig. 6.21 Adiabatic piston schematic drawing showing the Si_3N_4 head and Incoloy ring.

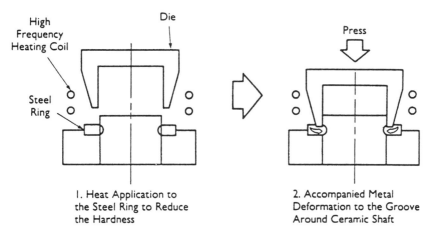

Fig. 6.22 New joining method.

schematically shown in Figure 6.22. The joining process is a combination of chemical bonding and mechanical joining whereby Si_3N_4 is joined to Incoloy. Initial test results of each process and the combination of processes are shown in Figure 6.23.

With the advent and commercial availability of Ag–Cu–Ti braze filler metals with the Ag–Cu–In–Ti family of filler metals, the one-step vacuum brazing of steels, superalloys, and ceramics became practical. Step

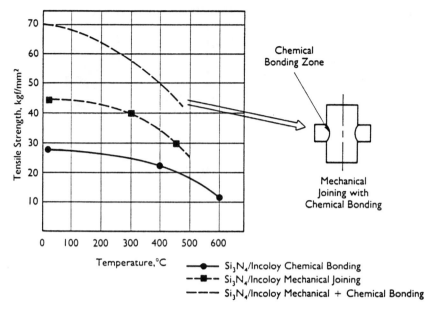

Fig. 6.23 Comparison of joining strength of chemical bonding, mechanical joining, and both. (kgf = 9.81 N: N/m² = Pa).

brazing, possible with the above braze filler metals, is a process in which the subsequent braze is made with a lower-melting filler metal such as Ag–Cu–In–Ti after an initial braze of Ag–Cu–Ti.

In selecting the filler metal composition, several criteria must be met. The filler metal must have suitable ductility, allow sufficient wetting on both ceramic and metal surfaces, and allow controlled flow to a preplaced position. Minimum filler metal flow (blushing) on both the ceramic and base surface is required. The degree of blushing depends on the filler metal and the metal-substrate compositions, brazing temperatures and atmosphere.

The thermal expansion of the base metal and the ceramic should follow similar behavior from room temperature to the solidus of the filler metal. Consequently, the typical brazing cycle is as followed:

1. Heat to about 50°C below the solidus temperature.
2. When melting begins, hold at this temperature for a given time until all parts, including braze fixtures, reach a uniform temperature.
3. Increase temperature above liquidus (25 to 50°C) to obtain complete melting.
4. Hold at this temperature up to 10 min, then cool.

Sandwich seal joining

Since the major problem in joining ceramics to metals is primarily their differences in thermal expansion, steps must be taken to alleviate the variation and differences.[37,38] In most cases, the metal parts to be joined are small (less than 12.7 mm in diameter) and the stress problems are minimized if a ductile filler metal is selected. If the parts being joined are larger (up to 127 mm in diameter or greater), the problem becomes more complicated. The combination of low brazing temperature, ductile filler metal, and proper engineering design is required for successful joining (Figure 6.24 and 6.25). The sandwich seal in Figure 6.24 increases joint reliability and uses a backup ceramic ring with two purposes: it eliminates the bending movement in the metal and distributes the shear stress equally between the two ceramic faces. In Figure 6.25 the metal part is attached edgewise to permit concentric distortion, with stress distributed across the total ceramic face by forming a full filler metal fillet.

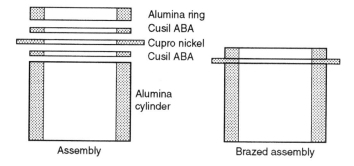

Fig. 6.24 Ceramic-to-ceramic sandwich seal.[35]

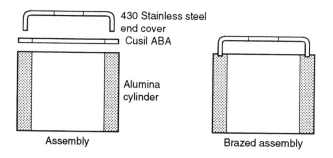

Fig. 6.25 Ceramic-to-430 stainless steel edge seal.[35]

6.2 FIXTURING AND TOOLING

Fixture design is one of the most important factors in ensuring high quality of brazed joints. Fixture design is also important in determining the efficiency and productivity of the brazing system.

A fixture is built to cradle, hold, or secure the assembly being joined. In assembling some components for brazing, the assembly may require additional positioning or support that cannot be provided by self-jigging alone, and so the use of auxiliary fixtures is unavoidable. These fixtures can take the form of simple baskets or wire stands, machined graphite blocks, clamps, or cast supports.

Depending on the brazing process, the fixture material will vary. Low-carbon steel is commonly used for fixtures for short runs in furnace or torch brazing, and sometimes in induction brazing. Steel has the advantage of low cost and the disadvantage of low strength at brazing temperatures. For long production runs, stainless steels and wrought and cast heat-resisting alloys are used. Stainless steels and Inconel are used for springs and clips in dip brazing fixtures. It is mandatory that these materials be used due to the corrosive nature of the flux bath and its attack on low-carbon steel. Figure 6.26 shows a method of deadweight loading for vacuum-furnace brazing which improves the quality of the joint. In the application for which it was developed – fabrication of stainless steel cold plates – the method decreased the rate of rejection from 57% to zero.

Previously, the plates to be brazed together were pressed together under a one-piece glide plate. The glide plate became distorted from the furnace heat and therefore did not apply its weight uniformly over the surfaces of the plates to be joined. In the improved method, the plates are weighted with heavy stainless steel blocks. The blocks act independently and are thus immune to distortion. Besides being uniformly distributed, the force they apply to the plates is larger, and is repeatable from one brazing operation to the next. Larger blocks are used on the edges of the plates, where more thermal mass is needed to reduce the differences between the temperatures of the interior and the edge as the temperature of the furnace is varied–a refinement that was not possible with the one-piece glide plate.

The cold plate layup consists of a stack of stainless steel sheets separated by brazing filler metal tape (0.1 mm) thick, all on a stainless-steel baseplate, Figure 6.26. A 'stopoff' coat of zirconium oxide powder is sprayed onto those surfaces that are in contact with each other but are not intended to be brazed together (e.g., between the layup and the baseplate), to ensure separation after brazing. The smaller blocks are placed on most of the top surface of the stack. The larger

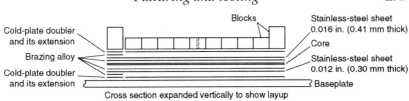

Cross section expanded vertically to show layup

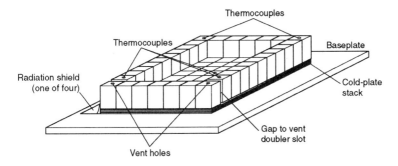

Typical layup with weighting blocks

Fig. 6.26 Stainless steel blocks in two sizes weigh a stack of layers to be brazed together to make a cold plate. Radiation shields, two of which have been removed to show the stack, prevent overheating at the edges. Thermocouples at selected locations monitor brazing temperatures.

blocks are placed at the periphery of the top surface. The blocks are separated from each other by gaps (0.4 mm) wide. Shields made of stainless-steel sheet are placed at the edges of the stack to prevent overheating of the edges by direct radiation. The shields are tack-welded in position. The assembly is brazed at a temperature between 1038 and 1085°C.

Although a number of different fixtures could be designed for any given assembly, there are specific considerations that determine the best type to use. Fixture design should adhere to the following principles:

- Fixtures should allow easy insertion of assembly components and easy removal of the brazed assembly.
- Fixtures should support assembly components to permit expansion and contraction during the heating and cooling cycles: external fixtures should expand more quickly, and internal fixtures more slowly, than the assembly; in applications where tight clamping is required, the reverse is true.
- The assembly should be supported at points away from the heat zone to prevent the fixture from becoming a heat sink.

- Fixtures should permit heat to be directed around the entire joint area so that the heating pattern designed for the system is free to cause flow of the brazing filler metal throughout the joint.
- Gravity should be used to assist capillary action wherever possible.
- Alignment and dimensional stability of assembly components should be maintained until the brazing filler metal solidifies.
- Fixtures should be sufficiently flexible to accommodate other similar assemblies where possible.
- Component pieces of the assembly should be self-locating so that the fixture only supports and cradles the components to the degree necessary to achieve good results (Figure 6.27).
- Fixtures should be designed for minimum surface contact with the assembly; point or line contact is preferable to overall surface contact – even with a minimum number of contact points between the fixture and the components to be brazed, it is sometimes difficult to prevent the fixture from being wetted by the filler metal and sticking to or being brazed to the assembly.

Fig. 6.27 Wave guide being assembled with AlSi–4 filler metal paste.[1]

- If the contact surface between the fixture and the components must be extensive, selection of a fixture material that resists wetting becomes critical.
- Sections of fixtures should be as thin as possible, consistent with required rigidity and durability.
- For efficient furnace brazing, the mass of the fixture should be held to a minimum; as the weight of the fixture is increased, the fuel efficiency of the furnace is reduced.
- In induction brazing, fixtures should be located well away from the work coil so that they will not act as heat sinks or interfere with the magnetic field.
- Fixtures should provide room for the coil to heat the joint area uniformly.

6.2.1 Specialized fixturing for vacuum brazing

In designing an assembly for vacuum furnace brazing, one must keep in mind how the assembly will be held together within the furnace and how it will be set up in the furnace so as to direct the flow of the brazing filler metal into the joints to best advantage and to allow minimum distortion or movement of the parts. These points are generally easy to determine by cut-and-try methods. When a proper procedure is found, the filler metal can be made to flow into all joints, leaving neat fillets and clean surrounding surfaces, and the job can usually be done without distortion.[39]

Some of the various methods of holding assemblies together within the furnace are as follows.

Laying parts together

Perhaps the simplest method of joining two parts is simply to lay one on top of the other, with filler metal either placed between the members or wrapped around one of the members near the joint. In this method, either the weight of the upper member must be sufficient to ensure good metal-to-metal contact, or a weight can be added to ensure such contact. This technique sometimes lacks the advantage of having a definite means of indexing or keeping the parts from moving in relationship to one another.

Pressing parts together

The most common method of assembling parts for vacuum furnace brazing is simply to press them together. In general, regardless of the degree of tightness, some scheme is usually employed to prevent slippage of the parts when they become heated in the furnace, particularly if the joint has a vertical axis. Figure 6.28 shows a shoulder formed on one member to accomplish this stability.

In pressing parts together, the usual tolerances used in machining the parts naturally result in variations in the amount of press fit, which cannot be avoided. An effort should be made, however, to have a snug fit at all times if possible. Sometimes a heavy press fit causes distortion of the parts by stretching them beyond their elastic limit when hot; this results in weakening of joints.

Spot welding and tack welding

Spot welding is frequently employed for maintaining definite relationships between parts assembled for vacuum furnace brazing. It is a fast, inexpensive, and generally neat operation.

Auxiliary fixtures

All of the foregoing methods of holding assemblies together are used for vacuum furnace brazing, the choice of any method depending on the characteristics of each individual product. In some instances, however, it

Fig. 6.28 Vacuum brazed ordnance projectiles.[1]

is impractical to use any of the suggested methods, and it is then necessary to resort to auxiliary fixtures to locate properly the members with respect to one another during furnace brazing. These fixtures sometimes take the form of graphite blocks, heat-resisting superalloy and/or refractory alloy supports, or clamps.

Auxiliary fixtures have several disadvantages. They constitute additional mass that must be heated, are subject to warpage that might make them unsuitable for repeated operations, and present an extra item of maintenance expense. However, two examples where auxiliary fixtures have been used to advantage are as follows:

1. *Blocks*. This type of tooling, consisting of weighted blocks, has been successfully utilized to produce brazed refractory and diffusion welded reactive metal components. The tools are usually made of the same material as the parts being brazed, thus minimizing thermal expansion problems (Figure 6.26).
2. *Pellets*. This new approach to tooling has resulted in a quantum step in removing the distortion problems associated with elevated-temperature brazing and diffusion welding (at 870 to 2480°C), and produces a fluid-type pressure over the assembly to be joined. Thus, control is attained over the mass of the fixturing, the weight of the fixturing, and continuous use of the fixturing without thermal warpage. Additionally, this flexweight tooling concept can be utilized in joining titanium alloys, steels, superalloys and refractory metals. The pellets used have been tungsten, graphite, and alumina. The versatility of this technique allows flat panels (Figure 6.29), curved panels, and cylindrical panels to be brazed.

It should be further noted that an alternative to pellets in some cases is the use of mesh materials in the form of netting or screen. Materials such as 0.10-mm molybdenum wrapped around a titanium honeycomb sandwich panel and a stainless steel tooling mandrel can also successfully produce 360° panels. The economic advantage of pellets over mesh is significant and usually governs one's choice.

Therefore, in selecting fixtures for assembling parts for vacuum brazing, the following factors should be considered:

(a) The mass of the fixture should be kept to the minimum value that will adequately accomplish the intended purpose. The fixture should be designed to provide minimum interference with even heating of the parts by removing heat by conduction from the brazing area. It is also important that the fixture not hamper the flow of the brazing filler metal.
(b) Vacuum is a determining factor in the selection of the material to be used in the fixture, and these materials must withstand the

Fig. 6.29 Steel container with graphite pellets surrounding titanium honeycomb sandwich panel prior to vacuum brazing.[1]

temperatures involved without being appreciably weakened, distorted, or vaporized.

(c) Consideration should be given to the expansion and contraction of the fixture in relation to the part being brazed to ensure a combination that will maintain proper joint clearance and alignment at the brazing temperature. Therefore, the coefficients of expansion of the fixture material and the parts should be considered.

Case history

A firm was brazing a nickel–chromium iron base metal with BNi–5 filler metal. The part was supported directly on a graphite fixture. Brazing was accomplished in a vacuum furnace at 10^{-3} and lower at 1177°C for 30 min at heat. On opening the furnace and inspecting the part, it was noted that the graphite fixture melted into the part. What went wrong?

The primary problem was that the part should not be supported directly on the graphite fixture. While some metals such as pure nickel, copper, copper alloys and copper–nickel will not pick up carbon: many others will readily pick up carbon. Carbon and alloy steels, 400-series stainless, 300-series stainless, titanium, zirconium and others, will readily alloy with the carbon to make a low-melting alloy. Once a liquid phase is present, alloying takes place at a faster rate and severe damage can be done to the part and the graphite fixture.

A second cause of the excessive melting of the part and fixture would come from the presence of the liquid BNi–5 brazing filler metal at the high brazing temperature. The chromium in this filler metal is a carbide former, thus, it reacts rapidly with the carbon. With the liquid brazing filler metal in contact with the base metal and graphite, the solution of the base metal and graphite takes place at a faster rate.

To prevent this problem, a thin ceramic fiber sheet or a suitable stopoff material should cover the graphite fixture so that no direct contact can be made between the graphite fixture and the base metal or brazing filler metal.

6.2.2 Graphite fixturing

Other problems can also occur with graphite fixtures. While graphite fixtures are excellent in retaining flatness and are readily machinable, the thermal expansion of graphite is much lower than most other materials being brazed like AISI 321 stainless steel. If a graphite fixture were to be used where the 321 stainless part was to be shouldered in a groove in the graphite fixture, the groove would have to be machined at least 7.6 mm larger than the flange (at room temperature) to allow for the larger growth of the AISI 321 part flange at brazing temperature.

Since we are discussing graphite fixtures, there is another important caution. Graphite fixtures should not be used in gas atmospheres containing hydrogen when brazing or processing stainless steels. Carbon monoxide and methane can be formed and these gases will carburize the low-carbon stainless steels, thus making the stainless steel susceptible to corrosion. In atmospheres of argon, nitrogen or vacuum, this carbon transfer does not occur. In all cases, however, the graphite fixture must not come in direct contact with the metal part.

There are certain other requirements for brazing fixtures that are peculiar to the vacuum brazing process. In vacuum brazing fixtures, materials should be selected that will not expel gas or otherwise contaminate the inner furnace environment. Whereas graphite and various steels and superalloy heat resisting materials are satisfactory at brazing temperatures up to 1095°C, the tooling materials for use at 1650°C are limited. Refractory metals and their alloys, ceramics, graphite, newly-developed carbides and borides and refractory-coated graphite are the only materials available for fixturing at such temperatures. Of these materials, ceramics such as alumina, zirconia, and beryllia have exhibited reactions with refractory metals, contamination, and problems of thermal expansion and contraction. Graphite is another unsatisfactory material. It has excellent dimensional stability but it embrittles the refractory metals that are usually used for the heating elements of the furnace. The newly developed carbides and borides (ZrB, ZrC, TiC) are still in limited use. This leaves only the refractory metals and their alloys.

6.2.3 Stopoff materials and parting agents

Frequently, it is necessary to prevent the brazing filler metal from wetting portions of assemblies, fixtures, and metallic supports. The materials customarily used for this purpose are refractory oxides such as levigated alumina, magnesium oxide, magnesium hydroxide, and titanium dioxide, which are used as extremely fine powders suspended in alcohol, lacquer, acryloid cement, water or acetone. Two types of stopoff mixtures are available. One is fast drying and behaves much like a commercial laquer. The second is a nonwicking type, composed of oxides in a gelled vehicle, that does not settle out on standing; this type dries slowly. Slurries usually are brushed on with a paint brush or roller. The use of an artist's brush is ideal for precision applications in fine areas, although this operation is time consuming and requires considerable skill.

Two developments in stopoff usage have recently been reported.[40,41] Ceramic stopoff materials, such as zirconium oxide, aluminum oxide, and boron nitride, are used to protect areas from unintentional brazing. However, if applied directly to a metal surface, these materials tend to spall during the brazing process. The problem is easily corrected by using a nickel alloy precoat (e.g., Nichrome or Inconel X, or equivalent) before the ceramic is applied. The precoat improves adhesion of the ceramic and makes the ceramic more tolerant to deformation.

The stopoff coating does not interfere with the brazing process and, when used in a vacuum, does not contaminate the pumping system. This process has been successfully used in brazing of tubes inside the Space Shuttle main engine nozzle.

In the second development, thin sheets of alumina-enriched paper prevent the workpiece from becoming attached to the tooling in brazing operations. Used in fluxless vacuum brazing of stainless steel parts, the paper acts as a barrier that prevents bonding of the filler metal to the tooling (Figure 6.30). Because of the high chemical stability of alumina, the paper does not react with the parts or the tooling, even at the high temperatures and pressures required for brazing.

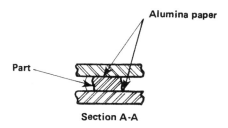

Section A-A

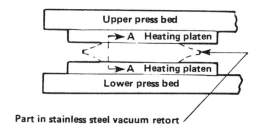

Part in stainless steel vacuum retort

Fig. 6.30 Sheets of alumina paper placed between the parts to be brazed and the heating platens of the press.[1]

The alumina barrier is especially useful in brazing of parts with perforated or otherwise irregular surfaces, because it prevents the brazing filler metal from extruding through the perforations and contacting the platens of the brazing press. Unlike other common barrier materials (various powders and solids), the alumina paper does not disintegrate in the press. Because the paper does not outgas, there is no contamination of the filler metal, ensuring uniform, reliable joints without voids.

In practice, flow may not stop when the joint is filled, and filler metal may flow onto areas where it is not wanted. For example:

1. In brazing of a threaded stud into a part, the filler metal is likely to follow the threads and render them out-of-tolerance.
2. Support points on fixtures used in furnace brazing may become wetted by the filler metal, producing an unwanted braze and perhaps resulting in loss of the fixture and assembly, because it may be impossible to separate them without damage.
3. Some parts, such as turbine and compressor brazements, are designed to close tolerances, and excess filler metal may be dimensionally objectionable.
4. Tubular assemblies, particularly small capillary tubes (1.57 mm ID, or less), can easily become partly or completely blocked with filler metal.

5. Excess filler metal may be unacceptable because of appearance.
6. In production brazing, where it may be necessary to use more filler metal than called for to allow for variations in fitup between parts, some joints will have excess filler metal, which will flow away from the joint area.

Although stopoff coatings can prevent flow/wetting by brazing filler metals on portions of assemblies that contact the coated surfaces, they will not necessarily form barriers against creep of filler metal on the assemblies, and filler metals often creep beneath the coatings. This problem becomes more acute in vacuum brazing. Because many stopoff materials are oxides, the vacuum removes them, leaving only a slight residue that is ineffective as a stopoff material. When zirconium oxide (ZrO_2) is mixed with a suitable nitrocellulose lacquer, it produces an effective parting agent for brazing in vacuum above 870°C.

In certain types of work it is necessary to confine the flow of brazing filler metal to definite areas. This may sometimes be accomplished by controlling the amount of filler metal used and its placement in the assembly.

Materials

For brazing of carbon and low-alloy steels in the more commonly used atmospheres, such as exothermic-based atmospheres, milk of magnesia painted on the appropriate areas is an effective stopoff. Also, painting of fixtures with a water solution of chromic acid and then heating them to the brazing temperature renders them resistant to wetting by the brazing filler metal, because a thin layer of chromium oxide forms.

For brazing in a hydrogen atmosphere or in a vacuum, commercial materials are used that are composed of graphite or oxides of aluminum, titanium and magnesium prepared in the form of a water slurry or organic binder mixture. More recently, a boron nitride additive has been added to various stopoff formulations for brazing applications above 1205°C.

Application

For large areas, the use of a brush or roller is satisfactory. This method is used to protect touch points of metal fixtures. Often it is advisable to repaint the fixture prior to each use, because some stopoff materials may crack off during each heating cycle.

Use of a medical syringe makes it possible to obtain extremely fine detail in stopoff application. Needles 0.25 mm in ID are often used. With a small needle, a drop of stopoff can be applied at a precise point.

With fast-drying stopoff, there is always danger that some of it will inadvertently run into the joint area. If this happens, the assembly must be taken apart and all stopoff removed.

A nonwicking stopoff can be applied by conventional equipment designed for application of liquid plastics, paste-type filler metals, and other organic compounds. It remains stable over long periods of time and does not clog the valves and tubing in the system.

Removal

Brazing stopoff materials of the 'parting compound' type can be removed by wire brushing, air blowing or water flushing. The "surface reaction" type can best be removed by a hot nitric acid hydrofluoric acid pickle, except when the brazed assemblies contain copper or silver. Solutions of sodium hydroxide or ammonium bifluoride can be used in all applications, including copper and silver. Other stopoff materials can be removed by dipping in a 5 to 10% solution of either nitric or hydrofluoric acid.

6.3 PRECLEANING, SURFACE PREPARATION AND POST-BRAZING TREATMENTS INCLUDING REPAIR

Cleaning of all surfaces that are involved in the formation of the desired brazed joint is necessary to achieve successful and repeatable braze joining. All obstructions to wetting, flow, and diffusivity of the molten brazing filler metal must be removed from both surfaces to be brazed prior to assembly. The presence of contaminants on one or both surfaces may result in formation of voids, restriction or misdirection of filler metal flow, and inclusion of contaminants within the solidified brazed area, all of which reduce the mechanical properties of the resulting brazed joint. Grease, oil, dirt, residual zyglo fluids, pigmented markings, residual casting or coring materials, and oxides prevent the uniform flow and bonding of the brazing filler metal, and they impair fluxing action resulting in voids and inclusions. With the refractory oxides or critical atmosphere brazing applications, precleaning must be more thorough and the cleaned components must be preserved and protected from contamination.

Therefore, precleaning must generally utilize a 'clean room' for the final handling and assembly of the cleaned parts. Although many fluxes have some cleaning effect, this is not the primary reason for their use, and complete reliance should not be placed on them for this function.

Fluxes are used primarily to prevent the formation of oxides during brazing, to reduce the surface tension of the brazing filler metal, and to form a protective covering of slag over the solidifying brazed joint. Their effectiveness in removing existing oxides is only slight and incidental (see Chapter 5).

6.3.1 Precleaning

It is important to realize that most metal surfaces actually consist of a thin layer of metal oxide crystals formed by reaction of the metal with oxygen in the air. It is only after penetration of the metal oxide layer that atoms of the metal itself are encountered. In addition to being covered with oxide crystals, metal surfaces are also characterized by the existence of absorbed moisture, oxygen and possibly other gases from the atmosphere. They also frequently are coated with various amounts of oil, grease, wax, perspiration, die lubricants and mill scale. Mill scale is a combination of salts and oils which form on the surface during the process of making the metal. Many of these substances are loosely held and therefore make a poor substrate for bonding. As a result, a large portion of the effort expended in surface preparation prior to bond formation is devoted to removing these materials.

The length of time that cleaning remains effective depends upon the metals involved, the atmospheric conditions, the amount of handling the parts may receive, the manner of storage and similar factors. It is recommended that brazing be done as soon as possible after the parts have been cleaned.

Maintaining part flow (first in, first out) is essential, particularly for atmosphere or vacuum brazing processes. Protection from contamination during storage is equally important. Maximum storage time and conditions must be determined for each product.

Solvent wiping, degreasing, hot alkaline cleaning and physical abrasion methods, or combinations thereof, are all intended to remove the loosely held contaminating materials and to expose a chemically inert and physically strong surface suitable for use as a base for adhesion. The key to success of brazing operations is the removal from the surface of extraneous substances by formation of relatively pure and nonreactive structures.

Cleaning is commonly divided into two categories: chemical and mechanical. Both chemical and mechanical cleaning methods are used to clean metal components for brazing, but chemical methods are the more widely used. Chemical cleaning methods vary from simple manual immersion to complex multistage operations. Chemical methods include alkaline cleaning, solvent cleaning, vapor degreasing, and acid pickling. The mechanical methods most commonly used are dry and wet abrasive blast cleaning. If warranted, machining or grinding may be used to obtain the necessary joint cleanliness and to ensure satisfactory wetting by the brazing filler metal.

Degreasing is generally done first. The following degreasing methods are commonly used, and their action may be enhanced by mechanical agitation or by applying ultrasonic vibrations to the bath:

1. *Solvent cleaning*: soak or spray operation with petroleum or chlorinated solvents.
2. *Vapor degreasing*: chlorinated and trichlorotrifluoroethane solvents that clean by soaking and condensation of the hot vapor on the work.
3. *Alkaline cleaning*: commercial mixtures of silicates, phosphates, carbonates, wetting agents and, in some cases, hydroxides.
4. *Emulsion cleaning*: mixtures of water, hydrocarbons, fatty acids and wetting agents.
5. *Electrolytic cleaning*: anodic, cathodic and periodic reversal.

Chemical cleaning methods

Alkaline cleaning methods, including soak, spray, and barrel cleaning, are widely used for removing oily, semisolid, or solid soils from metal components before brazing. They employ commercial mixtures of silicates, phosphates, carbonates, detergents, soaps, wetting agents, and in some cases hydroxides. They are generally satisfactory for removing most cutting and grinding fluids, grinding and polishing abrasives, and some pigmented drawing compounds.

Solvent cleaning is capable of removing oil, grease, loose metal chips, and other contaminants from metal components. Parts are immersed and soaked in a petroleum solvent or chlorinated hydrocarbon. Spray methods can also be employed.

Vapor degreasing, a cleaning process usually performed first, utilizes stabilized trichloroethylene or stabilized perchloroethylene. To supplement the cleaning action of the vapor, some degreasing units are equipped with facilities for immersing the work in the hot solvent or for spraying it with clean solvent.

Scale and oxide removal can be accomplished mechanically or chemically. Prior degreasing allows intimate contact of the pickling solution with the parts, and vibration aids in descaling with any of the following solutions:

1. *Acid cleaning*: phosphate-type acid cleaners;
2. *Acid pickling*: sulfuric, nitric, and hydrochloric acid;
3. *Salt bath pickling*: electrolytic and nonelectrolytic.

The selection of chemical cleaning agent will depend on the nature of the contaminant, the base metal, the surface condition, and the joint design.

For example, base metals containing copper and silver should not be pickled with nitric acid. In all cases, the chemical residue must be removed by thorough rinsing to prevent formation of other equally undesirable films on the joint surfaces, or subsequent chemical attack of the base metal.

Mechanical cleaning

Mechanical methods are less widely used than chemical methods in cleaning for brazing. However, they are usually preferred for removing heavy scale and may be indispensable in removing the more tenacious lubricants, such as pigmented drawing compounds. Mechanical abrading or roughening may be required on very smooth surfaces to promote filler metal wetting and flow. Chemical cleaning also roughens the mating surfaces to enhance capillary flow and wetting by the brazing filler metal.

Mechanical cleaning methods, such as grinding, filing, machining, blasting, and wire brushing, also are used to remove objectionable surface conditions and roughen faying surfaces in preparation for brazing. If a power-driven wire wheel is used, care should be exercised to prevent burnishing. Burnishing can result in surface oxide embedment which interferes with the proper wetting of the base metal by the filler metal. When rolling, fine grinding, or lapping has produced a base metal surface that is too smooth, the filler metal may not effectively wet the faying surfaces. In this case, the parts can be roughened slightly by rubbing with 30 to 40 grit emery cloth for improved wetting. Provision should be made to adjust and control joint clearances after machining or blasting operations. Cutting oils used in machining must be removed prior to brazing.

When faying surfaces of parts to be brazed are prepared by blasting techniques, there are several factors that should be understood and considered. The purpose of blasting parts to be brazed is to remove oxide films and to roughen the mating surfaces so that capillary attraction of the brazing filler metal will be increased. The blasting media must be clean and must not leave a deposit on the surfaces to be joined that will restrict filler-metal flow or impair brazing. The materials should be fragmented rather than spherical so that the blasted parts are lightly roughened rather than peened. The operation should be done in such a way that delicate parts are not distorted or otherwise harmed.

Recommended and nonrecommended blasting materials and methods, along with their advantages and disadvantages for use in preparing surfaces for brazing, are as follows.

Recommended blasting materials

Chilled cast iron and hardened steel fragmented shot are recommended because the roughening they produce and the residual iron they leave on the faying surfaces both promote the flow of brazing filler metal. However, the residual iron is also a disadvantage, because it may rust on standing. Also recommended are stainless steel grits and powders, modified nickel base braze filler metal grits and glass beads. The advantages of these materials are that they promote filler-metal flow on stainless steel surfaces and leave nonrusting residues. Their disadvantage is that they do not clean or roughen the surfaces as readily as do fragmented iron and steel shot.

Nonrecommended blasting materials and methods

Nonmetallic materials, such as alumina, zirconia, silica, silicon carbide, and other similar materials, are not recommended for prebraze preparation. Although excellent for cleaning and roughening, they can be undesirable because they embed themselves in the surfaces and may retard filler metal flow. Wet blasting methods, such as vapor blasting, generally are not recommended because they contaminate the surfaces with minerals from water, rust inhibitors, and refractory oxides.

Thermal treatments

Thermal treatments are utilized to clean and modify surfaces by heating in furnaces with specific atmospheres that reduce oxides and remove objectionable contaminants. Examples of this practice are bright annealing carbon steel parts in a controlled atmosphere and precleaning stainless steel in a dry hydrogen or vacuum atmosphere. Vacuum furnace operations are also conducted to remove objectionable contaminants from small capillary spaces and cracks that are to be repaired by filling with brazing filler metal.

Precoating and finishing

For some applications, parts to be brazed are precoated by electrodeposition, hot dip-coating, flame spraying and cladding methods. Precoatings and finishes frequently are used to ensure wetting and flow on base metals that contain constituents such as aluminum, titanium or other additions that are difficult to wet. Precoatings also protect clean surfaces and prevent the formation of oxides on base metals in storage and during the heating process. When brazing dissimilar metals, precoatings can reduce the tendency of a filler metal to wet and flow preferentially on one of the base metals. Occasionally, precoatings are used on refractory metals to prevent the rapid diffusion of filler metal constituents into the base metal and the subsequent formation of brittle intermetallic compounds.

The selection of the precoating depends on the base metal, filler metal and brazing technique. Electroplatings of copper on steel and low stress nickel on stainless steels are coatings that often are used. Brush plating is the technique used to apply electroplates locally at the joint areas. Proper procedures must be followed to activate the surface prior to applying the electrodeposit. Electroless deposition (autocatalytic) of nickel should not be employed to precoat materials that are to be brazed at temperatures above 870°C. The phosphorus content of the nickel plating provides a eutectic at 880°C and melting of the coating may interfere with wetting and flow.

In some brazing operations, the basic metals are clad with filler metal. Examples of this practice include aluminum alloys clad with BAl–Si filler, copper clad with BCuP filler and copper clad with BAg–3 for sandwich brazing of carbide cutting tool tips.

Specialized processes

Bucklow recently examined and evaluated the use of autodissolution and ion bombardment as a procedure for surface cleaning steels and high purity irons. The work demonstrated that ion bombardment was a more effective cleaning method than simple heating to 500°C, and was even more effective when a hot substrate was treated. It is therefore probable that surfaces ion bombarded at 700°C and immediately joined were much cleaner than the present analyses suggest, but in all the cases studied, steels appeared to be easier to clean high purity irons.

6.3.2 Surface preparation

For some applications, the parts to be brazed are precoated by electrodeposition, hot-dip or flame-spraying methods. Another technique utilizes electron-beam evaporation and the deposition of films on the substrate surfaces, which greatly enhances liquid metal wetting, spreading and brazing. The wetting of and bonding to the base metal take place by virtue of the liquid metal penetrating, displacing, and "tunneling" under the vapor-deposited film. Aluminum has been suitably used to wet and join beryllium.[42] Precoatings are used to retain clean surfaces and prevent the formation of oxides on base metals that contain refractory alloy additions. For brazing of dissimilar metals, precoatings reduce the tendency of the brazing filler metal to wet and flow preferentially on one of the base metals. Occasionally, precoatings are used on refractory metals to prevent the rapid diffusion of filler metal constituents into the base metal and the subsequent formation of brittle intermetallic compounds.

The use of nickel plating facilitates brazing of some alloys, but it may not always be feasible because of the manufacturing sequence, the size of the items involved, or the lack of suitable plating equipment. An alternative surface preparation method is plasma spraying of a 0.03 to 0.04 mm layer of fine nickel powder or brazing filler metal directly onto the surface. In some recently completed work it was found that this technique could be used for applying filler metals on A–286, Inconel 713 and 718, MAR–M246, Rene' 41, and others. These base metal alloys contain aluminum and titanium and are among the most difficult alloys to braze. Unlike wet methods, which allow little control and do not work with some difficult-to-wet alloys, the plasma-sprayed deposit can withstand handling under manufacturing conditions. Moreover, the sprayed filler metal does not contaminate furnace atmospheres.

Other techniques used to clean and prepare surfaces for brazing difficult-to-wet alloys include envelopment of parts in a chromium fluoride atmosphere or in a fluorocarbon gas atmosphere. The latter technique has been used extensively in preparing all types of brazeable γ' nickel-base superalloys and has been successfully used in repair of crack damage in gas turbine engine components.[43,44] Finally, selection of the precoating method depends on the base metal, the filler metal, and the brazing technique.

An example of the use of a combination of surface preparation methods is in braze repair of aircraft gas-turbine nozzles.[44] An activated diffusion healing (ADH) process, Figure 6.31, for maintaining aircraft gas-turbine engine parts was developed for GE aircraft engines. The first stage in the ADH process is a four-step cleaning operation that prepares parts for brazing:

1. An alkaline solution removes oxides and soils.
2. An acidic chemical solution strips parts of their protective aluminide coating.
3. Fluoride-ion cleaning (FIC) removes any remaining complex oxides, including those embedded in the smallest cracks.
4. A vacuum-cleaning cycle extracts any residual fluoride ions, resulting in an extremely clean part ready for ADH'ing.

The braze filler material typically contains chromium, aluminum, tantalum, and cobalt, and a 2.4% addition of boron that serves as a melting-point depressant. Vacuum brazing typically takes about 30 min at approximately 1205°C.[45] (More under Paragraph 6.3.4.)

The following section will cover cleaning procedures and compositions of cleaning solutions for some of the numerous brazeable metals and alloys. Many proprietary cleaning solutions are equally satisfactory.

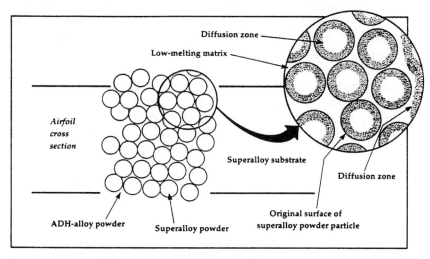

Fig. 6.31 In the ADH repair process, a superalloy powder is 'cast' into the crack using an assist from a lower-melting-point braze filler metal. The starting slurry also contains a standard brazing binder. Shown here: crack prior to 'casting' and, inset, after vacuum brazing. Note diffusion of braze filler metal into both powder and substrate.[44]

Aluminum and aluminum alloys

In the case of aluminum, precleaning requirements vary depending on surface condition, material thickness, the alloys to be brazed and finally the liquid or gas tightness required in the finished joint. Solvent-type cleaning operations for removal of surface lubricants have been quite satisfactory. Solvent cleaning is mandatory for fluxless brazing, particularly where a large surface area is exposed during brazing. Etchant-type cleaning may be done with a caustic or acid cleaner. Two cleaning procedures that are applicable to dip brazing and furnace flux brazing of aluminum alloys are as follows:

1. Caustic cleaning:
        Degrease (solvent or vapor)
        Dip in 5 wt % sodium hydroxide for up to 60 s at 60°C
        Rinse in cold water
        Dip in 50% vol cold nitric acid for 10 s
        Rinse in hot or cold water
        Dry
2. Acid cleaning:
        Degrease (solvent or vapor)

Dip in 10% vol cold nitric acid plus 0.25% vol hydrofluoric acid
for up to 5 min
Rinse in hot or cold water
Dry

In a new technique, aluminum is cleaned of its oxide film and is sealed
immediately with a polymeric material, making it suitable for vacuum
brazing. The time between cleaning and brazing is no longer a critical
factor. First, the surface of the aluminum is degreased with any common
degreaser, such as naphtha. After degreasing, the aluminum oxide is
removed by chemical cleaning with an alkali wash of sodium hydroxide
and sodium bicarbonate. A water rinse at 60 to 70°C and an acid wash
follow. After the acid treatment, the aluminum is rinsed in distilled
water. It is then immersed in an all-organic solvent miscible with water,
such as acetone, to remove all water from the surfaces. Immediately after
this step, the clean surfaces are coated with a sealer. One of the best
sealers comprises polystyrene in toluene and acetone. Sealed aluminum
surfaces can be stored for several days without appreciable surface
oxidation.

Beryllium

Beryllium, like titanium, reacts with oxygen at conventional brazing
temperatures. Beryllium also reacts with atmospheric nitrogen. Because
the presence of an oxidized or nitrided surface impairs the wetting and
flow properties of brazing filler metals, high-temperature brazing
usually is done in an argon atmosphere or in a vacuum after thorough
cleaning of the base-metal surfaces. Fluxes have been used to prevent
oxidation during low-temperature brazing of beryllium in air. Beryllium
surfaces are also sometimes plated with silver to improve filler metal
wetting and flow.

Cast iron

Special preparation methods are available that are designed to improve
wetting on the 'as cast' brazing surfaces of cast iron. One such method is a
proprietary electrochemical treatment that will remove graphite, silica, and
other oxides. The process utilizes a molten salt bath operating at 460 to
480°C. The bath composition consists of 75% sodium hydroxide, 5% sodium
chloride, 5% sodium fluoride, 14% sodium carbonate, and 1% potassium
carbonate.[46] Optimum cleaning is obtained by a 19-min direct-current
electrolysis treatment (reduction for 4 min, oxidation for 10 min, and
reduction for 5 min). Preparation is completed by rinsing in hot water to

remove the salt, and then drying. Once treated in this bath, machined surfaces rust rapidly. The rust film must be removed before brazing.

Another choice is **flame cleaning**, which results in strong brazements when done properly. Flame cleaning with an oxyacetylene torch, however, calls for expertise in playing the oxidizing flame over the entire surface being brazed to burn out all of the graphite. Burning with a reducing flame, to reduce surface iron oxide to elemental iron, is done next. **Grit blasting** of the faying surfaces may also be used. Finally, chemical treatment in fused sodium and potassium nitrate salts is not recommended, because it is as complex as electrolytic treatment and decreases brazement strength.

Copper and copper alloys

Standard solvent or alkaline degreasing procedures are suitable for cleaning of copper and copper-based metals, and mechanical methods (wire brushing, abrading, sanding, etc.) may be used to remove oxides. Complete chemical removal of oxides requires proper selection of pickling solution. Typical procedures used for chemical cleaning are as follows:

Copper. Immerse in cold 5 to 15% vol sulfuric acid.

Aluminum bronzes. Successively immerse in a cold mixture of 2% hydrofluoric acid and 3% sulfuric acid and then in a solution of 5% vol sulfuric acid at 27 to 49°C, and repeat until clean. Electroplating with copper at least 0.013 mm thick on surfaces to be brazed will aid wetting.

Copper–silicon alloys. Immerse in hot 5% vol sulfuric acid, then in a cold mixture of 2% vol hydrofluoric acid and 5% vol sulfuric acid.

Brass and Nickel–Silver Alloys. Immerse in cold 5% vol sulfuric acid.

Beryllium Copper. Immerse in 20% vol sulfuric acid at 70 to 80°C, water rinse, then quick dip (less than 30 s) in cold 30% vol nitric acid solution followed by immediate and thorough rinsing.

Chromium Copper. Immerse in hot 5% vol sulfuric acid, then in a cold mixture of 15 to 37gl^{-1} sodium bichromate with 3 to 5% vol sulfuric acid. Subsequent copper plating may facilitate wetting.

Copper Nickel. Standard solvent or alkaline degreasing procedures are used to remove sulfur or lead from the surface because these elements might cause cracking during the brazing cycle. Oxides are removed by abrading or by pickling in hot 5% vol sulfuric acid followed by immediate and thorough rinsing.

Magnesium and magnesium alloys

These can be satisfactorily cleaned by the mechanical method of abrading with aluminum oxide cloth or steel wool. A chemical cleaning method that has been successful consists of dipping for 5 to 10 min in hot alkaline cleaner followed by dipping for 2 min in ferric nitrate bright pickle solution.

Nickel and nickel alloys.

Precleaning of nickel alloys just prior to brazing is particularly important, because they are subject to attack by low-melting-point elements, particularly lead and sulfur, at elevated temperatures. Because grease, oil, paint, shop dirt, and other foreign materials usually contain these harmful elements, they must be entirely removed before brazing.

The oxide films formed on nickel alloys are tenacious, and wire brushing may not remove them; however, they may be removed with emery cloth or by grinding. Uniform oxide removal by pickling cannot be expected unless the high-nickel alloy being processed is first thoroughly cleaned of all foreign material.

Additional methods of chemical cleaning to remove oxides and other adherent metallic contaminants include immersion in phosphate acid cleaners.

Care must be taken in selecting time of exposure for both acid cleaning and pickling of heat-resistant nickel base alloys and superalloys. Overexposure during chemical cleaning can lead to excessive metal loss, grain boundary attack, and selective phase structure attack. As a last step in chemical cleaning, ultrasonic cleaning in alcohol or clean hot water is recommended.

Refractory metals

Surface oxide removal is mandatory before brazing of refractory metals. The cleaning operation should be performed immediately before brazing to prevent contamination. Degreasing should be used to remove oil, fingerprints and grease. Both mechanical and chemical cleaning methods are satisfactory. Sandblasting, liquid abrasive cleaning or abrasion may be used to remove oxide films from simple parts, but chemical cleaning is preferred, especially for complex assemblies.

Molybdenum and Its Alloys. For removing heavy oxide films from molybdenum and molybdenum alloys, molten salt baths, such as 70% sodium hydroxide and 30% sodium nitrite at 260 to 370°C or commercial martempering salt (mixture of sodium and potassium nitrates) at 370°C, have achieved good results. The former bath should be controlled closely, because it attacks molybdenum. Gross attack has not been noted with the latter bath. Light surface oxide films should be removed by appropriate means after the salt bath treatment.

Electrolytic etchants can be used to remove surface oxides from simple parts; however grain boundary attack of molybdenum by such etchants can be severe. Chemical etching is the most popular cleaning method, and three successful techniques are given in Reference 1.

Tantalum and Its Alloys. They can be cleaned by both mechanical and chemical methods. Hot chromic acid (glass-cleaning solution) is quite satisfactory. Hot caustic cleaning solutions will attack the metal and should not be used. Prior to chromic acid cleaning, tantalum can be blast cleaned. This procedure, however, should be followed by an immersion in a hydrochloric acid solution to dissolve the iron particles. The glass-cleaning solution then becomes more effective. Abrasion and other usual mechanical cleaning methods have been found to be acceptable. Tantalum has a tenacious oxide film that reforms immediately on exposure to air or vapor after any cleaning treatment. Another method of preparing tantalum is described below under niobium.

Niobium and Its Alloys. One method of preparing niobium (as well as tantalum) prior to brazing is to electroplate either copper or nickel onto an acid cleaned surface. The deposits are bonded to the niobium (or tantalum) by diffusion, and in the case of copper, melting actually occurs. Brazing is subsequently accomplished by using the plated surface as a base. Niobium can be cleaned by both mechanical and chemical methods (one chemical method is given in Reference 1).

Tungsten and Its Alloys. Thorough cleaning prior to brazing is essential for tungsten, and both mechanical and chemical cleaning can accomplish this purpose. Cleaning methods that are acceptable for tungsten are given in Reference 1. The most effective cleaning procedure will depend on the tenacity of the oxide film. In cases where wrought tungsten sheet has been mill cleaned, degreasing is sometimes the only cleaning operation necessary prior to brazing. However, the optimum conditions for preparation of tungsten should be determined for each particular application. In some cases, electroplating of tungsten with nickel and other elements has been used satisfactorily to stop diffusion of elements that form brittle intermetallic compounds with the base metal. A hydrogen-atmosphere furnace cleaning operation at 1065°C for 15 min has been effective in reducing light oxide films.

Low-carbon and stainless steels

For best results with low-carbon steels, faying surfaces should be cleaned mechanically or chemically to ensure essentially complete absence of oxides or organic matter. Stainless steels require more stringent precleaning than do carbon steels. The tenacious oxide films that impart corrosion resistance to stainless steels are more difficult to remove by fluxes or reducing atmospheres than the oxide films that form on carbon steels.

Precleaning of stainless steels for brazing should include a degreasing operation to remove any grease or oil films. The joint surfaces should also be cleaned mechanically or with an acid pickling solution. Wire brushing, however, should be avoided, especially with a carbon steel wire brush. The best practice is to braze parts immediately after cleaning. When this is not possible, it is desirable to enclose the cleaned parts in sealed polyethylene bags to exclude moisture and other contaminants until the parts can be brazed.

Carbides

Much of the difficulty in brazing carbides is the result of improper cleaning. The carbide surfaces should be grit blasted or ground on a silicon carbide or diamond wheel to remove any surface carbon enrichment, because such surfaces are not readily wetted by the brazing filler metal. The usual precaution of degreasing the surface prior to brazing also should be taken. Occasionally, some of the more difficult-to-wet carbides, such as titanium carbide, are coated with copper oxide or nickel oxide and then fired in a reducing atmosphere to fuse the copper or nickel onto the surface. This surface is readily wetted by the common brazing filler metals.

Ceramics

The inherent porosity associated with many ceramic bodies necessitates the use of very strict cleaning procedures before brazing. They are usually fired in air at a temperature from 800 to 1000°C to permit outgassing. Suitable alkaline cleaning solutions may be used, followed by immersion in dilute nitric acid and subsequent rinsing in a neutralizing solution.

6.3.3 Clean rooms

Within the past decade, more and more brazing is being performed in atmospheres without flux, and therefore the cleanness of the parts is of great importance. Although chemical fluxes are capable of removing greater amounts of residual oxides than are generated during the brazing cycle, atmospheres are not. Precleaning therefore must be more thorough, and the components must be preserved and protected in the clean condition. One of the more common techniques is to utilize a clean room for final handling and assembling of the cleaned parts.

Clean rooms are areas physically separated from the rest of the shop and provided with means for controlling the atmosphere and cleanness within the area. In almost all instances, clean rooms are air conditioned. The objective is to reduce or eliminate airborne contamination.

Workers in the clean room may be required to wear special lint-free clothing. Fresh clothing usually is provided daily. Special shoes also may be necessary. To avoid carrying outside contamination into the clean room, it may be advisable to provide dressing rooms between the shop area and the clean room where street clothes are exchanged for clean room clothes. Workers in clean rooms may be required to wear hair coverings, but this is not necessary in every case.

One of the most important functions of the clothing used in clean rooms is protection of the parts from contamination by handling with bare hands. Cleaned parts can become so contaminated by bare hands that the flow of brazing filler metal is impaired or prevented even in a satisfactory atmosphere. Therefore, workers in a clean room should be required to wear gloves. White, lint-free cotton gloves have proven most satisfactory. These gloves are changed several times a day depending on the amount of soil they accumulate. In some operations, the cotton gloves are supplemented by nylon gloves worn over them.

Clean rooms have been widely used as assembly areas for brazed honeycomb assemblies and brazed stainless steel heat exchangers used in connection with jet-engine fuel systems. They also are used as assembly areas for electronic components, such as vacuum tubes and solid-state devices.

6.3.4 Repair techniques with cleaning agents

Fluoride-ion cleaning, using hydrogen fluoride gas as the active agent, is rapidly becoming established as a cost-effective method of preparing nickel- and cobalt-based superalloys for braze repair in the aerospace industry. These alloys include Rene' 80, IN–100, IN–738, MAR–M–509, B–1900+Hf and MAR–M–200+Hf.

The method, as described,[47] can be practiced using relatively simple heat treat equipment, such as a hydrogen atmosphere retort in an air-atmosphere furnace.

An alternative fluoride-ion cleaning technique has been developed[48] and patented.[49] This process depends on the thermal decomposition of tetrafluoroethylene as the source of the HF gas.

Chemically, the hydrogen fluoride gas cleaning process is identical to the previously described processes, but has one significant difference: Instead of obtaining HF gas through secondary reactions, a small quantity of HF gas is introduced directly into the reactor system through

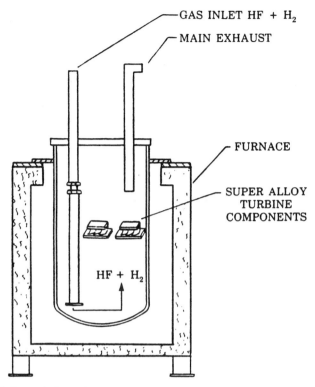

Fig. 6.32 CrF_2 pack fluoride-ion cleaning diagram. The fluoride ion is supplied by secondary chemical reaction.[47]

a precision electronic gas mass flowmeter. Figure 6.32 is a representation of the system configuration.

It therefore appears that the fluoride-ion cleaning technique, using HF gas, offers a simpler, more precise alternative to other available, more complex techniques for jet-engine superalloy components, Figures 6.33 and 6.34.

6.4 INSPECTION OF BRAZED JOINTS

Inspection of brazements should always be required to protect the ultimate user, but it's specified by regulatory codes and by the fabricator. Inspection of brazed joints may be conducted on test specimens or by tests of the finished assembly. The tests may be nondestructive or destructive.

Fig. 6.33 Gas turbine vanes before and after fluoride-ion cleaning (note characteristic bright appearance of cleaned parts).[47]

Generally, brazing discontinuities are of three general classes:

1. Those associated with drawing or dimensional requirements;
2. Those associated with structural discontinuities in the brazed joint;
3. Those associated with the braze metal or the brazed joint.

The inspection of a brazed assembly or subassembly is the last step in the brazing operation. This step is essential for assuring satisfactory and uniform quality of the brazement. It also indicates the adequacy with which the prior steps in the process were carried out with regard to the integrity of the brazed joints.

The design of the brazement is important to the inspection operations. During the design phase, inspection of the joints must be addressed to ensure that completed joints can be inspected by methods which show the required reliability for the service requirements. If a brazement is intended for a critical application, it must be designed so that it can be manufactured with the required quality and then properly inspected to assure that quality requirements have been met.

The inspection method chosen to evaluate a final brazed component should depend on the service requirements. When establishing the above-mentioned regulatory codes or standards of quality for brazed joints, an

Fig. 6.34 Vanes before (top) and after (bottom) gaseous HF fluoride-ion cleaning.[47]

approach similar to the one used in establishing standards for any other phase of manufacturing should be used. These standards should be based, if possible, upon requirements that have been established by prior service tests or history. The brazing process should be validated by destructive testing to assure the ability to produce brazements to the required quality standards and dependability.

Nondestructive inspection methods are applicable to brazements made by all of the brazing processes, and are essential to the quality control of brazements for critical applications. The size, shape, complexity, and degree of critical application will dictate the particular inspection method or methods that are most suitable. If no accurate and dependable method of inspecting a critical brazed joint can be found, either the part should be redesigned to permit inspection or another more inspectable joining technique should be used. Neither periodic destructive inspection nor process qualification requirements are completely acceptable substitutes for nondestructive inspection of the actual hardware entering critical service.

When the acceptance limit for any type of brazed defect is defined, the following must be considered: shape, orientation, and location in the brazement, including surface versus subsurface, and the relationship to other imperfections.

Judgements for disposition of discrepant components should be made by persons competent in the fields of brazing metallurgy and quality assurance who fully understand the function of the component. Such dispositions must be documented.

6.4.1 Nondestructive inspection methods (NDT)

The objectives of nondestructive inspection or testing of brazed joints should be:

1. To seek out discontinuities defined in quality standards or codes;
2. To obtain clues to the classes of irregularities in the fabricating process.

Visual inspection (VT)

Every brazed joint should be examined visually. It is a convenient preliminary test when other test methods are to be used. It will not be effective if the joint cannot be readily viewed.

A visual inspection is effective in evaluating external evidence of voids, porosity, surface cracks, fillet size and shape, noncontinuous fillets, base metal erosion, surface imperfections, roughness, liquation, and general braze appearance. Fillets on both sides of a joint, even if continuous, do not guarantee complete filling of the joint with filler metal. Visual inspection cannot reveal internal imperfections in a brazed joint, such as trapped flux, porosity, lack of fill, and internal cracks. To avoid misinterpretation, the inspector should be provided with samples, photos or sketches showing the precise visual conditions that are acceptable and unacceptable.

Proof testing

Proof testing is a method of inspection that subjects the completed joint to loads slightly in excess of those that will be experienced during its subsequent service life. These loads can be applied by hydrostatic methods, by tensile loading, by spin testing, or by numerous other methods. Occasionally, it is not possible to ensure a serviceable part by any of the other nondestructive methods of inspection, and proof testing then becomes the most satisfactory method.

This test does not evaluate braze quality but applies a one-time loading that may not closely simulate all the conditions encountered in service. It may not accurately predict service life, especially if cyclic loadings are encountered in service. A suitable inspection is required after proof testing to assure that the test itself did not cause cracks which could propagate in service.

Leak testing (LT)

Often called **pressure testing**, it is recommended when gas or liquid tightness is required. Pressure testing determines the gas or liquid tightness of a closed vessel. It may be used as a screening method to find gross leaks before adopting sensitive test methods. A low pressure air or gas test may be done by one of three methods (sometimes used in conjunction with a pneumatic proof test):

1. Submerging the pressurized vessel in water and noting any signs of leakage by rising air bubbles;
2. Pressurizing the assembly, closing the air or gas inlet source, and then noting any change in internal pressure over a period of time (corrections for temperature may be necessary);
3. Pressurizing the assembly and checking for leaks by brushing the joint area with a soap solution or a commercially available liquid and noting any bubbles and their source.

A method sometimes used in conjunction with a hydrostatic proof test is to examine the brazed joints visually for indications of the hydrostatic fluid escaping through the joint.

It should be emphasized that leak tightness when originally tested does not guarantee tightness of a joint subjected to various and repeated loadings in service.

Pressure testing with helium is frequently preferred because of the relative ease of finding very minute leaks while the leak testing of brazed assemblies with freon is extremely sensitive.

Vacuum testing (VT)

Evacuation of the brazed component generally is employed in checking assemblies such as refrigeration equipment, electronic devices, and other high vacuum systems. A mass spectrometer is used in this method of testing and helium is the sensing medium. If mass spectrometer helium leak testing is to be used, the brazed joints should not be exposed to liquid prior to the helium leak testing because of the possibility of plugging a leak with liquid.

Liquid penetrant inspection

This NDT method finds cracks, porosity, incomplete flow, and related surface flaws in a brazed joint. Commercially colored or fluorescent penetrants penetrate surface openings by capillary action. After the surface penetrant has been removed, any penetrant in a flaw will be drawn out by a white developer that is applied to the surface. Colored penetrant is visible under ordinary light. Fluorescent penetrant flaw indications will glow under an ultraviolet (black) light source. Since penetration of minute openings is involved, interpretation is sometimes difficult because of the irregularities in the braze fillets and residues of flux deposits. Inspection by another method must be used to differentiate surface irregularities from joint discontinuities. This type of inspection should not be used if subsequent repairs are contemplated because the penetrant often is difficult or impossible to remove completely.

Radiographic inspection

Radiographic inspection of brazements detects lack of bond or incomplete flow of filler metal. The joints should be uniform in thickness and the exposure made straight through the joint. The sensitivity of the method is generally limited to two percent of the joint thickness. X-ray absorption by certain filler metals, such as gold and silver, is greater than absorption by most base metals. Therefore, areas in the joint that are void of braze metal show much darker than the brazed area on the film or viewing screen.

Radiographic inspection may show the presence of filler metal in a joint, but it cannot verify a metallurgical bond between the base metal and filler metal. Metallurgical bonding must be assured by brazing process controls. A radiograph showing no indications does not guarantee that filler metal has flowed into the joint. Radiographic film readers should be aware of this, and be particularly suspicious of joints that have no discontinuities on the film. Joints in which the filler metal is preplaced as foil should not be inspected radiographically because filler metal will always be present whether or not it has wet both faying surfaces.

Toolmakers braze carbide tips to steel shanks by the millions throughout the world. High-volume brazing occasionally goes awry, so some brazes fail to bond soundly, allowing tips to break loose during rough service.

To detect poor bonds in carbide-tipped hammer drills, most firms use radiography. Specialists have found that X-rays could spot defects as small as 1 mm in diameter.

The use of X-rays in the testing of brazed seams in carbide-tipped hammer drills[50] revealed that they would reveal defects totalling 10% or more of the bond area. This area of no bond would cause a hammer drill to fail before drilling 10 m in concrete, the standard for good drills, Figure 6.35.

For production-line radiography of carbide-tipped hammer drills, it is recommended that automation should replace film radiography with a real-time X-ray intensifier. Using an intensifier, an inspector could examine X-ray images flashing by on a CRT monitor, as in fluoroscopy.

Thermal heat transfer inspection

Inspection by heat transfer will detect lack of bonding in specific brazed assemblies such as honeycomb and covered skin panel surfaces. Additionally, brazed aircraft propeller blades may be photographed while still hot, a few minutes after leaving the furnace. The covering skins appear to be bright red in areas where they are brazed to the reinforcing rib, but are a much darker red or black in areas where voids are present.

A method of inspecting brazed honeycomb panels utilizes powder or liquid materials with low melting points to indicate the differences in heat transfer characteristics when the honeycomb assembly is placed under infrared heat lamps. Temperature variations cause the liquid to be repelled from warm areas and to accumulate in cool spots. The core partitions act as heat sinks, causing the fluid to flow to the brazed areas.

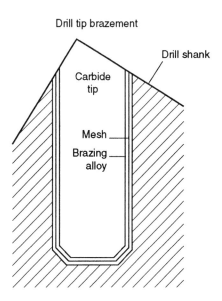

Fig. 6.35 Brazed carbide tips into shanks of hammer drills, assisted by ductile metal mesh. Radiography reveals areas of disbond.[50]

Other techniques use thermally sensitive phosphors, liquid crystals, and other temperature-sensitive materials. Infrared-sensitive electronic imaging devices with television readout are commercially available to monitor temperature differences produced by variations in heat transfer across a brazed joint. The resulting images, which can be recorded on video tape, show the brazed areas as light spots resulting from the rapid conduction of heat through to the opposite part. The void areas are seen as dark spots which do not emit infrared as intensely. These devices are able to monitor temperature differences less than (1°C) which indicate variations in braze quality.

Ultrasonic inspection

The ultrasonic testing method using low energy, high frequency mechanical vibration (sound waves) readily detects, locates or identifies discontinuities in brazed joints.[51] The applicability to brazements of this method depends largely on the design of the joint, surface condition, material grain size and the configuration of adjacent areas.

A transducer emits a pulse and then receives echoes from the surface. In the standard brazed lap joint, these echoes will come from the front surface and the rear surface of the part detail at areas of complete bonding. At defective areas, a third echo located between the first two will be caused by the unbrazed faying surfaces of the joint. These echoes may be displayed on an oscilloscope. The defect signal may be used to trigger the pen of a recording device to produce a facsimile of the joint showing bonded and unbonded areas.

There are two basic ultrasonic testing techniques, complete immersion and contour following. In the first method, flat panels are immersed in water, and a sonic transducer close to the surface of the facing sheet moves backward and forward across the panel in a scanning motion.

In contour following, water is emptied from the test tank. The sonic transducer, mounted on gimbels, describes a scanning motion identical to that of the immersion technique, but follows the curve of the panel. The sound waves travel to and from the panel through a stream of water jetted from the gimbal head. Very little water is required for effective coupling.

The angle of the sound waves transmitted to the panel and reflected back to the receiving head is important in both complete immersion and contour following in order to maintain full signal strength.

Ultrasonic inspection is sensitive to set-up variables, part configuration, and materials. A reference standard identical to the part being inspected and containing defects of known configuration should be used to set up and to calibrate the equipment at specified intervals.

The applicability of ultrasonic inspection depends largely on the design of the joint and the configuration of the adjacent areas of the brazed assembly. Ultrasonic techniques have been developed and used for a wide range of brazing applications and often provide one of the best methods of evaluating joint quality.[52–54]

The method is sensitive to both the presence of filler metal in the joint, and to the bonding between it, and the base metal. Thus, ultrasonic inspection is the preferred method for inspecting joints in which preplaced filler metal foils are used.

Other NDT methods

Inspection techniques, such as laser holography, ultrasonic scanning microscopy, acoustic emission and real time radiography, show promise of improving the ability to ensure quality. In all cases, however, a rational decision based on experimental evidence should be the basis for process selection and application.

6.4.2　Destructive test methods

Destructive methods of inspection clearly show whether a brazement design will meet the requirements of intended service conditions. Destructive methods must be restricted to partial sampling. It is used to verify the nondestructive methods of inspection, by sampling production material at suitable intervals. The methods and frequency are usually established by quality control procedures.

Destructive testing methods may be conducted on either test specimens or finished brazements. The three major types are (1) metallographic, (2) mechanical, and (3) chemical.[55]

Metallographic inspection

Requires the removal of sections from the brazed joints and preparing them for macroscopic or microscopic examination. This method detects flaws (especially porosity), poor flow of brazing filler metal, excessive base metal erosion, the diffusion of brazing filler metal, improper set-up of the joint, and it will reveal the microstructure of the brazed joint. When defects are found, it may be an indication that the brazing procedure is out of control or that improper techniques are being used.

Mechanical inspection

Peel tests are frequently employed to evaluate lap joints. One member is held rigid, as in a vise, while the other is peeled away from the joint. This test may be used as a means of production quality control to determine the general quality of the bond as well as the presence of voids and flux inclusions in the joint. The permissible number, size and distribution of these discontinuities will depend upon the service conditions of the joint and may be limited by applicable codes.

Tension or shear tests are used to determine quantitatively the strength of the brazed joint, or to verify the relative strengths of the joint and base metal. This method is widely used when developing a brazing procedure. Random sampling of brazed joints is used for quality control and verification of brazing performance.

The torsion test evaluates brazed joints with a stud, screw, or tubular member brazed to a base member. The base member is clamped rigidly and the stud, screw, or tube is rotated to failure which will occur in either the base metal or the brazing filler metal. Torsion testing is used on a sampling basis to validate the brazing procedure.

Impact tests, like fatigue tests, are generally limited to laboratory work in determining basic properties of brazed joints. As a general rule, the normal notch-type specimens do not appear to be best for evaluating brazed joints. Special types of specimens may be required to obtain accurate results.

Fatigue tests are employed to a limited extent under cyclic loading and in most cases are used to test the base metal as well as the brazed joint. As a general rule, fatigue tests require a long time to complete and for this reason are very seldom used for quality control.

Chemical inspection

It is used ordinarily to determine chemical composition or corrosion resistance. Tests for chemical composition are often required to ascertain whether the base metal and filler metal meet specifications or codes.

Corrosion tests have been devised to determine if the braze material will perform under service conditions. Such tests usually are accelerated laboratory-type tests. The brazed assembly is placed in a specific environment and is subjected to service-like conditions for a specific period of time.

6.5 SAFETY

Although brazing is well established throughout industry as a reliable and safe method of assembling metal components, attention to health and safety precautions is necessary, as with all operations involving molten metals and fluxes, and in particular to:

- burns;
- combustion products from brazing torches;
- fumes from fluxes and metals.

Additionally, good ventilation is essential, plus common sense, such as the operator not touching hot assemblies or learning over an assembly during brazing.

Other considerations concerning safety include the operation and maintenance of the brazing equipment or facility. The equipment should conform to the provisions of national standards established in all countries and governments.[56-61] See Appendix for typical types of figures and tables available on Safety. These standards provide detailed procedures and instructions for safe practices which will protect personnel from injury or illness and protect property and equipment from damage by fire or explosion arising from brazing operations.

Care must be exercised in the handling of hot objects. Braziers must wear clothing, including shoes, that will afford protection against falling objects, heat, and other hazards incidental to the brazing process. Goggles or spectacles, as dictated by the brazing operation, are also necessary.

Some brazing filler metals contain toxic materials such as cadmium or beryllium, and most fluxes contain fluorides and fluorine compounds. The use of such filler metals and fluxes require special precautions as detailed in the relevant safety standard.

You must try to prevent brazing fluxes from contacting skin. Occasional contact is not dangerous, but all flux should be thoroughly washed off before consuming food and drink. Cuts or breaks in the skin must be properly covered by a dressing. Flux, especially if it contains fluorides or chlorides, can delay the healing of wounds.

Fluxes produce fumes when heated, especially above the temperatures given as their maximum. Braze in work stations with large air space into which fumes can escape. Ventilate with fans or exhaust hoods to carry fumes away from workers, or equip operators with air-supplied respirators. Consult manufacturers' Material Safety Data Sheets for specific safety and health procedures connected with flux use, see Appendix Tables A.18 to A.21 and Figure A.2.

Ventilation

It is essential that adequate ventilation be provided so that personnel will not inhale gases and fumes generated during brazing. Confined areas must be ventilated. If questions arise, it is best to consult national regulations or other reference materials.[56, 57]

Cleaning

Base metals should be cleaned thoroughly. Surface contaminants of unknown composition may increase fume hazards and cause excessively rapid breakdown of flux, leading to overheating and fuming. Care should be exercised in using fluxes because they contain chemical compounds of fluoride, chlorine and boron, which are harmful if they are inhaled or if they contact the eyes or skin. Base metals should be heated broadly and uniformly. Intense localized heating uses up flux and increases the danger of fuming. Heat should be applied only to base metals, not to brazing filler metal. In torch brazing, direct impingement of the flame on the filler metal causes overheating and fuming.[58]

The commonest fumes likely to be encountered are fluorides from the fluxes, burnt gases from brazing torches and oxides arising from zinc, copper and especially cadmium. Zinc vaporizes from brasses and oxidizes readily to give white fumes which are unpleasant but clearly visible. Copper and copper oxide fumes are less frequently encountered but both zinc and copper may cause 'metal fume fever'. Cadmium oxide fume is particularly dangerous as it may lead to occupational exposure limits (OELs) for chemical substances in the atmosphere and should not be exceeded. The current limits are given in the Appendix Table A.20. When OELs are likely to be reached, ventilation must be improved locally or in general, applying expert advice if necessary.

Barrier cream may be applied before work, and gloves should be worn if skin irritation is a problem. Gloves required for handling hot metal should be properly treated chrome leather and asbestos-free. Eye irritation should be irrigated with water.

For salt bath or dip brazing the components must be completely dry before being immersed in the bath otherwise explosions may occur, ejecting molten salt possibly to cause severe burns. Temperature controls on salt or metal baths must operate effectively to avoid overheating.

REFERENCES

1. Schwartz M., *Brazing*, ASM International, Metals Park, OH, 438p, 1987.
2. Trimmer R. M., Kuhn A. T., The strength of silver-brazed stainless steel joints – a review, *Brazing and Soldering*, No. 2, Spring, 1982, pp. 6–13.
3. *Metals Handbook, 9th Ed, Vol 6, Welding, Brazing, and Soldering*, ASM, Metals Park, OH, 1983, pp. 941–944.
4. Zhuang H., Lugscheider E., Chen J., *Wide gap brazing of stainless steel with nickel-based brazing alloys*, Document SCIA–B–133, IIW Meeting, Strasbourg, France, Sept. 1985 (16 pages).
5. *Metal Construction*, Table 14, Data Sheet Series 3, pp. 166–168, March 1986.
6. Watson H. H. H., Fluid-tight joints for exacting applications, *Welding and Metal Fabrication*, Sept. 1976, pp. 491–495.
7. How to specify proper joint clearance for brazing dissimilar metals, *Industrial Heating*, Nov. 1961, pp. 2162–2164.

8. Biagi L. A., Koehler G. W., Patterson J. A., The fabrication and brazing of 15 A, 120 keV continuous duty accelerator grid assemblies, *Welding Journal*, 61(10), Oct. 1980, pp. 33–35.

9. Pattee H. E., Joining ceramics to metals and other materials, *WRC Bulletin* #178, Nov. 1972, 43 pages.

10. Schwartz M. M., *Ceramic Joining*, ASM International, Metals Park, OH, 1990, 185 p.

11. Loehman R. E., Tomsia A. P., Joining of ceramics, *Am. Ceram. Soc. Bull.*, 67(2), 1988, 375–80.

12. Bates C. H., Foley M. R., Rossi G. A., *et al.*, Joining of non-oxide ceramics for high-temperature applications. *Am. Ceram. Soc. Bull.*, 69(3), 1990, 350–56.

13. Johnson S. M., *The formation of high strength silicon nitride joints by brazing*, Rept. No. D88–1208, SRI International, Menlo Park, CA, Sept. 1987.

14. Nicholas M. G., Mortimer D. A., Ceramic metal joining for structural applications, *Mater. Sci. Technol.*, 1, 1985, 657–65.

15. Loehman R. E., Interfacial reactions in ceramic-metal systems, *Am. Ceram. Soc. Bull.*, 68(4), 1989, 891–96.

16. Nicholas M. G., Crispen R. M., Brazing ceramics with alloys containing titanium, *Ceram. Eng. Sci. Proc.*, 10(11–12), 1989, 1602–12.

17. Iwamoto N., Makino Y., Miyata H., Joining silicon carbide using nickel-active metal (or hydride) powder mixtures, *Ceram. Eng. Sci. Pro.*, 10(11–12), 1989, 1761–67.

18. Hare M., Keller R. F., Meneses H. A., *Electroformed ceramic-to-metal seal for vacuum tubes*, TR453–3, Cont DA–36–039–SC–73138, Menlo Park, CA, Stanford Univ., Nov. 17, 1958.

19. Bondley R., Metal-ceramic brazed seals, *Electronics*, 20(7), 1947, 97–99.

20. Pearsall C. S., Zingeser P. K., *Metal to nonmetallic brazing*, Tech. Rept. No. 104, MIT Res. Lab. of Elect., Cambridge, MA, April 5, 1949.

21. Kelly F. C., *Metallizing the bonding nonmetallic bodies*, U.S. Patent 2,570,248, Oct. 9, 1951.

22. Fox C. W., Slaughter G. M., Brazing of ceramics, *Welding Journal*, 43(7), Jul. 1964, 591–97.

23. Kutzer I. G., Joining ceramics and glass to metals, Carborundum Co., *Matls. Des. Engr.*, Jan. 1965, 106–10.

24. *Brazing Manual*, 3rd Ed, Amer. Weld. Soc., Miami, FL, 1976, 262–63.

25. Morrell R., Joining to other components, Part I, an introduction for the engineer and designer, *Handbook of Properties of Technical and Engineering Ceramics*, London, Her Majesty's Stationery Office, Section 3.5, 1985, 267–78.

26. Van Houten G. R., Ceramic-to-metal bonds, *Matls. Des. Engr.*, Dec. 1958, 112–14.

27. Weymueller C. R., Braze ceramics to themselves and to metals, *Weldg. Des. & Fab.*, Aug. 1987, 45–48.

28. Mizuhara H., Huebel E., Joining ceramic to metal with ductile active filler metal, *Welding Journal*, 65(10), Oct 1986, 43–51.

29. Hoop G. G., *Generator development, SPUR program, Part II, generator stator bore seal*, TDR–63–677–Part 2, Cont AF33(657)–10922 and AF33(615)–1551, Westinghouse Elect. Corp., Feb. 1967.

30. Yamada T. *et al.*, Development of ceramic exhaust valves, *16th CIMAC*, D-82, 1985.

31. Yamada T. *et al.*, Collected abstracts, *38th Meeting and Symposium of the Marine Engineering Soc.*, Japan.

32. Yamada T. *et al.*, Diffusion bonding SiC or Si_3N_4 to nimonic 80A, *Hi Temp Tech*, 15(4), Nov. 1987, 193–200.

33. Cassidy R. T., *et al. Bonding and fracture of titanium-containing braze alloys to alumina*, Monsanto Res. Corp., MLM-3431 (OP) and MLM-3394 and DE87002197 and DE87009195, U.S. Dept. of Energy Cont De–ACOE–76DP00053, Oct. 1987 and Oct. 1986.

34. Hammond J. P., David S. A., Santella M. L., Brazing ceramic oxides to metals at low temperatures, *Welding Journal*, 67(10), Oct. 1988, 227s–32s.

35. Mizuhara H., Vacuum brazing ceramics to metals, *Advanced Metals and Processes*, 131(2), Feb. 1987, 53–55.

36. Honnell R. E., Stoddard S. D., *Development of a high-temperature ceramic-to-metal seal for an air force weapons laboratory laser*, LA–10884–MS, UC-25, Los Alamos Nat. Lab., Los Alamos, NM, Mar. 1987.

37. Dalgleish B. J., Lu M. C., Evans A. G., The strength of ceramics bonded with metals, *Acta Metallurgica*, 36(8), Jan 1988, 2029–35.

38. Naka M. *et al.*, Influence of brazing on the shear strength of alumina-Kovar joints made with amorphous $Cu_{50}Ti_{50}$ filler metal, *Transactions of JWRI*, 12(2), 1983, 181–83.

39. Schwartz M. M., Brazing in a vacuum, *WRC Bulletin #244*, Dec. 1978, 23p.

40. Alumina barrier for vacuum brazing, MSC–18528, *NASA Tech Briefs*, Spring, 1980, p106.

41. A precoat prevents ceramic stopoffs from spalling, MFS–19495, *NASA Tech Briefs*, Spring, 1980, p114.

42. Weiss S., Adams Jr. C. M., The promotion of wetting and brazing, *Welding Journal*, 48(2), Feb. 1967, pp. 49s–57s.

43. Chasteen J. W., *Development and evaluation of wide clearance braze joints in gamma prime alloys*, Cont F33(615)–79–C–5033, AFWAL–TR–82–4016, Univ. of Dayton Res. Inst., Dayton, OH, 5/1/79–5/1/81, 194p.

44. Demo W. A., Ferrigno S. J., Brazing Method Helps Repair Aircraft Gas-Turbine Nozzles, *Advanced Metals and Processes*, 141(3), Mar. 1992,pp. 43–45.

45. Zhuang H., Li Y., Cleaning and repair brazing of cracks in turbine vanes, *Brazing, High Temperature Brazing and Diffusion Welding Symposium*, DVS 125, 19–20 Sept. 1989, Essen, GDR, pp. 103–06.

46. Riad S. M., El-Naggar A., Brazing of gray cast iron, *Welding Journal*, 62(10), Oct. 1981, pp. 22–24.

47. Clavel A. L., Kasperan J. A., Vapor-phase, fluoride-ion processing of jet engine superalloy components, *Plating & Surface Finishing*, pp. 52–57, Nov. 1991.

48. Chasteen J. W., *The U.D.R.I. fluorocarbon-cleaning process, 1980*.

49. Chasteen J. W., U. S. Patent 4,405,379, 1983.

50. *Schweissen und Schneiden*, pp. E72–E75, May 1987.

51. Cook K. V., McClung R. W., Development of Ultrasonic Techniques for the Evaluation of Brazed Joints, *Welding Journal*, 43(9), Sept. 1962, pp. 404s–408s.

52. Mirror checks braze joints, *Amer. Mach.*, Apr. 30, 1973, p23.

53. deSterke A., deBlieck T., Nondestructive examination of tube to tube plate connections, Technical Monograph A–2570, RTD, *International Symposium on Nondestructive Testing of Nuclear Power Reactor Components*, Rotterdam, Feb. 1970, 13p.

54. deRaad J. A., Ultrasonic and other nondestructive testing methods for tube joints used in heat exchangers, Technical Monograph A-2670, RTD, *International Symposium on Nondestructive Testing of Nuclear Power Reactor Components*, Rotterdam, Feb 1970, 18p.

55. Mock J. A., Guide to destructive testing, *Matls. Engineering*, Aug 1977, pp. 18–29.
56. *Safety and health in soldering and brazing*, U.S. Dept. of Health, Education and Welfare, DHEW (NIOSH) Publication No. 78–197, 1978, 42p.
57. *Soldering and brazing safety guide – a handbook on safe practice for those involved in soldering and brazing*, Enviro-Management and Research, Inc., Washington, PB82–151721, NIOSH, Cincinnati, OH, 1981.
58. *Safety in Welding and Cutting*, Amer. Nat. Std. Z49.1, AWS, Miami, FL.
59. *Environment and Facility Safety Signs*, ANSI Z535.2.
60. *Practices for Occupational and Educational Eye and Face Protection*, ANSI Z87.1.
61. *Metal Construction*, Sheets 14 and 15, May 1986.

7

Applications and the future of the process

The ability to produce reliable brazed metal/metal, ceramic/ceramic, ceramic/metal and matrix composite joints is a key enabling technology for many production, prototype and advanced developmental items and assemblies. Design and performance demands are now being made that Nicholas claims previously couldn't be met readily by existing processes and commercially developed metals, alloys and ceramics.[6] Considerable attention is now being paid to the potential usefulness of these new materials,[1] due to the improvement in atmospheres (fluxless),[2] cleaning methods,[3] modeling procedures, braze filler foils,[4] and their improved properties, non-toxic filler metals,[5] and NDE techniques to verify structural integrity, soundness and suitability. For example, ceramics are generally more refractory and less dense than metals and can be stronger. However, if these properties are to be exploited it is generally essential for an adequate joining technology to be available and that technology is brazing.[6–14]

7.1 AUTOMATION

The future of braze processing lies in automation. Automation of the brazing process is the key to maximizing production and quality and minimizing costs. Typical machines that are currently available are capable of automatic brazing by torch, furnace, induction, infrared and resistance processes. Rotary and shuttle machines automatically apply paste and filler metal, heat the parts, torch braze the assembly, cool the brazement and remove some or most of the flux.

Many engineers and manufacturing specialists are reluctant to try to automate brazing operations for fear of upsetting a well-established process and causing quality problems because of a processing and/or method change. Their attitude, in short, is "If it's working and producing parts, don't change it; don't upset it."

However, brazing processes, when properly understood, are just as adaptable to increased productivity through automation as any other process; on the other hand, of course, they are also subject to similar limitations.

Regardless of how attractive the idea of automating a brazing operation may be, the first step usually is to verify and compare the economics. If automation is economically justified, the next step is to investigate the mechanics of how to automate the operation.

In summary, the following steps are necessary for successful automation:

1. Make an economic analysis to justify automation.
2. Break down the total operation into its several parts[15,16] and determine which can be eliminated, which can be combined, and which can be carried on in conjunction with other operations.
3. Determine the required type of automated equipment (rotary indexing machine, constant rotary machine, in-line conveyor, in-line indexing conveyor, shuttle machine or racetrack-type conveyor). The types of parts to be brazed and the required production rate will normally determine what type of mechanized equipment will be most feasible.
4. Apply specialized skills and knowledge to the various design and engineering problems involved in realizing the objectives of automation: mechanics (in the selection of a mechanical movement and also in the design of the part-holding fixture); selection and application of flux and brazing filler metals; type and position of heat application; cooling; additional mechanics of automatic loading and unloading; and electrical control systems.

7.2 FLUXLESS BRAZING

7.2.1 Aluminum alloys

A fluxless bonding procedure[17] has been developed for the 8090 Al–Li alloy, based on an aluminum foil interlayer containing little or no lithium. This foils distorts during the bonding process; the oxide film thus mechanically disrupts, and lithium diffuses from the parent alloy into the foil, Figure 7.1.

The procedure requires little sophistication as the parts are simply mechanically abraded and degreased before assembly in the bonding equipment. Bonding parameters of 530°C, 8–10 MPa pressure and 20–60 min bonding time have been successfully applied, giving joint shear

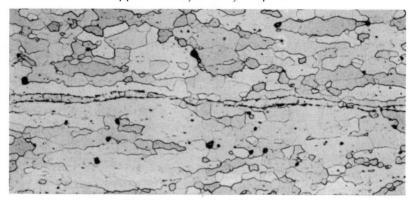

Fig. 7.1 Photomicrograph of 3% wt Li aluminum alloy joint interface.[17]

strengths of about 80 MPa. Further work has been done in examining different surface preparations, interlayer materials and testing techniques.

Due to their advantages, fluxless brazing methods for aluminum have increasingly come into common use within the last twenty years. Some aspects concerning the casting and rolling of the brazing filler material e.g., the purity of the alloys, the cleanliness and cladding tolerances of the brazing sheets has been examined as well as the essential brazing characteristics of the different filler metals e.g., cladding ratio, kind and purity of the brazing atmosphere, fillet formation, joint clearance and magnesium deposition in the furnace.

Braze filler metals in particular have been scrutinized for impurity levels and strict process compliance and material control to achieve reproducible sound brazed joints. In a recent study Schoer[18] evaluated the fluxless brazing of various braze filler metals which range from about 7–12% Si, but usually the Si content is about 10%. Currently three types of brazing filler metals are used worldwide with the following additional alloying constituents:

High Mg-type: 10AlSi–1.5 Mg
    4004: 1.5% Mg
    4104: 1.5% Mg and 0.1% Bi
Low Mg-type: 10AlSi–Mg
    modified 4045: 0.15% Mg
Mg-free type: XL–10AlSi
    low Bi content (GDR development)

With aluminum it is possible to clad the brazing filler metal directly to the base metal permitting complex parts to be formed and assembled without the need for replaced filler metal preforms. The use of filler wires and shims is not common. The base metals most frequently used for cladding are 3003, 3103, 3005, 3105, 6951 and 6063.

Aluminum has very favorable properties for heat-exchanger applications, and feasible designs for aluminum heat-exchangers and methods for their economical mass production have existed for at least 40 years. Despite these encouraging factors, no major automotive application of the aluminum radiator in the US occurred during the first 30 years of development and application efforts. It wasn't until the decade of the 1980s that mass production of aluminum radiators began in the US and it was expected that by 1992, aluminum radiators would be used in all Ford vehicle lines.

Processing

One of the most important factors determining the suitability of an aluminum substitution for copper to replace the standard radiator. The vacuum-braze process for aluminum requires no fluxing, and the braze is accomplished in one step. Assembly density of parts to be joined is not limited with this process, and the oxide barrier to brazing is overcome by the use of a solute addition.

Efforts during the 1990s and beyond will be focused on producing and improving reliability of aluminum radiators through improvements in aluminum materials and the brazing process. Use of alternate fluxes that pose no corrosion problems will compete with fluxless processes for reducing costs and improving reliability. Fluxless inert gas brazing will compete with the vacuum process for improving process rates and yields. Intelligent processing methods for controlling brazing will be developed and establish new standards for manufacturing reliability.

7.3 *IN-SITU* REACTION JOINING

Fiber-reinforced SiC composites

Rabin[19] and his associates have recently developed methods for *in situ* reaction joining of SiC–SiC composites. SiC fiber-reinforced SiC matrix composites (SiC–SiC) produced by chemical vapor infiltration are being developed for use in structural applications at temperatures approaching 1000°C.[20,22] These composites contain about 40% vol SiC fibers (Nicalon®) and are infiltrated to about 85% of the theoretical density with SiC. In order to fully realize the advantages of these materials, practical joining techniques are being developed. Successful joining methods will permit the design and fabrication of components with complex shapes and the integration of component parts into larger structures. These joints must possess acceptable mechanical properties and exhibit thermal and environmental stability comparable with the composite which is being joined.

Joining of SiC has been accomplished by a variety of techniques, including the following:

direct diffusion bonding;[23,24]
co-densification of interlayer and green bodies;[25]
diffusion welding or brazing with boride, carbide and silicide interlayers;[23]
hot pressing of sinterable SiC powder;[26]
bonding with polymeric precursors;[27]
brazing with oxide[28] or oxynitride materials;[29]
reactive metal bonding;[30]
active metal brazing.[31]

Although varying degrees of success have been achieved, these joining methods must be improved upon to withstand the intended service temperatures (1000°C).

The initial studies have identified two material systems with potential for joining SiC–SiC composites by reaction methods. Focus has been aimed on joints produced using TiC–Ni and SiC+ Si interlayers. The microstructures of joints have been characterized and the results appear promising. Further work is needed to optimize joint microstructures, understand interfacial reactions and to assess the mechanical properties of the joined components.[19]

7.4 METAL/CERAMIC JOINING: MICROELECTRONICS PACKAGING

Brazing processes have been previously confined to high temperature fired/cofired alumina substrates using either tungsten or molybdenum-based refractory metallizations. The metallized alumina packages are usually fired (cofired) at about 1600°C. Prior to brazing, the metallized pads and seal rings are normally nickel plated and heat treated. Conventional brazing of pins, leads and heat sinks to the metallized alumina substrate is usually carried out in a nitrogen–hydrogen atmosphere using a silver–copper eutectic filler metal in the 820–900°C temperature range.

Current developments cover the development of silver, gold and copper-based thick film paste compositions, brazing filler metal compositions, pin-lead windowframe and heatsink surface treatment and furnace conditions where brazing processes are accomplished in the 550–760°C temperature range in a nitrogen atmosphere, Figure 7.2.

The metal/ceramic joint strengths obtained using low-temperature fired thick films, were found to be comparable to those accom-

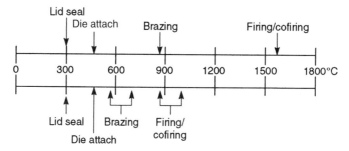

Fig. 7.2 Schematic of process hierarchies in high and low temperature cofired/fired packaging systems.[32]

plished using high-temperature fired tungsten- or molybdenum-based metallizations. The capabilities developed in the work by Keusseyan *et al.*[32], coupled with ceramic circuitization using existing (a) high electrical conductivity metallizations (Cu, Au or Ag-based), and (b) low dielectric constant and low TCE dielectric thick-film pastes and tapes, offer the package designer extended electrical, thermal, performance and reliability capabilities that were not available previously.

7.5 JOINING OF ABRASIVE TOOL MATERIALS

In recent years direct brazing of a monolayer of diamond crystals on a steel substrate with active filler metals has gained tremendous importance in the industry, with a view to developing tools which can out-perform the conventional galvanically-bonded diamond tools. An existing proprietary process uses a specially prepared Ni–Cr filler metal to facilitate its application on a steel substrate. The brazing is done either in a vacuum or a dry hydrogen furnace. Studies[33] have shown that a commercially-available Ni–Cr hardfacing alloy, flame-sprayed on a steel substrate with an oxyacetylene gun, could be used for direct brazing of diamond particles. During the induction brazing in an argon atmosphere the chromium present in the alloy segregated preferentially to the interface with diamond to form a chromium-rich reaction product promoting the wettability of the alloy. It has been further revealed that under a given set of brazing conditions, the wettability of the Ni–Cr hardfacing alloy towards diamond grits primarily depended on its layer thickness.[33]

7.6 JOINING WITH METALLIC-AMORPHOUS GLASS FOILS

Rapidly solidified (RS) amorphous and microcrystalline brazing filler metals are currently used in a wide variety of brazing applications. The RS materials, typically cast to foil form for direct use in metal joining, offer superior purity and chemical and microstructural homogeneity when compared with conventionally-formed brazing filler metals. This homogeneity, in turn, manifests itself in uniform melting, flow in the joint area and solidification during the brazing process. Accurate control of brazing in this manner permits the production of uniform joint microstructures that are free of voids and macroscopic segregation. The results are dramatic reductions in reject rates and superior joint properties. Over a broad range of base metal/filler metal combinations, the use of RS filler metals yields joints with superior mechanical properties and improved resistance to thermal fatigue and corrosion. Moreover, the use of RS technology uniquely permits the formation of foils in many filler metal systems which are brittle and unformable in the crystalline state.

The basic difference between crystalline and glassy metals is in their atomic structures. Crystalline metals are composed of regular, three-dimensional arrays of atoms which exhibit a long-range order. Metallic glasses do not have long-range structural order. Atoms are packed in a random arrangement similar to that of a glass or a liquid metal.

Despite this vast structural difference, crystalline and glassy metals of the same composition will have nearly identical densities. Typically, a metallic glass will be a few percent less dense than its crystalline counterpart.

Metallic glasses also lack the microscopic structural features common in crystalline metals. In the absence of crystallinity, grains, grain boundaries, grain orientations, and additional phases do not exist. The glassy state is essentially one phase, possessing complete chemical homogeneity.[34]

A whole family of RS brazing filler metals have now been produced, including Cu–Ni–Sn–P (78Cu–10Ni–4Sn–8P and 77Cu–6Ni–10Sn–7P), RSNi–2 (82.5Ni–7Cr–3Fe–3B–4.5Si) and RSNi–3 (92Ni–0.5Fe–3B–4.5Si), RSNi–Pd, and Al–Mg–Si filler metals.[35,36] The applications for these RS materials range from high-temperature brazing of superalloys in critical assemblies within gas-turbine engines to low-temperature soldering of semiconductors and lead frames in microelectronic devices.

An example are the corrosion-resistant, leak-free seals in pressure regulators which have been made by brazing with clean nickel-based foils. In the regulator, Figure 7.3, a 3 mm thick convoluted diaphragm (316L stainless or Carpenter Custom 450) is joined to an actuator of 316L stainless steel. The continuous joint must be void-free to eliminate leaks under vacuum. Joints cannot be welded because the temperatures required (up to 1371°C) can deform the diaphragm, and change its base metal properties.

As a result brazing with amorphous foil brazing filler metal yields a clean, leak-free joint, accomplished at lower temperature. Melting at 1024°C, this 100% metal, amorphous nickel-based foil seals the joint completely without contamination. In the bonding process, 3 mm thick brazing foil preform rings are placed between the parts to be joined. The components are fixtured together by tack welding, and the entire assembly is heated in a vacuum furnace to 1066°C for 5 min. Brazing at this temperature assures complete liquefaction, wetting, and capillary flow of the brazing filler metal. When cooled, the solidified filler metal produces a uniform, corrosion-resistant joint. Hundreds of assemblies have been brazed at the same time, and no cleaning is required.

Other applications include the brazing of honeycomb Inconel 625 exhaust nozzles, cones, plugs, fan ducts, and tail pipes for jet engines. In all cases the foil is applied to the face sheets using resistance tack welding with roller wheel electrodes. For complex contours, precut foil is chosen. The braze filler metal foil chosen is the nickel-based AWS BNi2 or MBF–20 or AMS–4777B.

The metallic glass foil is really metal with an amorphous glass-like atomic structure. It is produced by very rapid quenching of a stream of liquid metal into a highly ductile ribbon. The supercooling bypasses the nucleation and grain growth stages altogether, producing an extremely strong material.

The future of RS technology lies in the joining of ceramics to metals in structural applications such as internal-combustion engines and gas-turbine components. Two approaches have been adopted in the development of ceramic brazing systems: alteration of the ceramic

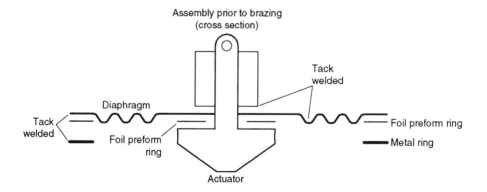

Fig. 7.3 Pressure regulator brazed with RSNi–3 amorphous foil.

substrate by metallization, and use of Ti–Zr filler metals which react with ceramic materials and promote wetting.

7.7 DIFFUSION OF BORON IN DUCTILE FOILS

Another form of brazing filler metal that has recently been developed, is used and is produced by diffusion of boron into the surface of ductile foil (Figure 7.4) or wire (Figure 7.5). The filler metal is produced in its final configuration prior to the introduction of boron, thereby eliminating the ductility problem which accompanies the presence of boron. Boron is then added by diffusion into the surface to produce a composition suitable for brazing.

As with foil, nickel-based wire can be drawn in the boron-free, ductile state. Preforms can be produced in any desired configuration. Boron is subsequently diffused into the surface to achieve the proper finished brazing composition.

Ductile-based nickel foils including (but not limited to) alloys such as AMS 4775, 4776, 4777, 4778 and 4779 have been developed.

With the diffusion of boron directly into the surfaces to be joined, the surface assumes the composition of the base metal plus boron, the melting point depressant. The surface to be bonded is thus converted to a braze-like alloy, permitting diffusion brazing.

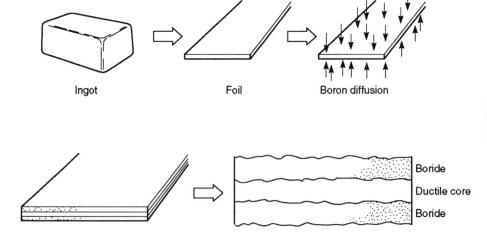

Fig. 7.4 Schematic representation of the basic steps used to produce ductile filler metal foil.

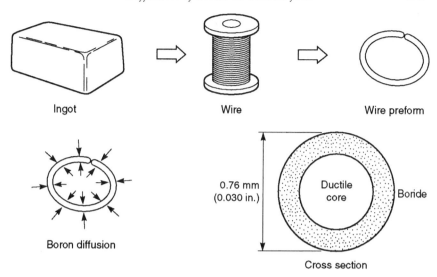

Fig. 7.5 Basic steps followed in producing filler metal wire preforms.

Although a number of applications for these materials are based on the substitution of foil or wire for conventional powder-based filler metals, many areas of interest have developed which are not currently utilizing brazing techniques. For example, conventional brazing processes are not suitable for the production of air-cooled axial-flow turbine wheels, Figure 7.6.

With the diffusion brazing approach, a turbine wheel (Astroloy) of the configuration shown in Figure 7.6 is produced by stacking a series of laminates which have been photoetched to develop the appropriate

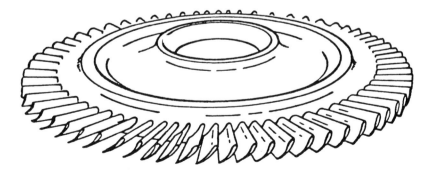

Fig. 7.6 Air-cooled turbine integral wheel.

pattern (e.g., cooling passages, etc.). The laminates have boron added to the bonding surfaces; the stack is then fixtured and exposed to the proper bonding cycle. The bonded assembly therefore contains the internal cooling passages, and no blade casting and attachment is required.

The wire and foil preforms described above are being used in the TLP® process in which specifically tailored foil compositions and bonding cycles are utilized to produce high quality bonds in INCO 713C turbine vanes in a turbofan engine used for commercial wide-body transports.[37]

TLP bonding[37] or Activated Diffusion Bonding (ADB)[38] is a bonding process that combines the manufacturing ease of brazing with the high efficiency of solid-state diffusion bonding. TLP bonding is applicable to nickel-based superalloys that are difficult to bond by conventional fusion welding, because of their fusion cracking troubles. This process is carried out with vacuum furnace brazing using filler metals with a specific composition, usually nickel-based filler metals containing boron. These filler metals temporarily melt and then resolidify at the bonding temperature according to the boron diffusion into the base metals. Moreover, it is said that by the post-bond heat treatments, the elements composing base metal diffuse into the TLP bonds and make a similar bond as the base metal.[39] Nakahashi[40] and his colleagues examined the TLP bonding of nickel-based superalloys Mar–M247 and IN939, using filler metals specially designed and fabricated into flexible coils by RS processes.

TLP bonds, using those specially-designed filler metals for Mar–M247 and IN939 reflect the following results:

1. Microstructures of Mar–M247 TLP bonds were almost the same as those for the base metal and were attained with the filler metal containing adequate quantities of strengthener 10.8Co–8.8Cr–3.9W–3.0Ta–3.0Al–2.5B.
2. Sufficient bonding pressure was also indispensable for the Mar–M247 TLP bonds to achieve high stress rupture properties.
3. IN939 TLP bonds showed better bonding efficiency than Mar–M247 TLP bonds.

Transient Liquid Insert Metal Diffusion Bonding (TLIM bonding) consists of three processes, viz., a dissolution process of base metal, an isothermal solidification process and a homogenizing process.

Advanced TLIM bonding process uses an amorphous filler and a metal powder sheet. In this new process, both the time necessary to complete the isothermal solidification process and that for the homogenizing process are shortened compared with those of conventional processes. The mechanism to shorten the TLIM bonding process was the use of a powder sheet with an insert metal. The morphology and size of the powder, kind of powder and the thickness of powder sheet were also factors to control the mechanical properties of the bonded joints.

Nakao[41] and his associates have applied the above theories on their TLIM process successfully to the ODS alloy, MA754, Mar–M247 and Alloy 713C. The MBF–80 (Ni–15.5Cr–3.7B) amorphous foil insert material was used as the intermediate filler metal.

The process was developed especially to join hot cracking susceptible nickel-based cast superalloys, DS alloys, single crystal alloys and ODS alloys and is similar to the above TLP and ADB processes.[37,38]

7.8 MODELING BEHAVIOR OF BRAZING PROCESSES AND MATERIALS

In the past several years especially with the assistance of computational equipment and software, a variety of modeling techniques evaluating solidification, dissolution, filler metal concentrations, etc. have taken place. These studies as well as others will produce the finite differences in the many brazing processes that currently exist and a better understanding of the melting, wetting and solidification of braze filler metals and base metal effects.[42] The future appears to offer an exciting time for the braze specialists and technicians.

7.9 EVALUATION OF STRUCTURAL DEFECTS

In the past four decades NDE methods to detect braze defects grew from visual means to X-ray methods to ultrasonic to eddy current examination, and even several heating methods so today all of these nondestructive techniques are available in production processes.

Now and in the future metals and ceramics will be joined to each other in various fields of industry and structural defects and integrity of the joints will require more sophisticated means of verification. Due to the difference between the thermal expansion coefficients of ceramics and metals high thermal stresses can nucleate cracks which then propagate either in the ceramic or along the ceramic/braze interface. Since applied stresses are added to the residual stresses in the joints it is of great importance to detect reliably defects in the ceramic-to-metal joints before use using nondestructive test methods. As a result new equipment has been developed and applied in the evaluation of the integrity of the brazed ceramic-to-metal joints. Equipment includes SAM (scanning acoustic microscopy), C-SAM (C-mode scanning acoustic microscopy), [43] X-ray CT (X-ray computer tomography).[44] SLAM (scanning laser acoustic microscopy)[44] and SPAM (scanning photoacoustic microscopy).[44]

The results of the comparative C-SAM and SEM (scanning electron microscope) studies conducted by Kauppinen[43] show that structural defects in brazed ceramic-to-metal assemblies can be reliably detected by using a reflection-type C-SAM operating at frequencies up to 100 MHz. For the detection of disbonding of the interface and laminar and surface opening cracks in the tested nitride ceramic, different focusing techniques must be used.

X-ray CT is a bulk characterization technique that can display real-time, two-dimensional X-ray sections of complex parts, such as turbocharger rotors and engine valve components. Current X-ray systems take several minutes for the beam to scan a part, but the use of a cone-beam X-ray system to collect all the data at once, coupled with powerful computer processing, will make future systems much faster.

Conventional X-ray can only detect 1 to 2% density variations, whereas X-ray CT is 100 times more sensitive and can detect density variations ranging from 0.01 to 0.02%. In addition, X-ray CT can be used to determine the quality of a ceramic part from the beginning to the end of processing.

The SLAM scans the surface perturbations continuously with a focused laser beam. The use of a scanning laser beam to detect surface displacements (namely, the sound field) allows images of the transmitted and scattered-mode converted sound fields to be visualized independently. This type of acoustic microscope provides images through the entire thickness of a part and for a typical advanced ceramic, a 10 MHz beam will penetrate a few millimeters of material with resolution of 250 μm, whereas a 500 MHz beam generates a resolution of 5 μm.[44]

SPAM shows excellent potential for detecting surface and near-surface flaws in opaque ceramics.[44]

In conclusion, at the present time 5 μm flaws are not detectable, 50 μm flaws are always detectable, and flaws of intermediate size may be detectable. In addition, surface flaws as small as 1 μm are detectable.

There is no single technique that will detect all flaws. A number of techniques must be used, and these must be carefully optimized for the material, part, and application. X-ray CT is a much more widely useful technique than neutron radiography since the latter requires a nuclear reactor to provide enough flux to examine ceramic parts properly. However, ultrasonic techniques may be preferred over X-ray methods since ultrasonic waves present no hazard to the operator.

7.10 CERAMIC/METAL/GRAPHITE JOINING

The application of ceramics in structural components such as turbine engines has received extensive attention in recent decades due to their excellent high temperature strength and resistance to corrosion and wear.

However, because of their brittle nature, joining of ceramics to metals is frequently required. As a consequence, the lack of joining techniques has in many cases limited their use. Normally, conventional fusion welding is not performed due to the risk of brittle fracture initiation as a result of the high concentration stresses formed on cooling. Hence, solid-state bonding and various types of brazing are currently applied to maintain the excellent base-metal properties of ceramics.

Brazing possesses a major advantage compared with conventional welding, as the base metals do not melt. This allows brazing to be applied in the joining of dissimilar materials which cannot be joined by fusion processes due to metallurgical incompatibility. In general, brazing produces less thermally-induced stress and distortion since the entire component is subjected to heat treatment, thus preventing the localized heating which may cause distortion in welding. In addition, it is possible to maintain closer assembly tolerances without costly secondary operations. Moreover, brazing can be easily adopted for mass production.[45,46]

Therefore, the scientific principles involved in the brazing of ceramics have been explained, discussed and accepted and in general, wetting seems to be the limiting factor to obtain sufficient adherence. This fact has led to the development of various techniques to metallize the ceramic surfaces prior to brazing (indirect brazing).[44,47] Recent advances have, however, resulted in new types of brazing filler metals, which provide sufficiently low contact angles due to the addition of active elements. These are essentially Ag–Cu, Ag or Cu brazes with additions of Ti, and Sn-based solders.[48]

Sueyoshi and his university associates[49] evaluated the joining of Si_3N_4 with thin sputter-deposited Ti and Ni films to 304 stainless steel using metallic buffers in a series of $Si_3N_4/Ni/Mo/Ni/304$. Calculations using a finite-element method (FEM) indicated a marked reduction in thermal stress induced in the joined Si_3N_4 with increasing thickness of the Mo buffer. The conclusion showed that the strong interfacial bond inducing the fracture of the joined Si_3N_4 was interpreted in terms of a good interfacial reaction, the interdiffusions and the reduction for thermal stress being due to the insertion of the Mo buffer.

A-P Xian *et al.*[50] verified that direct brazing of Si_3N_4–steel joint using Ag–28Cu brazing filler metal with Ti interlayer could probably solve the main difficulties in brazing ceramic to metal (wetting[51,52] and residual stresses due to thermal expansion mismatch[53–55]). Future efforts will evaluate direct brazing of ceramic to metal with a reactive interlayer material such as Nb, Ta, Zr or Hf.

Nb as an interlayer has been used in joining SiC to 304 stainless steel.[56] Niobium as an interlayer material for joining ceramics possesses two superior properties. First, niobium easily reacts with ceramics, alumina[57] and SiC.[58] Secondly, soft niobium with a low expansion coefficient can relax the thermal stress that arises from the difference between ceramics and metal.

In the work Naka and Saito[56] used reaction-sintered SiC (RS–SiC) with 13% wt Si and pressureless sintered SiC (PLS–SiC) with a few per cent alumina, and Nb with a purity of 99.9% wt.

Figure 7.7 shows the change in fracture shear stress for the RC–SiC–304 joint with the thickness of the Nb interlayer. The use of an Nb interlayer with a thickness of 2 mm gives a lower strength for the SiC–304 joint, because cracks were observed in SiC near the interface after joining. The stress arising from the difference in thermal expansion between SiC and 304 stainless steel causes the cracks in the joint. An increase in the thickness of the Nb interlayer from 2 to 4 mm increases the strength of the SiC–304 joint and removes the cracks in the SiC. The thick Nb interlayer of 4 mm thickness relaxes the stress in the joint. This leads to an increase in the strength of the SiC–304 joint.

In the case of an Nb thickness of 4 mm, PLS–SiC gives the higher strength of 126 MPa for the SiC–304 joint versus that of 49 MPa for the RS–SiC as shown in Figure 7.8. The mechanical properties of SiC themselves affect the strength of the SiC–304 stainless steel joint. The sound joint of SiC to Nb or Nb to 304 stainless steel are the result of the formation of Nb_5Si_3 at the interface.[58]

Using the Ti–Ag–Cu braze filler material graphite can be joined to 95% alumina ceramics. This will provide an important basis for the widening application of graphite according to Qui and Xia.[59]

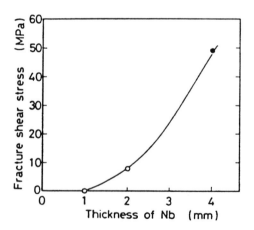

Fig. 7.7 Change in the strength of the RS–SiC–SUS304 joint with the Nb thickness; (○) crack in SiC after joining and (●) no crack in SiC after joining.

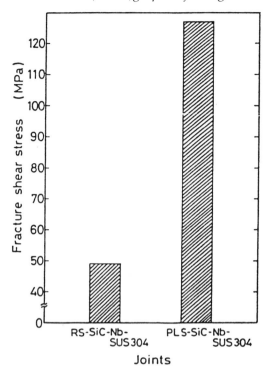

Fig. 7.8 Comparison of the strength of RS–SiC–SUS304 and PLS–SiC–SUS304 joints with a 4 mm thick Nb interlayer.

For the existence of C, the Ti–Ag–Cu method of brazing is used to make $AlAg_3$ and Ti_3Al exist in the physical phase of the sealed region, which is an essential factor in the possible forming of gas-tight seals between graphite and 95% alumina.

For the existence of $AlAg_3$ and Ti_3Al, the use of Ti–Ag–Cu method is a significant advance in the mechanism research of seals between 95% alumina ceramics and oxygen-free copper,[60], Kovar, beryllium, beryllium oxide,[61] tungsten, molybdenum, and stainless steel.[62]

REFERENCES

1. Dunford D. V., Partridge P. G., Strength and fracture behaviour of diffusion-bonded joints in Al-Li (8090) alloy, Part 1, *J. Mater. Sci.*, 25, pp. 4957–64, 1990.
2. Crabtree G. E., Atmospheres and surface chemistry effects on copper brazing of Kovar, Joining Subgroup of the Interagency Mechanical Operation Groups

— US Dept. of Energy, Paper B2B, *22nd AWS International Brazing & Soldering Conf.*, Detroit, MI, Apr 16–18, 1991.

3. Manente D., Stoute P., Immarigeon J. P., Development and evaluation of gaseous cleaning and diffusion braze repair techniques for cobalt and nickel base superalloys, Vac-Aero Inter., Canada, Paper B4B, *23rd AWS International Brazing & Soldering Conf.*, Chicago, IL, March 24–26, 1992.

4. Miyazawa Y., Asiga T., Brazing of a nickel base alloy and stainless steel by nickel base brazing foils, Tokai Univ., Japan, Paper B1D, ibid.

5. Timmons P. F., The development of non-toxic Ag-based brazing, Husky Oil, Canada, Paper B3B, ibid.

6. Nicholas M. G., *Material aspects of ceramic-ceramic and ceramic-metal bonding*, Harwell Lab., IIW, Madrid, Spain, Sept. 1992.

7. Santella M. L., High strength silicon nitride braze joints, ORNL, TN, *22nd AWS International Brazing & Soldering Conf.*, Paper B3C, Detroit, MI, Apr. 16–18, 1991.

8. Xu R., Indacochea J. E., Brazing silicon nitride to metals using active filler metal, Univ. of ILL., Paper B2A, *23rd AWS International Brazing & Soldering Conf.*, Chicago, IL, March 24–26, 1992.

9. Dunkerton S., Brazing of sialon to steel for automotive applications, TWI, Abington, U.K., Paper B2E, ibid.

10. Lee H. -K., Lee J. -Y, Decomposition and interfacial reaction in brazing of SiC by copper-based active alloys, *J. Mater. Sci. Lett.*, 11, pp. 550–53, 1992.

11. Santella M. L., Pak J. J., Fruehan R. J., Analysis of microstructures formed by brazing Al_2O_3 and ZrO_2 with Ag-Cu-Ti alloys, ORNL, TN, Paper B1C, *20th AWS International Brazing & Soldering Conf.*, Wash. D.C., Apr. 3–5, 1989.

12. Palm B., in *Ceramics in Advanced Energy Technologies* (ed. Kröckel, H., Merz, E., Van der Biest, O., Reidel Publ., Dordrecht, p. 178, 1984.

13. Mizuhara H., Huebel E., Joining ceramic to metal with ductile active filler metal, *Welding Journal*, 65(10), Oct. 1986, pp. 43–51.

14. Joy T., Crispin R. M., Nicholas M.G., *High Technology Joining*, B.A.B.S., Abington, UK, p 23/1, 1987.

15. Williams G., Increased productivity through brazing automation, *Welding Journal*, 54(10), Oct. 1973, pp. 640–43.

16. Rabinkin A., Fundamental aspects of the brazing process, Allied Signal, NJ, Paper B4A, *20th AWS International Brazing & Soldering Conf.*, Wash. D.C., Apr. 3–5, 1989.

17. *Connect*, TWI, Abington, UK, p. 6, Dec. 1989.

18. Schoer H., Aluminum brazing alloys for fluxless brazing of aluminum, VAW, Bonn, FRG, *Light Metals 1989*, TMS, pp. 699–702.

19. Rabin B. H., Joining of fiber-reinforced SiC composites by *in situ* reaction methods, EG&G, ID, *Mat. Sci. & Engr.*, A130, pp. L1–L5, 1990.

20. Caputo A. J., Lackey W.J., Fabrication of fiber-reinforced ceramic composites by chemical vapor infiltration, ORNL Tech. Memo 9235, ORNL, TN, 1984.

21. Stinton D. P., Caputo A. J., Lowden R. A., Synthesis of fiber-reinforced SiC composites by chemical vapor infiltration, *Am. Ceram. Soc. Bull.*, 65(2), pp. 347–50, 1986.

22. Caputo A. J., *et al.* Fiber-reinforced SiC composites with improved mechanical properties, *Am. Ceram. Soc. Bull.*, 66(2), pp. 368–72, 1987.

23. Moore T. J., Feasibility study of the welding of SiC, *J. Am. Ceram. Soc.*, 68(6), pp. C151–C153, 1985.

24. Nicholas M. G., Crispin R. M., *Diffusion Bonding* (ed. Pearce, R.), Cranfield Inst. Tech., Cranfield, p. 173, 1988.

25. Bates C. H., *et al.*, Joining of non-oxide ceramics for high-temperature applications, *Am. Ceram. Soc. Bull.*, 69(3), pp. 350–56, 1990.
26. Iseki T., Arakawa K., Suzuki H., Joining of dense silicon carbide by hot pressing, *J. Mater. Sci. Lett.*, 15, pp. 1049–1050, 1980.
27. Yajima S., *et al.*, Joining of SiC to SiC using polyborosilozane, *Am. Ceram. Soc. Bull.*, 60(2), p. 253, 1982.
28. Gehris J. A. P., High temperature bonding of silicon carbide, M.S. Thesis, New Mexico Inst. of Mining and Technology, Socorro, NM, 1989.
29. Tamari N., *et al.*, Joining of silicon carbide ceramics with Si_3N_4–Y_2O_3–La_2O_3–MgO mixture, Yogyo-Kyokai-Shi, *Japan Ceram. Society*, 94(10), pp. 1087–91, 1986.
30. Morozumi S., *et al.*, Bonding mechanism between silicon carbide and thin foils of reactive metals. *J. Mater. Sci.*, 20, pp. 3976–82, 1985.
31. Boadi J. K., Yano T., Iseki T., Brazing of pressureless sintered SiC using Ag-Cu-Ti alloy, *J. Mater. Sci.*, 22, pp. 2431–34, 1987.
32. Keusseyan R. L., Goeller P. T., Page J. P., *et al.*, A new method of metal-ceramic joining for advanced microelectronic packaging, Dupont Electronics, Res. Triangle Park, NC, *IEEE/ISHM '90 IEMT Symposium*, Italy, pp. 186–95.
33. Chattopadhyay A. K., Chollet L., Hintermann H. E., Experimental investigation on induction brazing of diamond with Ni-Cr hardfacing alloy under argon atmosphere, *J. Mater. Sci.*, 26, pp. 5093–5100, 1991.
34. DeChristofaro N., Henschel C., Metglas brazing foil, *Welding Journal*, 57(7), July 1978, pp. 33–38.
35. Datta A., Rabinkin A., Bose D., Rapidly solidified copper–phosphorus base brazing foils, *Welding Journal*, 63(10), Oct. 1984, pp. 14–21.
36. Datta A., DeChristofaro N. J. Rapidly solidified filler metals in *Brazing and Soldering Applications: Rapidly Quenched Metals*, Elsevier Sci. Publishers B.V., 1985, pp. 1715–22.
37. Duvall D. S., Owczarski W. A., Paulonis D. F., TLP Bonding: A new method for joining heat resistant alloys, *Welding Journal*, 53(4), Apr. 1974, pp. 203–14.
38. Hoppin III G. S., Berry T. F., Activated diffusion bonding, *Welding Journal*, 49(11), Nov. 1970, pp. 505s–509s.
39. *Brazing, High Temperature Brazing and Diffusion Welding*, 2nd International Conf., Essen, FRG, IBSN3-87155-430-8, 164p, 19–20 Sept. 1989.
40. Nakahashi M., Suenaga S., Shirokane M., *et al.*, Transient liquid phase bonding for Ni-base superalloys, Mar–M247 and IN939, *Matls. Transactions, JIM*, Vol. 33, No. 1, pp. 60–65, 1992.
41. Nakao Y., Nishimoto K., Shinozaki K., *et al.*, *Transient liquid insert metal-diffusion bonding of nickel-base superalloys*, Osaka Univ. and Pusan Nat. Univ. IIW, Madrid, Spain, Sept. 1992.
42. Nakagawa H., Lee C. H., North T. H., Modeling of base metal dissolution behavior during transient liquid-phase brazing, *Metall. Transactions A*, Vol. 22A(2), pp. 543–55, Feb. 1991.
43. Kauppinen P., Kivilahti J., Evaluation of structural defects in brazed ceramic-to-metal joints with C-mode scanning acoustic microscopy, *NDT&E International*, Vol. 24(4), pp. 187–90, Aug. 1991.
44. Schwartz, M. M., *Handbook of Structural Ceramics*, McGraw-Hill, ISBN 0-07-055719-5, 666p, 1992.
45. Yoshimi N., Nakae, H., Fujii, H., New approach to estimating wetting in a reaction system, *Mater. Trans. JIM*, 30, pp. 41–147, Feb. 1990.
46. Akselsen, O. M., Review advances in brazing of ceramics, *J. Mater. Sci.*, 27, pp. 1989–2000, 1992.

47. Schwartz M. M., *Brazing*, ASM International, Materials Park, OH, 437p, 1987.
48. Xian A. -P., Si Z. -Y., An improvement of the oxidation resistance of Ag–Cu eutectic-5 at% Ti brazing alloy for metal/ceramic joints, *Matls. Lett.*, 12, pp. 84–88, North Holland, 1991.
49. Sueyoshi H., Tabata M., Nakamura Y., Joining of silicon nitride to SUS 304 stainless steel using a sputtering method, *J. Mater. Sci.*, 27, pp. 1926–32, 1992.
50. Xian A. -P., Si Z. -Y., Direct brazing of Si_3N_4–steel joint using Ag–28 Cu brazing filler metal with Ti interlayer, *J. Mater. Sci. Lett.*, 10, pp. 1381–83, 1991.
51. Kapoor R. R., Eager T. W., Oxidation behavior of silver- and copper-based brazing filler metals for silicon nitride/metal joints, *J. Am. Ceram. Soc.*, 72, pp. 448–54, Mar. 1989.
52. Kapoor R. R., Eager T. W., Tin-based reactive solders for ceramic/metal joints, *Metall. Transactions*, 20B(6), pp. 919–24, Dec. 1989.
53. Evans A. G., Ruhle M., Turwitt M., On the mechanics of failure in ceramic/metal bonded systems. *J. de Physique Colloque*, 46, No. C4, pp. 613–26, Apr. 1985.
54. Naka M., Tanaka T., Okamoto I., Amorphous alloy and its application to joining, *Trans. JWRI*, 14(2), pp. 185–95, Dec. 1985.
55. Suganuma K., Okamoto T., Koizumi M., *et al.*, *J. Mater. Sci. Lett.*, 4, p. 648, 1985.
56. Naka M., Saito T., Niobium interlayer for joining SiC to stainless steel, *J. Mater. Sci. Lett.*, 10(6), pp. 339–40, Mar. 1991.
57. Morozumi S., Endo M., Kikuchi M., *et al.*, Bonding mechanism between silicon carbide and thin foils of reactive metals, *J. Mater. Sci.*, 20(11), pp. 3976–82, Nov. 1985.
58. Naka M., Saito T., Okamoto I., *Proc. of the 4th International Symposium on the Sci. and Technology of Sintering*, Elsevier Applied Science, Barking, Essex, (1988) p. 1373.
59. Qiu C., Xia. H., A research on the mechanism of joining graphite to ceramics of 95% Al_2O_3, in *Advances in Joining Newer Structural Materials*, Montreal, Canada, 23–25 July 1990, Pergamon Press, Headington Hill Hall, UK, Section 1.5, pp. 129–35.
60. Qiu, C., Mechanism research on seal between 95% alumina ceramics and activated metal of oxygen-free copper, *Proc. of 4th Ann. Meeting of PRC Elec. Soc.*, E-Vacu. Assoc., pp. 155–156, 1982.
61. Qiu C., Technical improvements on the weld of beryllium and beryllium oxide ceramics, *Proc. of the Exp. Exch. Meeting on Brazing Matl. & its Applications*, 1, p. 8, 1984.
62. Qiu C., Advantages of titanium and its alloy in welding applications, *Proc. of 5th National Weld Symposium*, 8, 1989.

Appendix

Table A.1 Commercially available silver-based brazing filler metals

				Chemical composition, (%)							Solidus (melting point)		Liquidus (flow point)		
Ag	Cu	Li	Zn	Pd	Ni	Sn	Mn	In	Cd	Other	(°C)	(°F)	(°C)	(°F)	Class
98	—	2	—	—	—	—	—	—	—	—	699	1290	760	1400	—
97	—	3	—	—	—	—	—	—	—	—	601	1115	668	1235	—
95	—	—	—	5	—	—	—	—	—	5 Al	782	1440	821	1510	—
95+	—	—	—	10	—	—	—	—	—	—	971	1780	1010	1850	—
90+	—	—	—	—	—	—	—	—	—	—	1001	1835	1066	1950	—
85	—	—	—	—	—	—	15	—	—	—	960	1760	970	1778	BAg–Mn
84.6+	7.5	0.2	—	2.2	—	—	—	5.5	—	—	768	1415	877	1610	—
84.5+	7.4	0.2	—	2.4	—	—	—	5.5	—	—	760	1400	871	1600	—
80	16	—	4	—	—	—	—	—	—	—	729	1345	810	1490	—
80+	—	—	—	20	—	—	—	—	—	—	1071	1960	1177	2150	—
77	21	—	—	—	2	—	—	—	—	—	780	1435	830	1525	—
75	24.5	—	—	—	0.5	—	—	—	—	—	780	1435	801	1475	BAg–II
75	22	—	3	—	—	—	—	—	—	—	740	1365	788	1450	—
75	20	—	5	—	—	—	—	—	—	—	732	1350	774	1425	—
75	—	—	25	—	—	—	5	—	—	—	704	1300	729	1345	—
75+	—	—	—	20	—	—	—	—	—	—	999	1830	1121	2050	—
71.5	28	—	—	—	0.5	—	—	—	—	—	780	1435	788	1450	—
71.15	28.10	—	—	—	0.75	—	—	—	—	—	795	1463	795	1463	—

Table A.1 (Continued)

Ag	Cu	Li	Zn	Pd	Ni	Sn	Mn	In	Cd	Other	Solidus (melting point) °C	Solidus (melting point) °F	Liquidus (flow point) °C	Liquidus (flow point) °F	Class
70+	—	—	—	30	—	—	—	—	—	—	1160	2120	1232	2250	—
68+	27	—	—	5	—	—	—	—	—	—	804	1480	810	1490	—
65+	20	—	—	15	—	—	—	—	—	—	851	1565	899	1650	—
65	28	—	—	—	2	—	5	—	—	—	749	1380	788	1450	—
64+	—	—	—	33	—	—	3	—	—	—	1149	2100	1199	2190	—
63	27	—	—	—	—	—	—	10	—	—	685	1265	729	1345	—
62.5	12.5	—	—	—	5	—	—	—	—	—	780	1435	866	1590	—
61.5	24	—	—	—	—	—	—	14.5	—	—	624	1155	710	1310	—
60	25	—	15	—	—	—	—	—	—	—	674	1245	718	1325	—
60	27	—	—	—	—	—	—	13	—	—	607	1125	743	1370	—
58+	32	—	—	10	—	—	—	—	—	—	827	1520	851	1565	—
57.5	32.5	—	—	—	—	7	3	—	—	—	604	1120	729	1345	—
57	33	—	—	—	—	7	3	—	—	—	604	1120	729	1345	—
54+	21	—	—	25	—	—	—	—	—	—	899	1650	949	1740	—
52+	28	—	—	20	—	—	—	—	—	—	880	1615	899	1650	—
50	50	—	—	—	—	—	—	—	—	—	780	1435	857	1575	—
50	28	—	22	—	—	—	—	—	—	—	677	1250	727	1340	—
50	22	—	20	—	—	1	—	—	7	—	618	1144	666	1231	—
50	15	—	25	—	—	—	—	—	10	—	627	1160	640	1185	—
45	30	—	12	—	—	—	13	—	—	—	699	1290	702	1298	—
44	27	—	13	—	—	—	—	—	15	1 P	607	1125	619	1145	—
41	17	—	18	—	—	—	—	—	24	—	613	1135	643	1190	—
40	18	—	15	—	—	—	—	—	27	—	613	1135	651	1205	—
40	36	—	24	—	—	—	—	—	—	—	668	1235	768	1415	—
40	30	—	25	—	5	—	—	—	—	—	660	1220	860	1580	—
40	30.5	—	29.5	—	—	—	—	—	—	—	621	1150	707	1305	—
50	20	—	28	—	2	—	—	—	—	—	677	1250	727	1340	BAg–6a

Table A.2 Various noble metal base and other brazing filler metals

Chemical composition, (%)												Solidus (melting point)		Liquidus (flow point)		Class
Ag	Cu	Li	Zn	Pd	Ni	Sn	Mn	In	Cd	Au	Other	(°C)	(°F)	(°C)	(°F)	
40	30	—	30	—	—	—	—	—	—	—	—	674	1245	727	1340	—
30	27	—	23	—	—	—	—	—	20	—	—	607	1125	710	1310	—
30	38	—	32	—	—	—	—	—	—	—	—	743	1370	766	1410	—
25	52.5	—	22.5	—	—	—	—	—	—	—	—	677	1250	857	1575	—
20	45	—	30	—	—	—	—	—	5	—	—	616	1150	816	1500	—
20	45	—	35	—	—	—	—	—	—	—	—	713	1315	816	1500	—
—	92.75	—	—	—	—	—	—	—	—	—	7.25 P	732	1350	732	1350	BCuP-2
10	52	—	38	—	—	—	—	—	—	—	—	788	1450	851	1565	—
9	53	—	38	—	—	—	—	—	—	—	—	766	1410	851	1565	—
7	85	—	—	—	—	8	—	—	—	—	—	663	1225	985	1805	—
6	86.5	—	—	—	—	—	—	—	—	—	7.5 P	643	1190	716	1320	BCuP-4
15	80	—	—	—	—	—	—	—	—	—	5 P	643	1190	801	1475	BCuP-5
5	89	—	—	—	—	—	—	—	—	—	6 P	643	1190	813	1495	BCuP-3
6	88	—	—	—	—	—	—	—	—	—	6 P	643	1190	793	1460	—
5	58	—	37	—	—	—	—	—	—	—	—	857	1575	871	1600	—
2	91	—	—	—	—	—	—	—	—	—	7 P	643	1190	785	1445	—
50	34	—	16	—	—	—	—	—	—	—	—	677	1250	774	1425	—
60	—	—	—	—	40	—	—	—	—	—	—	1238	2260	1238	2260	—
—	55	—	—	20	15	—	10	—	—	—	—	1060	1940	1104	2020	—
—	—	—	—	21	48	—	31	—	—	—	—	1121	2050	1121	2050	—
—	—	0.2	—	59.7	40	—	—	—	—	—	0.05 B	1210	2210	1210	2210	—
—	—	—	—	54	36	—	—	—	—	—	10 Cr	1232	2250	1260	2300	—
—	52	—	—	—	8.5	—	39.5	—	—	—	—	880	1615	927	1700	—

Table A.2 (Continued)

Ag	Cu	Li	Zn	Pd	Ni	Sn	Mn	In	Cd	Au	Other	Solidus (melting point) (°C)	(°F)	Liquidus (flow point) (°C)	(°F)	Class
40	30	—	28	—	—	2	—	—	—	—	—	650	1202	716	1320	—
65	20	—	15	—	—	—	—	—	—	—	—	693	1280	718	1325	—
50	20	—	28	—	2	—	—	—	—	—	—	660	1220	707	1305	—
25	41	—	32	—	—	2	—	—	—	—	—	685	1265	760	1400	—
56	42	—	—	—	2	—	—	—	—	—	—	771	1420	893	1640	—
63	28.5	—	—	—	2.5	6	—	—	—	—	—	690	1275	801	1475	—
45	19.85	—	10	—	—	—	—	—	—	25.15	—	707	1305	729	1345	—
30.8	19.90	—	1	—	—	—	—	—	19	29.3	—	693	1280	760	1400	—
25.7	19	—	0.9	—	—	—	—	—	16	38.4	—	635	1175	704	1300	—
24	16.18	—	9	—	—	—	—	—	9	41.817	—	643	1190	701	1295	—
41	—	—	17.183	—	—	—	—	—	—	41.817	—	701	1295	732	1350	—
45	—	—	—	—	—	—	—	—	—	38	17 Ge	510	950	535	995	—
25	21.85	—	3	—	—	—	—	—	—	50.15	—	732	1350	788	1450	—
18	11.82	—	11.7	—	6	—	—	—	—	54.48	—	721	1330	754	1390	—
15	5.52	—	15	—	—	—	—	—	—	58.48	—	704	1300	746	1375	—
17.75	75	—	—	—	—	—	—	—	—	—	7.25 P	643	1190	644	1191	—
—	91.75	—	—	—	—	—	—	—	—	—	8.25 P	710	1310	716	1320	—
—	27.25	—	64.75	—	—	7.5	—	—	—	—	0.5 Pb	751	1385	782	1440	—
4.5	51.5	—	44	—	—	—	—	—	—	—	—	766	1410	890	1635	—
—	55	—	44.75	—	—	—	0.25	—	—	—	—	877	1610	890	1635	—
—	80	—	—	—	—	20	—	—	—	—	—	799	1470	890	1635	—
—	90	—	—	—	—	10	—	—	—	—	—	954	1750	999	1830	—
—	90	—	—	—	—	—	—	—	—	—	10 Cu$_2$O	1082	1980	1082	1980	—
—	95	—	—	—	—	—	—	—	—	—	5 Fe$_2$O$_3$	1082	1980	1082	1980	—
—	90	—	—	—	—	—	—	—	—	—	7 CuO/ 3 Fe$_2$O$_3$	1082	1980	1082	1980	—

Table A.2 (Continued)

Ag	Cu	Li	Zn	Pd	Ni	Sn	Mn	In	Cd	Au	Other	Solidus (melting point) (°C)	(°F)	Liquidus (flow point) (°C)	(°F)	Class
—	—	—	—	27	22	—	—	—	—	41	10 Cr	1054	1929	1110	2030	—
—	—	—	—	60	40	—	—	—	—	—	—	1238	2260	1238	2260	—
—	60	—	—	—	—	—	—	—	—	—	40 Pt	1185	2165	1216	2221	—
20	20	—	—	—	—	—	—	—	—	60	—	835	1535	845	1553	—
5	20	—	—	—	—	—	—	—	—	75	—	885	1625	895	1643	—
—	37	—	—	—	—	—	—	3	—	60	—	860	1580	900	1652	—
—	16.5	—	—	—	2	—	—	—	—	81.5	—	910	1670	925	1697	—
—	50	—	—	—	—	—	—	—	—	50	—	955	1751	970	1778	—
—	55	—	—	—	—	—	—	—	—	45	—	954	1749	971	1780	—
—	60	—	—	—	10	—	—	—	—	40	—	980	1796	1000	1832	—
—	40	—	—	—	—	—	—	—	—	50	—	971	1780	1004	1839	—
—	65	—	—	—	—	—	—	—	—	35	—	990	1814	1010	1850	—
—	—	—	—	8	22	—	—	—	—	70	—	1005	1841	1037	1899	—
—	90	—	—	—	—	—	—	—	—	10	—	1063	1945	1078	1972	—
—	—	—	—	25	25	—	—	—	—	50	—	1102	2016	1121	2050	—
—	—	—	—	8	—	—	—	—	—	92	—	1200	2192	1240	2264	—
—	45.28	—	—	7	10	—	37.5	—	—	—	0.02 C, 0.2 La	882	1620	904	1660	—
—	52.28	—	—	—	10	—	37.5	—	—	—	0.02 C, 0.2 La	882	1620	904	1660	—
92.5	—	—	—	3.5	—	—	—	—	—	—	4 Al	966	1770	982	1800	—
88	—	—	—	4	—	—	—	—	—	—	8 Al	816	1500	843	1550	—
—	4.5	—	—	—	69.8	—	19.7	—	—	—	5.9 Si	1043	1910	1066	1950	—
—	86.2	—	—	—	—	7	—	—	—	—	6.75 P	650	1202	700	1292	—

Table A.2 (Continued)

				Chemical composition, (%)								Solidus (melting point)		Liquidus (flow point)		
Ag	Cu	Li	Zn	Pd	Ni	Sn	Mn	In	Cd	Au	Other	(°C)	(°F)	(°C)	(°F)	Class
8.5	50.5	—	36	—	—	5	—	—	—	—	—	750	1381	850	1561	—
78	—	—	—	—	—	—	—	2	—	20	—	975	1787	1025	1877	—
80	—	—	—	—	—	—	—	—	—	20	—	993	1820	1027	1880	—
—	—	—	—	34	36	—	—	—	—	30	—	1135	2075	1169	2136	—
65	—	—	—	—	—	—	—	—	—	35	—	990	1814	1010	1850	—
60	—	—	—	8	22	—	—	—	—	40	—	965	1769	1000	1832	—
—	—	—	—	—	22	—	—	—	—	70	—	1005	1841	1037	1899	—
—	—	—	—	—	3	—	—	—	—	72	6 Cr	975	1787	1000	1832	—
—	15.5	—	—	—	—	—	—	—	—	81.5	—	900	1652	910	1670	—
—	6	—	—	—	—	—	—	—	—	94	—	965	1769	990	1814	—
—	75	—	—	—	—	—	25	—	—	—	—	871	1600	890	1635	—
—	30.9	—	—	10	14.5	—	9.5	—	—	35	0.1 La	1010	1850	1051	1925	—
—	47.1	—	—	2.5	4.3	—	8	—	—	38	0.1 La	927	1700	1010	1850	—
—	—	—	—	—	66.2	—	—	—	—	20.5	3.4 Si, 5.3 Cr, 2.3 B, 2.3 Fe	1010	1850	1079	1975	—
—	—	—	—	—	55.75	—	—	—	—	41	1.75 Si, 1.0 B, 0.5 Fe	1010	1850	1079	1975	—
—	58.5	—	—	—	—	—	31.5	—	—	—	10 Co	1010	1850	1079	1975	—
—	—	—	—	36	48.7	—	—	—	—	—	11 Cr, 2.2 Si, 2.1 B	1004	1840	1051	1925	—
—	—	—	—	25	53	—	—	—	—	—	6 Cr, 14.5 U, 1.5 B	1010	1850	1051	1925	—

Table A.3 Vacuum-grade brazing filler metals

Ag	Cu	Ni	Sn	Au	Pd	Zn	Cd	Pb	P	C	Other	Solidus (melting point) (°C)	(°F)	Liquidus (flow point) (°C)	(°F)	Class(a)
60	30	—	10	—	—	0.001	0.001	0.002	0.002	0.005	—	600	1115	720	1325	BVAg-18*
61.5	24	—	—	—	—	0.001	0.001	0.002	0.002	0.005	14.5 In	625	1155	705	1300	BVAg-29*
5	20	—	—	75	—	—	—	—	—	—	—	885	1625	895	1640	—
—	65	—	—	35	—	—	—	—	—	—	—	1000	1832	1020	1865	—
—	62	3	—	35	—	—	—	—	—	—	—	975	1787	1030	1885	—
—	—	18	—	82	—	—	—	—	—	—	—	950	1742	950	1740	BVAu-4
72	28	—	—	—	—	0.001	0.001	0.002	0.002	0.005	—	780	1435	780	1435	BVAg-8*
71.5	28	0.5	—	—	—	0.001	0.001	0.002	0.002	0.005	—	754	1390	795	1465	BVAg-8b*
50	50	—	—	—	—	0.001	0.001	0.002	0.002	0.005	—	780	1436	855	1571	BVAg-6b*
68	27	—	—	—	5	0.001	0.001	0.002	0.002	0.005	—	805	1481	810	1490	BVAg-30*
58	32	—	—	—	10	0.001	0.001	0.002	0.002	0.005	—	825	1517	852	1566	BVAg-31
65	20	—	—	—	15	—	—	—	—	—	—	852	1566	900	1652	—
52	28	—	—	—	20	—	—	—	—	—	—	880	1616	900	1652	—
54	21	—	—	—	25	—	—	—	—	—	—	900	1652	950	1742	BVAg-32
95	—	—	—	—	5	—	—	—	—	—	—	970	1778	1010	1850	—
90	—	—	—	—	10	—	—	—	—	—	—	1000	1832	1065	1949	—
80	—	—	—	—	20	—	—	—	—	—	—	1070	1958	1175	2147	—
—	20	—	—	80	—	0.001	0.001	0.002	0.002	0.005	—	890	1634	890	1634	BVAu-2
—	50	—	—	50	—	—	—	—	—	—	—	950	1742	975	1787	—
—	62.5	—	—	37.5	—	—	—	—	—	—	—	990	1814	1015	1859	—
—	65	—	—	35	—	—	—	—	—	—	—	1000	1832	1020	1868	—
—	6	—	—	94	—	—	—	—	—	—	—	965	1769	990	1814	—
—	15.5	3	—	81.5	—	—	—	—	—	—	—	900	1652	910	1670	—

364

Table A.3 (Continued)

Ag	Cu	Ni	Sn	Au	Pd	Zn	Cd	Pb	P	C	Other	Solidus (melting point) (°C)	(°F)	Liquidus (flow point) (°C)	(°F)	Class(a)
99.95	0.05	—	—	—	—	0.001	0.001	0.002	0.002	0.005	—	961	1761	961	1761	BVAg–0
0.05	99.95	—	—	—	—	0.001	0.001	0.002	0.002	0.005	—	1083	1981	1083	1981	BVCu–1x
—	25	—	—	50	24	0.001	0.001	0.002	0.002	0.005	0.06 Co	1102	2015	1121	2050	BVAu–7
—	—	—	—	92	8	0.001	0.001	0.002	0.002	0.005	—	1200	2190	1240	2265	BVAu–8
—	—	0.06	—	—	65	0.001	0.001	0.002	0.002	0.005	35 Co	1230	2245	1235	2255	BVPd–1

(a) All vacuum-grade filler metals for which AWS classifications are given are considered grade 1 filler metals (i.e., BVAg–18, grade 1). A grade 2 variety is also available for each vacuum-grade filler metal with an AWS classification. Each grade 2 filler metal has the exact same composition as the corresponding grade 1 variety, except that its zinc and cadmium contents are 0.002%. Grade 2 varieties also have 0.002% phosphorus except for those marked with an asterisk (*), which contain 0.02% P.

Table A.4 Commercially-available nickel-based brazing filler metals

Ni	Cr	C	P	Fe	B	Si	Co	Mn	Cu	Other	Solidus (melting point) (°C)	(°F)	Liquidus (flow point) (°C)	(°F)
99.26	0.04	0.04	0.30	—	—	—	—	—	—	—	1082	1980	1121	2050
98.88	—	0.015	1.10	—	—	—	—	—	—	—	1038	1900	1204	2200
97.25	—	—	—	—	0.75	2.00	—	—	—	—	1021	1870	1066	1950
96.00	—	—	—	—	1.50	2.50	—	—	—	—	1010	1850	1066	1950
95.50	—	—	—	—	1.50	3.00	—	—	—	—	1010	1850	1066	1950
95.20	—	—	—	—	1.80	3.00	—	—	—	—	1004	1840	1066	1950
94.60	—	0.03	—	—	1.90	3.50	—	—	—	—	982	1800	1066	1950
93.72	—	0.03	—	1.25	1.50	3.50	—	—	—	—	982	1800	1066	1950
93.60	2.20	—	—	1.00	1.60	2.60	—	—	—	—	954	1750	1038	1900
92.80	2.30	—	—	1.00	1.50	2.40	—	—	—	—	954	1750	1038	1900
92.60	2.70	—	—	1.10	1.90	2.80	—	—	—	—	954	1750	1038	1900
92.54	—	0.06	—	—	2.90	4.50	—	—	—	—	982	1800	1024	1875
92.40	2.90	—	—	1.30	1.90	2.80	—	—	—	—	954	1750	1038	1900
92.30	2.25	—	—	1.00	1.70	2.75	—	—	—	—	954	1750	1038	1900
91.70	2.30	—	—	1.20	1.80	3.00	—	—	—	—	954	1750	1038	1900
91.60	2.50	—	—	1.10	1.80	3.00	—	—	—	—	954	1750	1038	1900
91.30	2.70	—	—	1.20	1.80	3.00	—	—	—	—	954	1750	1038	1900
91.22	—	0.03	—	1.25	3.00	4.50	—	—	—	—	954	1750	1024	1875
90.80	3.00	—	—	1.30	1.90	3.00	—	—	—	—	954	1750	1038	1900
90.66	0.20	0.14	—	1.50	3.00	4.50	—	—	—	—	982	1800	1038	1900
88.95	5.00	0.25	—	1.00	1.80	3.00	—	—	—	—	971	1780	1182	2160
88.80	3.80	—	—	1.70	2.20	3.50	—	—	—	—	954	1750	1038	1900
88.40	5.60	—	—	2.40	2.80	3.20	—	—	—	—	954	1750	1038	1900
87.10	5.00	—	4.00	0.60	1.20	2.10	—	—	—	—	954	1750	1038	1900
85.90	5.60	—	—	2.40	2.50	3.60	—	—	—	—	954	1750	1038	1900
84.39	9.00	0.35	—	2.10	1.66	2.50	—	—	—	—	1010	1850	1066	1950
84.08	5.50	0.12	1.50	2.50	2.50	3.80	—	—	—	—	982	1800	1038	1900

Table A.4 (Continued)

Ni	Cr	C	P	Fe	B	Si	Co	Mn	Cu	Other	Solidus (melting point) (°C)	(°F)	Liquidus (flow point) (°C)	(°F)
83.40	8.00	—	6.00	0.40	0.80	1.40	—	—	—	—	982	1800	1049	1920
82.97	6.50	0.03	—	3.00	3.00	4.50	—	—	—	—	971	1780	999	1830
82.50	11.50	—	—	—	—	6.00	—	—	—	—	1082	1980	1138	2080
82.10	7.00	0.03	—	3.00	2.90	5.00	—	—	—	—	1038	1900	1065	1950
81.50	11.40	—	—	—	0.30	6.80	—	—	—	—	1082	1980	1138	2080
81.50	15.00	—	—	—	3.50	—	—	—	—	—	1054	1930	1054	1930
81.10	11.50	—	—	—	0.40	7.00	—	—	—	—	1082	1980	1138	2080
80.85	10.00	0.50	7.50	0.25	0.50	0.90	—	—	—	—	954	1750	1038	1900
79.25	11.00	0.35	—	3.50	2.25	3.50	—	—	—	—	971	1780	1160	2120
79.20	10.00	—	—	4.00	2.10	4.35	—	—	—	—	971	1780	1149	2100
78.87	13.30	—	—	—	0.23	7.60	—	—	—	—	1082	1980	1138	2080
78.65	13.00	—	—	—	0.35	8.00	—	—	—	—	1082	1980	1138	2080
78.40	13.00	0.70	—	3.50	2.90	1.50	—	—	—	—	1082	1980	1138	2080
77.00	15.00	—	—	—	—	8.00	—	—	—	—	1082	1980	1138	2080
76.80	15.20	—	—	—	—	8.00	—	—	—	—	1082	1980	1138	2080
76.30	7.00	0.15	—	3.00	3.20	4.50	—	—	—	6.00 W	977	1790	1038	1900
75.60	13.50	0.15	—	3.50	3.25	4.00	—	17.00	—	—	976	1789	1052	1925
74.85	—	—	—	—	—	8.00	—	—	—	—	1024	1875	1065	1950
74.60	—	0.10	—	—	3.15	4.25	18.00	—	—	—	982	1800	1024	1875
74.40	15.00	0.70	—	3.00	3.00	4.50	—	—	—	—	971	1780	1077	1970
73.90	14.80	—	—	3.80	3.00	3.80	—	—	—	—	977	1790	1038	1900
73.62	17.10	—	—	—	0.08	9.20	—	—	—	—	1082	1980	1138	2080
73.40	17.00	0.06	—	—	0.10	9.50	—	—	—	—	1082	1980	1138	2080
72.44	15.00	—	—	4.00	3.50	5.00	—	—	—	—	982	1800	1024	1875
72.20	—	—	—	—	3.30	4.50	20.00	—	—	—	982	1800	1024	1875
72.00	—	0.03	—	—	3.60	4.50	20.00	—	—	—	982	1800	1024	1875
71.90	15.00	0.60	—	4.00	3.50	5.00	—	—	—	—	977	1790	1052	1925

Table A.4 (Continued)

Ni	Cr	C	P	Fe	B	Si	Co	Mn	Cu	Other	Solidus (melting point) (°C)	(°F)	Liquidus (flow point) (°C)	(°F)
70.00	16.00	—	—	2.00	3.50	3.50	—	—	2.50	2.50 Mo	1093	2000	1149	2100
69.20	17.50	0.80	—	4.50	3.50	4.50	—	—	—	—	949	1740	1065	1950
68.59	18.00	—	—	4.50	—	9.00	—	—	—	12.00 W	1082	1980	1138	2080
68.10	10.00	0.40	—	3.50	2.50	3.50	—	—	—	—	971	1780	1093	2000
67.00	20.00	—	—	3.00	—	10.00	—	—	—	—	1079	1975	1135	2075
66.85	19.00	0.15	—	3.00	—	10.00	0.50	0.50	—	—	1082	1980	1138	2080
65.50	—	0.01	—	—	—	7.00	—	23.00	4.50	—	982	1800	1010	1850
65.00	—	0.07	—	—	—	7.00	—	23.00	5.00	—	982	1800	1010	1850
63.00	—	—	—	5.00	2.00	—	—	—	—	30.00 Mo	1066	1950	1138	2080
61.95	12.00	0.55	—	3.50	2.50	3.50	—	—	—	16.00 W	971	1780	1104	2020
61.50	19.00	—	—	4.50	—	9.80	—	—	—	—	1082	1980	1138	2080
61.40	19.50	—	—	—	—	9.60	5.20	9.50	—	—	1082	1980	1107	2025
60.35	11.50	0.55	—	3.40	2.50	3.20	2.50	—	—	16.00 W	1010	1850	1121	2050
59.15	19.50	—	—	2.25	—	9.50	—	9.50	—	—	1082	1980	1121	2050
57.10	3.50	—	—	1.00	0.90	2.50	—	35.00	—	—	987	1800	1066	1950
50.20	4.00	—	—	—	0.80	—	—	45.00	—	—	996	1825	1079	1975
50.10	—	—	3.50	30.00	—	11.00	—	—	—	5.40 Mo	954	1750	1001	1835
36.00	—	—	—	—	—	—	—	—	59.00	5.00 In	1038	1900	1204	2200
17.00	21.00	0.80	—	—	3.25	3.00	44.95	—	—	10.00 W	1038	1900	1121	2050
17.00	19.00	0.40	—	—	0.80	8.00	50.80	—	—	4.00 W	1107	2025	1149	2100
16.00	—	—	0.10	66.90	1.00	—	16.00	67.00	—	—	1004	1840	1021	1870
12.00	18.00	—	—	—	—	1.00	—	2.00	—	—	1121	2050	1163	2125
9.00	—	0.03	—	—	—	—	—	23.50	67.50	—	910	1670	932	1710
9.00	—	—	—	—	—	—	—	38.50	52.50	—	880	1615	927	1700
9.00	—	0.03	—	—	—	—	—	37.00	52.00	—	871	1600	927	1700
5.00	—	—	—	—	—	—	10.00	22.00	63.00	—	943	1730	957	1755
2.00	16.00	0.20	—	81.80	—	—	—	—	—	—	1149	2100	1204	2200

Table A.5 Nickel-based brazing filler metals

AWS A5.8 Class.	AMS Spec.	British Spec. BS 1845	UNS NO.	Ni	Cr	B	Si	Fe	C	P	Co	W
BNi-1	4775	HTN1	N99600	73	14	3.1	4.5	4.5	0.75	—	—	—
BNi-1a	4776	HTN1A	N99610	73	14	3.1	4.5	4.5	.06 Max	—	—	—
BNi-2	4777	HTN2	N99620	82	7.0	3.1	4.5	3.0	.06 Max	—	—	—
BNi-3	4778	HTN3	N99630	92	—	3.1	4.5	—	.06 Max	—	—	—
BNi-4	4779	HTN4	N99640	94	—	1.8	3.5	—	.06 Max	—	—	—
BNi-5	4782	HTN5	N99650	71	19	—	10	—	.06 Max	—	—	—
BNi-6	—	HTN6	N99700	89	—	—	—	—	.06 Max	11	—	—
BNi-7	—	HTN7	N99710	76	14	—	—	—	.06 Max	10	—	—
*	—	HTN9	N99612	81	15	3.6	—	1.5 Max	.06 Max	—	—	—
*	—	HTN10	N99622	63	11.5	2.5	3.5	3.5	.48	—	—	16
*	—	HTN11	N99624	67	10.4	2.7	3.8	3.3	.40	—	—	12.1
BCo-1	4783	HTN12	R30040	17	19	0.80	8.0	—	.06 Max	—	50	4.0

* Possible additions

Table A.6 Brazing filler metals for brazing of graphite, ceramics, and refractory and reactive metals

Chemical composition, (%)						Solidus (melting point)		Liquidus (flow point)	
Ti	Zr	Be	Ni	Al	Cu	(°C)	(°F)	(°C)	(°F)
43	43	2	12	—	—	799	1470	816	1500
45	45	2	8	—	—	899	1650	899	1650
48	47	5	—	—	—	893	1640	904	1660
43	47	5	—	5	—	927	1700	927	1700
70	—	—	15	—	15	950	1742	1000	1832
48	48	4	—	—	—	1000	1832	1050	1922

Table A.7 Nominal compositions and melting ranges of common brazeable aluminum alloys

Commercial designation	Aluminum Assoc. No.	Brazeability rating (a)	Nominal composition (b), (%)						Approximate melting range	
			Cu	Si	Mn	Mg	Zn	Cr	(°C)	(°F)
1350	1350	A	99.45 min Al						646–657	1195–1215
1100	1100	A	99 min Al						643–657	1190–1215
3003	3003 (c)	A	—	—	1.2	—	—	—	643–654	1190–1210
3004	3004	B	—	—	1.2	1.0	—	—	629–652	1165–1205
3005	3005	A	0.3	0.6	1.2	0.4	0.25	0.1	638–657	1180–1215
5005	5005	B	—	—	—	0.8	—	—	632–654	1170–1210
5050	5050	B	—	—	—	1.2	—	—	588–649	1090–1200
5052	5052	C	—	—	—	2.5	—	—	593–649	1100–1200
6151	6151	C	—	1.0	—	0.6	—	0.25	643–649	1190–1200
6951	6951 (d)	A	0.25	0.35	—	0.65	—	—	616–654	1140–1210
6053	6053	A	—	0.7	—	1.3	—	—	596–652	1105–1205
6061	6061	A	0.24	0.6	—	1.0	—	0.25	593–652	1100–1205
6063	6063	A	—	0.4	—	0.7	—	—	615–652	1140–1205
7005	7005	B								
7072	7072	A	—	—	—	—	1.0	—	607–646	1125–1195
Cast 43	Cast 443.0	B	—	5.0	—	—	—	—	574–632	1065–1170
Cast 356	Cast 356.0	B	—	7.0	—	0.3	—	—	557–613	1035–1135
Cast 406	Cast 406	A	99 min Al						643–657	1190–1215
Cast A612	Cast 710	B	—	—	—	0.7	6.5	—	596–646	1105–1195
Cast C612	Cast 711	A	—	—	—	0.35	6.5	—	604–643	1120–1190

(a) A = alloys readily brazed by all commercial methods and procedures. B = alloys that can be brazed by all techniques with a little care. C = alloys that require special care in brazing. (b) Aluminum and normal impurities constitute remainder. (c) Used both plain and as the core of brazing sheet. (d) Used only as the core of brazing sheet.

Table A.8 Brazeable magnesium alloys and recommended brazing filler metals

ASTM alloy designation	Solidus (°C)	(°F)	Liquidus (°C)	(°F)	Brazing range (°C)	(°F)	Suitable brazing filler metal BMg–1 or AZ92A	BMg–2 or AZ125A
Base metals								
AZ10A	632	1170	643	1190	582–616	1080–1140	X	X
AZ31B	566	1050	627	1160	582–593	1080–1100		X
K1A	649	1200	650	1202	582–616	1080–1140	X	X
M1A	648	1198	650	1202	582–616	1080–1140	X	X
ZK21A	626	1159	642	1187	582–616	1080–1140	X	X
Brazing filler metals								
AZ92A (a)	443	830	599	1110	604–616	1120–1140	—	—
AZ125A (b)	410	770	566	1050	582–610	1080–1130	—	—

(a) Available as wire or rod. (b) Available as wire, rod, strip, or powder.

Table A.9 Brazing filler metals for elevated-temperature service

Nickel-base filler metals

AWS classification	Cr	B	Si	Fe	C	P	S	Al	Ti	Mn	Cu	Zr	Ni	Others (total)	Solidus (°C)	(°F)	Liquidus (°C)	(°F)	Brazing range (°C)	(°F)
BNi-1	13.0–15.0	2.75–3.50	4.0–5.0	4.0–5.0	0.6–0.9	0.02	0.02	0.05	0.05	—	—	—	Rem	0.50	977	1790	1038	1900	1066–1204	1950–2200
BNi-1a	13.0–15.0	2.75–3.50	4.0–5.0	4.0–5.0	0.06	0.02	0.02	0.05	0.05	—	—	—	Rem	0.50	977	1790	1077	1970	1077–1204	1970–2200
BNi-2	6.0–8.0	2.75–3.50	4.0–5.0	2.5–3.5	0.06	0.02	0.02	0.05	0.05	—	—	—	Rem	0.50	971	1780	999	1830	1010–1177	1850–2150
BNi-3	—	2.75–3.50	4.0–5.0	0.5	0.06	0.02	0.02	0.05	0.05	—	—	0.5	Rem	0.50	982	1800	1038	1900	1010–1177	1850–2150
BNi-4	—	1.5–2.2	3.0–4.0	1.5	0.06	0.02	0.02	0.05	0.05	—	—	—	Rem	0.50	982	1800	1066	1950	1010–1177	1850–2150
BNi-5	18.5–19.5	0.03	9.75–10.50	—	0.10	0.02	0.02	0.05	0.05	—	—	—	Rem	0.50	1079	1975	1135	2075	1149–1204	2100–2200
BNi-6	—	—	—	—	0.10	10.0–12.0	0.02	0.05	0.05	—	—	—	Rem	0.50	877	1610	877	1610	927–1093	1700–2000
BNi-7	13.0–15.0	0.01	0.10	0.2	0.08	9.7–10.5	0.02	0.05	0.05	0.04	—	—	Rem	0.50	888	1630	888	1630	927–1093	1700–2000
BNi-8	—	—	6.0–8.0	—	0.10	0.02	0.02	0.05	0.05	21.5–24.5	4.0–5.0	0.05	Rem	0.50	982	1800	1010	1850	1010–1093	1850–2000

AWS Classification	Au	Cu	Pd	Ni	Others (total)	Solidus (°C)	(°F)	Liquidus (°C)	(°F)	Brazing range (°C)	(°F)
Precious metals											
BAu-1	37.0–38.0	Rem	—	—	0.15	990	1815	1016	1860	1016–1093	1860–2000
BAu-2	79.5–80.5	Rem	—	—	0.15	890	1635	890	1635	890–1010	1635–1850
BAu-3	34.5–35.5	Rem	—	2.5–3.5	0.15	974	1785	1030	1885	1030–1090	1885–1995
BAu-4	81.5–82.5	—	—	Rem	0.15	949	1740	949	1740	949–1004	1740–1840
BAu-5	29.5–30.5	—	33.5–34.5	35.5–36.5	0.15	1135	2075	1166	2130	1166–1232	2130–2250

AWS classification	Cr	Ni	Si	W	Fe	B	C	P	S	Al	Ti	Zr	Co (total)	Others	Solidus (°C)	(°F)	Liquidus (°C)	(°F)	Brazing range (°C)	(°F)
Cobalt-base filler metals																				
BCo-1	18.0–20.0	16.0–18.0	7.5–8.5	3.5–4.5	1.0	0.7–0.9	0.35–0.45	0.02	0.02	0.05	0.05	0.05	Rem	0.05	1121	2050	1149	2100	1149–1232	2100–2250

Table A.10 Brazing filler metals for brazing TD–Ni and TD–NiCr

Base metal	Brazing filler metal	Brazing temperature (°C)	(°F)	Nominal composition
TD–Ni	TD–20	1302	2375	Ni–16Cr–25Mo–4Si–5W
	TD–6	1302	2375	Ni–22Cr–17Mo–4Si–5W
	J8600	1177	2150	Ni–33Cr–25Pd–4Si
	Ni–Pd	1246	2275	Ni–60Pd
TD–NiCr	TD–6	1302	2375	Ni–22Cr–17Mo–4Si–5W
	CM50	1066–1121	1950–2050	Ni–3.5Si–1.9B
	NX77	1177–1191	2150–2175	Ni–5Cr–7Si–1B–1W–4Co
	NSB	1288	2350	Ni–2Si–0.8B

Table A.11 Filler metals for brazing of refractory metals

Brazing filler metal (a)	Liquidus temperature (°C)	(°F)	Brazing filler metal (a)	Liquidus temperature (°C)	(°F)
Nb	2416	4380	Mn–Ni–Co	1021	1870
Ta	2996	5425			
Ag	960	1760	Co–Cr–Si–Ni	1899	3450
Cu	1082	1980	Co–Cr–W–Ni	1427	2600
Ni	1454	2650	Mo–Ru	1899	3450
Ti	1816	3300	Mo–B	1899	3450
Pd–Mo	2127	3860	Cu–Mn	871	1600
Pt–Mo	1774	3225	Nb–Ni	1191	2175
Pt–30W	2299	4170			
Pt–50Rh	2049	3720	Pd–Ag–Mo	1316	2400
			Pd–Al	1177	2150
Ag–Cu–Zn–Cd–Mo	618–702	1145–1295	Pd–Ni	1204	2200
Ag–Cu–Zn–Mo	718–788	1325–1450	Pd–Cu	1204	2200
Ag–Cu–Mo	779	1435	Pd–Ag	1316	2400
Ag–Mn	971	1780	Pd–Fe	1316	2400
			Au–Cu	885	1625
Ni–Cr–B	1066	1950	Au–Ni	949	1740
Ni–Cr–Fe–Si–C	1066	1950	Au–Ni–Cr	1038	1900
Ni–Cr–Mo–Mn–Si	1149	2100	Ta–Ti–Zr	2093	3800
Ni–Ti	1288	2350			
Ni–Cr–Mo–Fe–W	1304	2380	Ti–V–Cr–Al	1649	3000
Ni–Cu	1349	2460	Ti–Cr	1482	2700
Ni–Cr–Fe	1427	2600	Ti–Si	1427	2600
Ni–Cr–Si	1121	2050	Ti–Zr–Be (b)	999	1830
			Zr–Nb–Be (b)	1049	1920
			Ti–V–Be (b)	1249	2280
			Ta–V–Nb (b)	1816–1927	3300–3500
			Ta–V–Ti (b)	1760–1843	3200–3350

(a) Not all the braze filler metals listed are commercially available. (b) The liquidus temperature (and therefore the brazing temperature) depends on the specific composition.

Table A.12 Brazing filler metals for torch brazing of low-carbon and low-alloy steels

AWS classification	Product form	Nominal composition, (%)										Solidus		Liquidus		Brazing temperature	
		Ag	Cu	Zn	Cd	Ni	Sn	Fe	Mn	Si	P	(°C)	(°F)	(°C)	(°F)	(°C)	(°F)
Silver-base filler metals																	
BAg–1	Strip, wire, powder	45	15	16	24	—	—	—	—	—	—	607	1125	618	1145	618–760	1145–1400
BAg–1a	Strip, wire, powder	50	15.5	16.5	18	—	—	—	—	—	—	627	1160	635	1175	635–760	1175–1400
BAg–2	Strip, wire, powder	35	26	21	18	—	—	—	—	—	—	607	1125	701	1295	701–843	1295–1550
BAg–2a	Strip, wire, powder	30	27	23	20	—	—	—	—	—	—	607	1125	710	1310	710–843	1310–1550
BAg–3	Strip, wire, powder	50	15.5	15.5	16	3.0	—	—	—	—	—	632	1170	688	1270	688–816	1270–1500
BAg–4	Strip, wire, powder	40	30	28	—	2.0	—	—	—	—	—	671	1240	779	1435	779–899	1435–1650
BAg–5	Strip, wire, powder	45	30	25	—	—	—	—	—	—	—	677	1250	743	1370	743–843	1370–1550
BAg–6	Strip, wire, powder	50	34	16	—	—	—	—	—	—	—	688	1270	774	1425	774–871	1425–1600
BAg–7	Strip, wire, powder	56	22	17	—	—	5.0	—	—	—	—	618	1145	651	1205	651–760	1205–1400
BAg–20	Strip, wire, powder	30	38	32	—	—	—	—	—	—	—	677	1250	766	1410	766–871	1410–1600
BAg–27	Strip, wire, powder	25	35	26.5	13.5	—	—	—	—	—	—	607	1125	746	1375	746–857	1375–1575
BAg–28	Strip, wire, powder	40	30	23	—	—	2	—	—	—	—	649	1200	710	1310	710–816	1310–1500
Copper–zinc filler metals																	
RBCuZn–A (a)	Strip, rod, wire, powder	—	59	40	—	—	0.61	—	—	—	—	888	1630	899	1650	910–954	1670–1750
RBCuZn–D (a)	Strip, rod, wire, powder	—	48	41	—	10.0	—	—	—	0.15	0.25	921	1690	935	1715	938–982	1720–1800
RBCuZn–B (b)	Rod	—	58	38	—	0.5	0.95	0.7	0.25	0.08	—	866	1590	882	1620	—	—
RBCuZn–C (b)	Rod	—	58	39	—	—	0.95	0.7	0.25	0.08	—	868	1595	882	1620	—	—

(a) Classified for braze welding and brazing. (b) Classified for braze welding.

Table A.13 Typical salts used for dip brazing of carbon and low-alloy steels with various brazing filler metals

Filler metal	Type of salt	Brazing temperature range (a) (°C)	(°F)
BAg–l through BAg–8	Neutral	621–871	1150–1600
and BAg–18	Cyaniding-fluxing	649–871	1200–1600
RBCuZn-A	Neutral	732–871	1350–1600
	Neutral	913–940	1675–1725
	Fluxing	913–940	1675–1725
	Carburizing-fluxing (water soluble)	913–940	1675–1725
RBCuZn–D	Carburizing and		
BCU–1 and –1a	self-fluxing	816–927	1500–1700
	Neutral	1038–1051	1900–1925
	Neutral	1093–1149	2000–2100
	Neutral	1093–1149	2000–2100

(a) Temperatures shown are those of the salt bath.

Table A.14 Typical composition and properties of standard filler metals for brazing stainless steels (silver based)

Filler metal	Composition, (%)								Other elements (total)	Solidus temperature		Liquidus temperature		Brazing temperature range	
	Ag	Cu	Zn	Cd	Ni	Sn	Li	Mn		(°C)	(°F)	(°C)	(°F)	(°C)	(°F)
BAg-1	44.0–46.0	14.0–16.0	14.0–18.0	23.0–25.0	—	—	—	—	0.15	607	1125	618	1145	618–760	1145–1400
BAg-1a	49.0–51.0	14.5–16.5	14.5–18.5	17.0–19.0	—	—	—	—	0.15	627	1160	635	1175	635–760	1175–1400
BAg-2	34.0–36.0	25.0–27.0	19.0–23.0	17.0–19.0	—	—	—	—	0.15	607	1125	701	1295	701–843	1295–1550
BAg-2a	29.0–31.0	26.0–28.0	21.0–25.0	19.0–21.0	—	—	—	—	0.15	607	1125	710	1310	710–843	1310–1550
BAg-3	49.0–51.0	14.5–16.5	13.5–17.5	15.0–17.0	2.5–3.5	—	—	—	0.15	632	1170	688	1270	688–816	1270–1500
BAg-4	39.0–41.0	29.0–31.0	26.0–30.0	—	1.5–2.5	—	—	—	0.15	671	1240	779	1435	779–899	1435–1650
BAg-5	44.0–46.0	29.0–31.0	23.0–27.0	—	—	—	—	—	0.15	677	1250	743	1370	743–843	1370–1550
BAg-6	49.0–51.0	33.0–35.0	14.0–18.0	—	—	—	—	—	0.15	688	1270	774	1425	774–871	1425–1600
BAg-7	55.0–57.0	21.0–23.0	15.0–19.0	—	—	4.5–5.5	—	—	0.15	618	1145	651	1205	651–760	1205–1400
BAg-8	71.0–73.0	Rem	—	—	—	—	—	—	0.15	779	1435	779	1435	779–899	1435–1650
BAg-8a	71.0–73.0	Rem	—	—	—	—	0.25–0.50	—	0.15	766	1410	766	1410	766–871	1410–1600
BAg-9	64.0–66.0	19.0–21.0	13.0–17.0	—	—	—	—	—	0.15	671	1240	713	1325	713–843	1325–1550
BAg-10	69.0–71.0	19.0–21.0	8.0–12.0	—	—	—	—	—	0.15	690	1275	738	1360	738–843	1360–1550
BAg-13	53.0–55.0	Rem	4.0–6.0	—	0.5–1.5	—	—	—	0.15	713	1325	857	1575	857–969	1575–1775
BAg-13a	55.0–57.0	Rem	—	—	1.5–2.5	—	—	—	0.15	771	1420	893	1640	871–982	1600–1800
BAg-18	59.0–61.0	Rem	—	—	—	9.5–10.5	—	—	0.15	601	1115	713	1325	713–843	1325–1550
BAg-19	92.0–93.0	Rem	—	—	—	—	0.15–0.30	—	0.15	760	1400	885	1635	877–982	1610–1800
BAg-20	29.0–31.0	37.0–39.0	30.0–34.0	—	—	—	—	—	0.15	677	1250	766	1410	766–871	1410–1600
BAg-21	62.0–64.0	27.5–29.5	—	—	2.0–3.0	5.0–7.0	—	—	0.15	690	1275	801	1475	801–899	1475–1650
BAg-22	48.0–50.0	15.0–17.0	21.0–25.0	—	4.0–5.0	—	—	7.0–8.0	0.15	682	1260	699	1290	699–830	1290–1525
BAg-23	84.0–86.0	—	—	—	—	—	—	Rem	0.15	960	1760	971	1780	971–1038	1780–1900
BAg-24	49.0–51.0	19.0–21.0	26.0–30.0	—	1.5–2.5	—	—	—	0.15	660	1220	707	1305	707–843	1305–1550
BAg-25	19.0–21.0	39.0–41.0	33.0–37.0	—	—	—	—	4.5–5.5	0.15	738	1360	790	1455	790–846	1455–1555
BAg-26	24.0–26.0	37.0–39.0	31.0–35.0	—	1.5–2.5	—	—	1.5–2.5	0.15	707	1305	801	1475	801–871	1475–1600
BAg-27	24.0–26.0	34.0–36.0	24.5–28.5	12.5–14.5	—	—	—	—	0.15	607	1125	746	1375	746–857	1375–1575
BAg-28	39.0–41.0	29.0–31.0	26.0–30.0	—	—	1.5–2.5	—	—	0.15	649	1200	710	1310	710–843	1310–1550

Table A.15 Compositions and properties of standard filler metals for brazing steels (copper, gold, cobalt and nickel based)

Copper alloys

Filler metal	Cu	Zn	Sn	Fe	Mn	Ni	P	Pb	Al	Si	Others (total)	Solidus temperature (°C)	(°F)	Liquidus temperature (°C)	(°F)	Brazing temperature range (°C)	(°F)
BCu-1	99.90 min	—	—	—	—	—	0.075	0.02	0.01	—	0.10	1082	1980	1082	1980	1093–1149	2000–2100
BCu-1a	99.0 min	—	—	—	—	—	—	—	—	—	0.30	1082	1980	1082	1980	1093–1149	2000–2100
BCu-2	86.5 min	—	—	—	—	—	—	—	—	—	0.50	1082	1980	1082	1980	1093–1149	2000–2100

Precious-metal alloys

Filler metal	Au	Cu	Pd	Ni	Others (total)	Solidus temperature (°C)	(°F)	Liquidus temperature (°C)	(°F)	Brazing temperature range (°C)	(°F)
BAu-1	37.0–38.0	Rem	—	—	0.15	990	1815	1016	1860	1016–1093	1860–2000
BAu-2	79.5–80.5	Rem	—	—	0.15	890	1635	890	1635	890–1010	1635–1850
BAu-3	34.5–35.5	Rem	—	2.5–3.5	0.15	974	1785	1030	1885	1030–1090	1885–1995
BAu-4	81.5–82.5	—	—	Rem	0.15	949	1740	949	1740	949–1004	1740–1840
BAu-5	29.5–30.5	—	33.5–34.5	35.5–36.5	0.15	1135	2075	1166	2130	1166–1232	2130–2250
BAu-6	69.5–70.5	—	7.5–8.5	21.5–22.5	0.15	1007	1845	1046	1915	1046–1121	1915–2050

Cobalt alloys

Filler metal	Cr	Ni	Si	W	Fe	B	C	P	S	Al	Ti	Zr	Co	Others (total)	Solidus temperature (°C)	(°F)	Liquidus temperature (°C)	(°F)	Brazing temperature range (°C)	(°F)
BCo-1	18.0–20.0	16.0–18.0	7.5–8.5	3.5–4.5	1.0	0.7–0.9	0.35–0.45	0.02	0.05	0.05	0.05	0.05	Rem	0.50	1121	2050	1149	2100	1149–1232	2100–2250

Table A.15 (Continued)

Nickel alloys

Filler metal	Composition (%)														Solidus temperature (°C) (°F)		Liquidus temperature (°C) (°F)		Brazing temperature range (°C) (°F)	
	Cr	B	Si	Fe	C	P	S	Al	Ti	Mn	Cu	Zr	Ni	Others (total)	(°C)	(°F)	(°C)	(°F)	(°C)	(°F)
BNi-1	13.0–15.0	2.75–3.50	4.0–5.0	4.0–5.0	0.6–0.9	0.02	0.02	0.05	0.05	—	—	—	Rem	0.05	977	1790	1038	1900	1066–1204	1950–2200
BNi-1a	13.0–15.0	2.75–3.50	4.0–5.0	4.0–5.0	0.06	0.02	0.02	0.05	0.05	—	—	—	Rem	0.05	977	1790	1077	1970	1077–1204	1970–2200
BNi-2	6.0–8.0	2.75–3.50	4.0–5.0	2.5–3.5	0.06	0.02	0.02	0.05	0.05	—	—	—	Rem	0.05	971	1780	999	1830	1010–1177	1850–2150
BNi-3	—	2.75–3.50	4.0–5.0	0.5	0.06	0.02	0.02	0.05	0.05	—	—	—	Rem	0.05	982	1800	1038	1900	1010–1177	1850–2150
BNi-4	—	1.5–2.2	3.0–4.0	1.5	0.06	0.02	0.02	0.05	0.05	—	—	—	Rem	0.05	982	1800	1066	1950	1010–1177	1850–2150
BNi-5	18.5–19.5	0.03	9.75–10.50	—	0.10	0.02	0.02	0.05	0.05	—	—	—	Rem	0.50	1080	1975	1135	2075	1149–1204	2100–2200
BNi-6	—	—	—	—	0.10	10.0–12.0	0.02	0.05	0.05	—	—	—	Rem	0.50	877	1610	877	1610	927–1093	1700–2200
BNi-7	13.0–15.0	0.01	0.10	0.2	0.08	9.7–10.5	0.02	0.05	0.04	0.04	—	—	Rem	0.50	888	1630	888	1630	927–1093	1700–2000
BNi-8	—	—	6.0–8.0	—	0.10	0.02	0.02	0.05	0.05	21.5–24.5	4.0–5.0	—	Rem	0.05	982	1800	1010	1850	1010–1093	1850–2000

Table A.16 Composition of brazing atmospheres

Brazing atmosphere number	Source (a)	Maximum dew-point of incoming gas (°C)	Composition of atmosphere, (%)				Brazing filler metals	Base metals	Remarks
			H_2	N_2	CO	CO_2			
1	Combusted fuel gas (low hydrogen)	20	5–1	87	5–1	11–12	BAg, BCuP RBCuZn	Copper, brass (b)	—
2	Combusted fuel gas (decarburizing)	20	14–15	70–71	9–10	5–6	BCu, BAg(b) RBCuZn, BCuP	Copper (c), brass, low-carbon steel, nickel, Monel, medium-carbon steel (d)	Decarburizes
3	Combusted fuel gas, dried	−40	15–16	73–75	10–11	—	Same as 2	Same as 2 plus medium- and high-carbon steels, Monel, nickel alloys	—
4	Combusted fuel gas, dried (carburizing)	−40	38–40	41–45	17–19	—	Same as 2	Same as 2 plus medium- and high-carbon steels	Carburizes
5	Dissociated ammonia	−54	75	25	—	—	BAg(b), BCuP, RBCuZn(b), BCu, BNi	Same as 1, 2, 3, 4 plus alloys containing chromium (e)	—
6A	Cryogenic or purified N_2	−68	1–30	70–99	—	—	Same as 5	Same as 3	—
6B	Cryogenic or purified N_2	−29	2–20	70–97	1–10	—	Same as 5	Same as 4	—
7	Deoxygenated and dried hydrogen	−59	100	—	—	—	Same as 5	Same as 5 plus cobalt, chromium, tungsten alloys and carbides (e)	—

Table A.16 (Continued)

Brazing atmosphere number	Source (a)	Maximum dew-point of incoming gas (°C)	Composition of atmosphere, (%)				Brazing filler metals	Base metals	Remarks
			H₂	N₂	CO	CO₂			
8	Heated volatile materials (inorganic vapors – zinc, cadmium, lithium, volatile fluorides)	—	—	—	—	—	BAg	Brasses	Special purpose. May be used in conjunction with 1 thru 5 to avoid use of flux.
9	Purified inert gas (e.g., helium, argon)	—	—	—	—	—	Same as 5	Same as 5 plus titanium, zirconium, hafnium	Special purpose. Parts must be very clean and atmosphere must be pure.
10	Vacuum above 266.6 Pa (2 Torr)	—	—	—	—	—	BCuP, BAg	Copper	—
10A	Vacuum from 66.65 to 266.6 Pa (0.5 to 2 Torr)	—	—	—	—	—	BCu, BAg	Low-carbon steels, copper	—
10B	Vacuum from 0.13 to 66.65 Pa (0.001 to 0.5 Torr)	—	—	—	—	—	BCu, BAg	Carbon and low-alloy steels, copper	—
10C	Vacuum of 0.13 Pa (10⁻³ Torr) and lower	—	—	—	—	—	BNi, BAu, BAlSi, titanium alloys	Heat- and corrosion-resisting steels, aluminum, titanium, zirconium, refractory metals	—

(a) Types 6, 7, and 9 include reduced pressures down to 266.6 Pa. (b) Flux required in addition to atmosphere when alloys containing volatile components are used. (c) Copper should be fully deoxidized or oxygen-free. (d) Heating time should be minimized to avoid objectionable decarburization. (e) Flux must be used in addition to the atmosphere if appreciable quantities of aluminum, titanium, silicon or beryllium are present.

Table A.17 Brazing joint clearances

Filler metal group	Joint clearance (a) (mm)	Comments
BAlSi	0.05 to 0.20	For laps less than 6.4 mm long
	0.20 to 0.25	For laps more than 6.4 mm long
BCuP	0.03 to 0.13	
BAg	0.05 to 0.13	Flux brazing (mineral fluxes)
	0.00 to 0.05 (b)	Atmosphere brazing (gas-phase fluxes)
BAu	0.05 to 0.13	Flux brazing (mineral fluxes)
	0.00 to 0.05 (b)	Atmosphere brazing (gas-phase fluxes)
BCu	0.00 to 0.05 (b)	Atmosphere brazing (gas-phase fluxes)
BCuZn	0.05 to 0.13	Flux brazing (mineral fluxes)
BMg	0.10 to 0.25	Flux brazing (mineral fluxes)
BNi	0.05 to 0.13	General applications (flux or atmosphere)
	0.00 to 0.05	Free-flowing types, atmosphere brazing

(a) Values given are radial clearances when rings, plugs, or tubular members are involved. For some applications it may be necessary to use recommended values as diametral clearances to prevent excessive clearance when the entire joint gap is on one side. Excessive clearances will produce voids, particularly when brazing is performed in a high-quality atmosphere (gas-phase fluxing). (b) For maximum strength, a press fit of 0.001 mm/mm of diameter should be used.

382

Table A.18 Fume information sheet: copper and copper alloys; brazing and soldering

COPPER AND COPPER ALLOYS

These include the brasses, bronzes, and beryllium-hardened copper

Potential health hazards

Shielded-metal-arc welding (SMAW) and gas-metal-arc welding (GMAW) with copper-alloy consumables generates fumes containing high concentrations of copper. Exposure to copper fume may cause respiratory irritation, coughing, and bronchial spasm and can lead to the temporary flu-like symptoms of metal-fume fever – aches, nausea, and fever.

Some components (e.g., electrical contacts, springs, gears, bearings) are hardened by alloying with beryllium, which is extremely toxic. Exposure to even low concentrations of beryllium fume can cause severe respiratory distress and possibly death.

BRAZING AND SOLDERING

Potential health hazards

Brazing generates fumes that are typically rich in zinc and may contain cadmium and small amounts of irritant fluorides, chlorides, and boron compounds from the flux. Amount of fume generated and concentration of cadmium in the fume increase substantially with brazing temperature.

Exposure to zinc fumes may cause metal-fume fever, with its flu-like symptoms – aches, chills, nausea, and fever.

Precautions

Position the welder's head out of the fume plume. General ventilation may not adequately protect personnel; use local-exhaust equipment or respiratory protection. In confined spaces and poorly ventilated areas, use local-exhaust and respiratory protection.

Determine whether the copper-alloy components to be joined or the consumables to be used contain beryllium. If they do, use both local-exhaust ventilation and respiratory protection.

Precautions

Use local-exhaust ventilation or respiratory protection adequate to minimize exposure to fumes generated during brazing.

Avoid overheating. Hold brazing temperatures within the range recommended for the specific braze alloy in use.

Table A.18 (Continued)

Cadmium-containing fumes are highly toxic, even in low concentrations. Immediate symptoms include chills, heavy cough, and nausea. In the long term, serious damage to kidneys and lungs may result.	Use cadmium-free braze alloys where possible. If cadmium-bearing alloys must be used, apply local-exhaust ventilation or respiratory protection. In confined spaces or poorly ventilated areas, use local-exhaust ventilation and respiratory protection.
In confined spaces, extensive brazing with oxyfuel torch can generate elevated levels of nitrogen oxides.	Employ general precautionary measures for fume control (See Fume-Information Sheet 1, WELDING DESIGN & FABRICATION, July 1991, page 38). Provide adequate clean air to the workplace, using air-supplied respirator or hood in confined spaces or poorly ventilated areas. Extinguish the torch flame when it is not in use.
Soldering fluxes can generate irritating fumes, causing wheezing, cough, and respiratory irritation. Some workers may exhibit an allergic reaction.	Solder in a well-ventilated area, and avoid inhalation of fumes. In poorly ventilated areas or where several soldering stations are in operation, use a tool-mounted or bench-top fume extractor.
High-temperature brazing or flame soldering with lead-based solders can generate high levels of lead fume.	Avoid overheating. Use local-exhaust ventilation or respiratory protection in poorly ventilated areas.

Adapted from FIS-10 and FIS-12, International Institute of Welding

Table A.19 Precautions for fluoride-containing fluxes

FUMES AND GASES CAN BE DANGEROUS TO YOUR HEALTH. BURNS EYES AND SKIN ON CONTACT. CAN BE FATAL IF SWALLOWED.

- Before use, read and understand the manufacturer's instructions, Material Safety Data Sheets (MSDSs), and your employer's safety practices.
- Keep your head out of the fumes.
- Use enough ventilation, exhaust at the work, or both, to keep fumes and gases from your breathing zone and the general area.
- Avoid contact of flux with eyes and skin.
- Do not take internally.
- Keep out of reach of children.
- See American National Standard Z49.1, *Safety in Welding and Cutting*, published by the American Welding Society, 550 N.W. LeJeune Rd., P.O. Box 351040, Miami, Florida 33135; OSHA Safety and Health Standards. 29 CFR 1910, available from U.S. Government Printing Office, Washington, DC 20402.

First Aid: If flux comes in contact with eyes, flush immediately with clean water for at least 15 minutes. If swallowed, induce vomiting. Never give anything by mouth to an unconscious person. Call a physician.

Table A.20 Threshold limits for exposure to hazardous brazing environments

Substance	Long term exposure limit (8hr TWA value) (mg/m³)
Fluxes	
Hydrogen fluoride (as F)	2.5
Boron trifluoride	3.0
Fumes (gases)	
Nitric oxide	30
Nitrogen dioxide	9.0
Metal fume	
Copper	0.2
Zinc oxide	5.0
Cadmium oxide (as Cd)	0.05

Table A.21 Warning for use with cadmium-based brazing materials

DANGER: CONTAINS CADMIUM. Protect yourself and others. Read and understand this label.
FUMES ARE POISONOUS AND CAN KILL.

- Before use, read and understand the manufacturer's instructions, Material Safety Data Sheets (MSDSs), and your employer's safety practices.
- Do not breathe fumes. Even brief exposure to high concentrations should be avoided.
- Use enough ventilation, exhaust at the work, or both, to keep fumes and gases from your breathing zone and the general area. If this cannot be done, use air supplied respirators.
- Keep children away when using.
- See American National Standard Z49.1, *Safety in Welding and Cutting*, published by the American Welding Society, 550 N.W. LeJeune Rd., P.O. Box 351040, Miami, Florida 33135; OSHA Safety and Health Standards, 29 CFR 1910, available from U.S. Government Printing Office, Washington, DC 20402.

If chest pain, shortness of breath, cough, or fever develop after use, obtain medical help immediately.

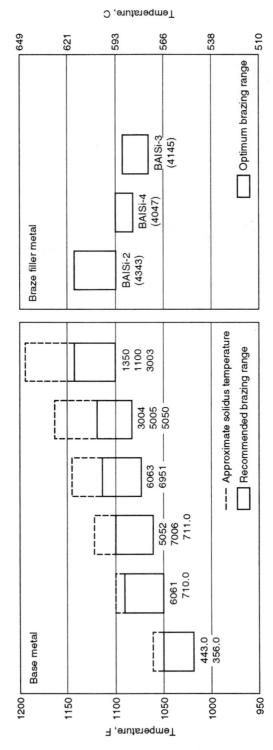

Fig. A.1 Comparison of brazing temperature ranges of aluminum alloy base metals and aluminum brazing filler metals.

Thin metals are frequently joined by brazing. Obey the safety rules when you enjoy the simplicity of torch brazing.

1. **Don't braze on unknown plated materials.** Cadmium or zinc may be present. They form toxic fumes during heating. Remove platings before heating for brazing.

2. **Don't braze on dirty parts.** Parts and brazing filler metal must be clean. Unknown dirt may add to the fume hazard. Don't expect flux to do the cleaning job.

3. **Don't braze without flux coverage.** Completely cover joint area. Dip filler metal in flux. Flux prevents oxidation of base metal and filler metal, reducing formation of hazardous fumes, while aiding flow of molten filler.

4. **Heat joint area uniformly before applying brazing filler metal.** Don't heat too fast or in one spot. Localized overheating increases fumes from flux and filler metal.

5. **Use flux as a temperature guide.** When flux melts and is fluid, the work is near the brazing temperature range. Further rapid or localized heating will end fluxing and cause fuming.

6. **Never apply flame directly to filler metal.** Use heat in work to melt filler metal. Filler metals flow freely.

7. **Especially avoid overheating filler metals that contain cadmium.** Look for warning notice. Four cadmium filler metals are: BAg-1, BAg-2, BAg-2a, BAg-3 (see AWS A5.8)

8. **Fumes? Overheated? Stop!** Reclean work and restart.

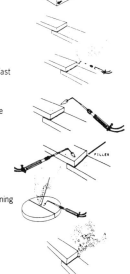

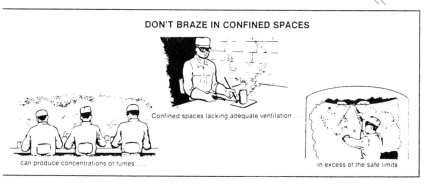

DON'T BRAZE IN CONFINED SPACES

Confined spaces lacking adequate ventilation . . .

can produce concentrations of fumes . . .

in excess of the safe limits

Fig. A.2 Safety precautions for joining thin metals by brazing.

INDEX

Page numbers appearing in **bold** refer to figures and page numbers appearing in *italic* refer to tables